The PHYSICAL WORLD

PREDICTING MOTION

Edited by Robert Lambourne

IoP

Institute of Physics Publishing
Bristol and Philadelphia
in association with

The Open University

The Physical World Course Team

Course Team Chair	Robert Lambourne
Academic Editors	John Bolton, Alan Durrant, Robert Lambourne, Joy Manners, Andrew Norton
Authors	David Broadhurst, Derek Capper, Dan Dubin, Tony Evans, Ian Halliday, Carole Haswell, Keith Higgins, Keith Hodgkinson, Mark Jones, Sally Jordan, Ray Mackintosh, David Martin, John Perring, Michael de Podesta, Ian Saunders, Richard Skelding, Tony Sudbery, Stan Zochowski
Consultants	Alan Cayless, Melvin Davies, Graham Farmelo, Stuart Freake, Gloria Medina, Kerry Parker, Alice Peasgood, Graham Read, Russell Stannard, Chris Wigglesworth
Course Managers	Gillian Knight, Michael Watkins
Course Secretaries	Tracey Moore, Tracey Woodcraft
BBC	Deborah Cohen, Tessa Coombs, Steve Evanson, Lisa Hinton, Michael Peet, Jane Roberts
Editors	Gerry Bearman, Rebecca Graham, Ian Nuttall, Peter Twomey
Graphic Designers	Steve Best, Sue Dobson, Sarah Hofton, Pam Owen
Centre for Educational Software staff	Geoff Austin, Andrew Bertie, Canan Blake, Jane Bromley, Philip Butcher, Chris Denham, Nicky Heath, Will Rawes, Jon Rosewell, Andy Sutton, Fiona Thomson, Rufus Wondre
Course Assessor	Roger Blin-Stoyle
Picture Researcher	Lydia K. Eaton

The Course Team wishes to thank the following individuals for their contributions to this book: Chapter 1: David Martin; Chapter 2: Keith Higgins; Chapter 3: Stan Zochowski; Chapter 4: Derek Capper and Alice Peasgood; Chapter 5: David Broadhurst and Stuart Freake; Chapter 6: Andrew Norton. The book made use of material originally prepared for the S271 Course Team by John Bolton, Keith Meek, Shelagh Ross and Milo Shott. For the multimedia package *Stepping through Newton's laws* thanks are due to Carole Haswell.

The Open University, Walton Hall, Milton Keynes MK7 6AA

First published 2000

Written, edited, designed and typeset by the Open University.

Published by Institute of Physics Publishing, wholly owned by The Institute of Physics, London.
IoP Publishing, Dirac House, Temple Back, Bristol BS1 6BE, UK.

US Office: Institute of Physics Publishing, The Public Ledger Building, Suite 1035, 150 South Independence Mall West, Philadelphia, PA 19106, USA.

Printed and bound in the United Kingdom by the Alden Group, Oxford.

ISBN 0 7503 0716 1

Library of Congress Cataloging-in-Publication Data are available.

This text forms part of an Open University course, S207 *The Physical World*. The complete list of texts that make up this course can be found on the back cover. Details of this and other Open University courses can be obtained from the Course Reservations Centre, PO Box 724, The Open University, Milton Keynes MK7 6ZS, United Kingdom: tel. +44 (0) 1908 653231; e-mail ces-gen@open.ac.uk

Alternatively, you may visit the Open University website at http://www.open.ac.uk where you can learn more about the wide range of courses and packs offered at all levels by the Open University.

To purchase this publication or other components of Open University courses, contact Open University Worldwide Ltd, The Berrill Building, Walton Hall, Milton Keynes MK7 6AA, United Kingdom: tel. +44 (0) 1908 858785, fax +44 (0) 1908 858787, e-mail ouwenq@open.ac.uk; website http://www.ouw.co.uk

1.1

s207book3i1.1

PREDICTING MOTION

Introduction

The previous book in *The Physical World* series was concerned with kinematics – the description of motion. Quantities such as position, displacement, velocity and acceleration were defined and relations between them established for several types of motion, including uniform linear motion, projectile motion, uniform circular motion, simple harmonic motion and orbital motion. However, the causes of these particular types of motion were not discussed.

In this book our concern is with **dynamics** — the study of forces and their effect upon motion. In Chapter 1 you will learn that the motion of a body is determined by its initial position and velocity, and by the forces that subsequently act upon it. In other words, the forces that act on a body determine the way in which the motion of the body changes with time. The laws that embody this deterministic relationship between forces and changes in motion are Newton's three laws of motion. They are the central laws of dynamics, and it is our knowledge of them that makes it possible, at least in principle, to predict how a body will move.

In practice, the detailed consequences of Newton's laws can sometimes be very difficult to work out. In such circumstances, two other quantities, energy and momentum, can often be used to determine important features of motion without requiring that a fully detailed prediction be made. Chapters 2 and 3 will explain the relation between energy, momentum and force and will also show why energy and momentum are important in their own right.

Chapter 4 concerns the dynamics of rotating bodies, such as spinning tops and rotating planets, while Chapter 5 deals with chaotic motion. This is the kind of motion that arises when the behaviour of a system is so sensitive to minute details of its initial condition that it is, in practice, impossible to predict the motion of the system, even though such a prediction is allowed in principle.

Open University students should view Video 3, *The Mother of All Collisions*, at some stage in their study of this book. The video can be viewed at any stage, but might be most effective at the end of Chapter 3.

Chapter 1   Forces and Newton's laws

1   *Voyager's* odyssey — an example of Newton's laws

The *Voyager 2* space probe is one of the most distant human artefacts in the Universe (Figure 1.1). *Voyager 2* was launched from the Earth in the summer of 1977. It passed the giant planet Jupiter in 1979, Saturn in 1981, Uranus in 1986 and Neptune in 1989 (Figure 1.2). By January 2000 it was about 9 billion kilometres from the Sun, and still travelling outwards, into interstellar space, at more than 15 kilometres per second.

Figure 1.1 *Voyager 2.*

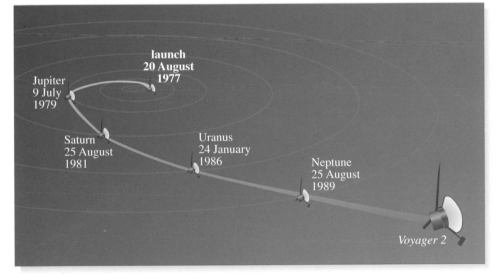

Figure 1.2   The planetary tour of *Voyager 2.*

Almost every step in the epic journey of *Voyager 2* provides an excellent example of one or more of Newton's three laws of motion, which will be introduced in Section 2. First, the space probe had to be launched from the Earth and set on an elliptical transfer orbit that would take it from the Earth to Jupiter. This was achieved by using a 700 tonne Titan-Centaur rocket that provided enough thrust to accelerate the 825 kilogram probe to the required speed (about 11 kilometres per second) to get it into the Earth–Jupiter transfer orbit. The acceleration wasn't instantaneous; it occurred in five distinct stages over a period of about an hour, and each was a separate illustration of Newton's second law.

The ability of rocket motors to supply the force needed to accelerate the space probe illustrates another of Newton's laws, the third law, which says that whenever a body exerts a force on some other body, a force of equal magnitude but acting in the opposite direction acts on the first body in reaction to the original force. Burning fuel in a rocket motor exerts a force on the rocket that accelerates it forward. The reaction acting on the fuel has the effect of accelerating some of the burnt fuel rearward and expelling it from the rocket's combustion chamber.

Once set on its course for Jupiter, *Voyager 2* continued to be acted upon by the gravitational pull of the Sun and, to a lesser extent, by the attraction of the various planets. (Incidentally, these forces are well described by another of Newton's laws; his law of universal gravitation.) It was the acceleration caused by the sum of these gravitational forces that, in accordance with Newton's second law, determined the precise shape of the curved orbital path that *Voyager 2* followed.

An even more spectacular demonstration of Newton's laws occurred as *Voyager 2* flew by Jupiter. Using a subtle manoeuvre that effectively played off the gravitational force due to the Sun against that due to Jupiter, *Voyager 2* was able to gain a substantial 'gravity assist' from its first planetary encounter. This had the effect of very slightly slowing Jupiter (by about one foot per trillion years) but it increased *Voyager 2*'s speed to nearly 16 kilometres per second relative to the Sun and redirected the probe towards its next goal, Saturn. The increase in speed thus achieved was so great that it ensured the space probe would eventually escape completely from the gravitational grip of the Sun. Although there were to be subsequent gravity assists at Saturn, Uranus and Neptune, each segment of *Voyager 2*'s post-Jupiter trajectory would be part of an open curve called a hyperbola, rather than part of a closed orbital ellipse.

Technically, *Voyager 2* is still following a hyperbolic trajectory, but it is now so far from the Sun that the gravitational force acting on it is very slight and the consequent acceleration almost negligible. Under such circumstances, *Voyager 2* is now behaving very much like a particle that is free of any net force. It thus illustrates another of Newton's laws, the first law, according to which a free particle will travel with uniform speed along a straight line. In the case of *Voyager 2*, the almost linear trajectory it is now following will eventually lead it into the vicinity of Sirius, the brightest star in the night sky. The distance to be travelled is so great that it will be millions of years before *Voyager 2* makes its closest approach to Sirius, but with Sir Isaac Newton in the driving seat, its eventual arrival seems assured!

2 Newton's laws and the definition of force

2.1 Underlying concepts: frames of reference and observers

Before giving a detailed account of Newton's laws it is necessary to define some underlying concepts that underpin those laws. You should already be aware that any description of motion requires the use of a coordinate system so that we can describe the position of a particle quantitatively. Typically we will use a Cartesian coordinate system, in which a particle located at the point with position coordinates (x, y, z) may be said to have the position vector $r = (x, y, z)$. Establishing such a system of coordinates involves choosing a point that can act as the origin, and specifying the orientation of the three mutually perpendicular axes which meet at that origin. In principle it also involves choosing appropriate units of measurement, but we shall generally adopt SI units for this purpose, as is the convention in scientific work.

Since we are interested in motion — the progressive change of position with time — we also need, implicitly at least, a clock or some equivalent method of timekeeping to allow us to interpret phrases such as 'at some particular instant' or 'at time t'. The combination of a coordinate system to fix positions, and a clock to fix times, constitutes a **frame of reference**. One such frame is shown schematically in Figure 1.3, along with an **observer** who uses that frame to record positions and times. If a certain observer

Figure 1.3 A frame of reference that is stationary relative to the observer. (Note that the coordinate system is right-handed, as is conventional.)

says that an object is moving, what that observer really means is that the position of the object is changing with time as measured in the frame of reference that the observer has chosen to use. An observer may generally be supposed to be at rest relative to the frame of reference he or she chooses to use, though they may be located at any point within that frame.

It is important to realize that the way in which the motion of a body is described will depend upon the frame of reference from which it is observed. For instance, an observer standing on the platform of a railway station might well choose to use a frame in which the station is at rest, so that a train drawing away from the station would be described as moving. However, an observer sitting on the train, would be more likely to use a frame of reference that was attached to the train. According to this second observer, the train would always be at rest; it would be the station and the surrounding town that were moving, since it is their positions that would be changing with time. Neither of these views is wrong. They are equally valid, but they are different, and one may well be simpler or more convenient than another for some particular purpose. In any case the conclusion is clear:

> The description of a body's motion depends on the frame of reference from which the motion is observed.

Question 1.1 A person standing on a railway platform observes that a train passing through the station is in uniform motion, that is to say its velocity is constant — the train is travelling with constant speed in a fixed direction in the person's frame of reference. Assuming that each of the following observers uses a frame of reference in which they themselves are at rest, which would agree that the train is moving uniformly?

(a) A painter who had just fallen from the roof of the station and was accelerating downwards towards a consignment of mattresses on the platform below.

Note that a body may accelerate by changing its direction of motion without changing its speed.

(b) A holidaymaker accidentally trapped in the station's revolving door who was being forced to walk around rapidly in small circles as the door continued to turn.

(c) A would-be passenger running at top speed after the train in a futile attempt to catch up with it. ∎

The laws of physics are ultimately justified by their ability to predict the outcome of experiments. However, as you have just seen, descriptions of motion are frame dependent. Two different observers, using different frames of reference, may well describe the motion resulting from an experiment differently. It follows that the laws of motion might also be expected to take different forms in different frames of reference. Any particular formulation of a law of motion might hold true in a specific frame of reference or, perhaps, in a restricted range of reference frames, but it seems unlikely that it could be written down in such a way that it would be true in all frames of reference. Determining the extent to which physical laws may be formulated in a way that allows them to take the same form in all frames of reference is an important matter, and considering it in detail would eventually lead us to a discussion involving Einstein's special theory of relativity. We shall not pursue that subject here, but we should note that Newton's three laws, as usually formulated, will only hold true in a particular class of reference frames called *inertial frames*. Just what constitutes an inertial frame, and how such a frame may be established will be discussed in the next subsection.

2.2 Newton's first law: inertia and inertial frames of reference

Newton's great contribution to the science of mechanics was to establish quantitative relationships which accounted for the changes in the motion of an object in terms of the forces acting on that object. But what do we mean by **force**? In everyday life we can think of a force as a push or a pull. It is a *vector* quantity characterized by a certain strength or *magnitude*, and a definite *direction* along which it acts. It follows that force is a quantity that can be represented diagrammatically by an arrow, with a length that represents the magnitude of the force, and an orientation that indicates the direction of the force. We will soon develop a more precise notion of force, but for the moment this simple idea of a push or pull that may be represented by an arrow will suffice.

Everyday experience seems to provide plenty of evidence that a force must be continuously applied to an object if it is to move at constant velocity. For example, to make a book slide at a given speed across a horizontal surface you have to push the book with a constant force; if you stop pushing, the book will quickly slow down and come to rest. Such experiences make it appear that force is required to *sustain* motion. However, appearances can be deceptive. If you carried out the same experiment on a variety of surfaces: wood, plastic, concrete, ice, etc., you would find that the strength of the force you needed to apply to maintain the given speed would vary from one surface to another. You would also find that when you stopped pushing the book, the time taken for it to come to rest would depend on the nature of the surface involved. On concrete the book would stop almost immediately; on ice it might continue sliding for some time. These observations suggest that what really happens when a book slides across a surface is that the surface exerts its own force on the book, a force that opposes the motion of the book, and that some surfaces exert a greater force than others.

The phenomenon that causes resistance to the relative motion of two surfaces in contact is called **friction**. The force on the sliding book that arises because of friction is called a **frictional force** (see Figure 1.4). A book sliding across a horizontal surface will move at constant speed provided it is pushed with a force that is equal in magnitude, but opposite in direction to the frictional force that opposes its motion. If the applied force (i.e. the push) is removed, the frictional force opposing the motion will continue to act and this will slow the book and bring it to rest. Viewed in this way — taking into account the frictional force as well as the applied force — it becomes clear that the effect of the *total* horizontal force on the book is not that of *sustaining* motion, but rather that of *changing* motion. While the total horizontal force is zero, the book moves with uniform velocity; when the total force is non-zero the book ceases to move uniformly, either speeding up or slowing down.

Figure 1.4 The horizontal forces on a sliding book. The total horizontal force on the book is the vector sum (i.e. the resultant) of the individual forces.

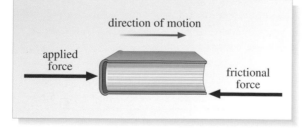

Thus, it is *not* the case that bodies have a natural tendency to be at rest and that forces are required to move them. Rather, the natural tendency of a moving body is to continue moving with the same speed and in the same direction, and the effect of an unbalanced force is to change the motion of a body by changing its speed or direction, or both, i.e. causing the body to accelerate. It is this principle that is enshrined in Newton's first law of motion:

Newton's first law of motion

A body remains at rest or in a state of uniform motion unless it is acted on by an unbalanced force.

The phenomenon that causes a body to continue in its state of uniform motion (including the possibility of remaining at rest) is called **inertia**, and Newton's first law, which implicitly recognizes the existence of inertia, is sometimes referred to as the law of inertia. Newton was not the first scientist to appreciate the need for a law of inertia. The principle had already been pretty well understood by Galileo Galilei (1564–1642), and had been even more accurately comprehended by the French philosopher René Descartes (1596–1650). However, by adopting the law of inertia as the first of his laws of motion, Newton did reveal his appreciation of the significance of force, and the need for some sort of test that would reveal when an unbalanced force was acting. The test offered by the first law is very simple: when a body ceases to move uniformly (i.e. when it accelerates) it is being acted on by an unbalanced force.

But wait a moment. In the last subsection we argued that the way a body moves depends on the frame of reference from which we choose to observe it. You can make any body accelerate simply by observing it from a frame of reference in which either its speed, or its direction of motion, or both, change with time. How can we reconcile this with Newton's first law? The simple fact is that we can't. Rather, we have to accept that Newton's first law, as stated above, is not true in all frames of reference.

To emphasize the limited validity of Newton's first law, imagine a suitcase on the frictionless luggage rack in a coach which is travelling at constant velocity relative to a road. If the driver of the coach steps on the brakes you know what will happen; the suitcase will slide along the luggage rack towards the front of the vehicle as shown in Figure 1.5. An observer standing by the roadside would see the suitcase continue to move with the same constant velocity it had before the brakes were applied, whereas the coach itself would be seen to slow down and stop. In a reference frame fixed to the road, these observations are consistent with Newton's

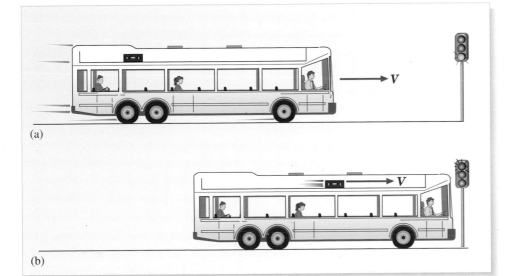

Figure 1.5   A suitcase on the frictionless luggage rack of a coach. In (a) the coach and suitcase travel with constant velocity V relative to the road. In (b) the coach slows down and eventually stops as its brakes are applied, but the suitcase continues to move with velocity V relative to the road.

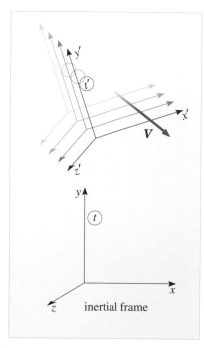

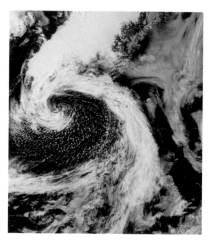

Figure 1.6 Any frame that moves with constant velocity relative to an inertial frame, while maintaining a fixed orientation, will also be an inertial frame. In this figure the clocks are represented by the circled *t* and *t'*.

Figure 1.7 A cyclonic depression; an anticlockwise circulation of air around a centre of low pressure in the Northern Hemisphere.

first law, since there is no unbalanced force acting on the suitcase, but there is a braking force acting on the coach. However, from the viewpoint of an observer sitting on the coach, using a reference frame fixed in the coach, the situation is very different. Such an observer would see the suitcase accelerate towards the front of the coach, even though no unbalanced force acted on it. To the observer in the coach, it would be clear that Newton's first law did not work.

The lesson to draw from the example of the suitcase is this; in order to apply Newton's first law (or in fact any of Newton's laws of motion) it is vital that observations are made from an appropriate frame of reference. Frames of reference in which Newton's first law holds true are called **inertial frames**. An important aspect of Newton's first law is that it enables us to identify inertial frames.

In principle it is fairly easy to identify an inertial frame, and once you have done so it's easy to set up infinitely many others. One way to establish whether a given frame is inertial is to use it to observe three force-free particles moving in mutually perpendicular directions. If all three particles move with uniform velocity (constant speed in their respective fixed directions) then the frame is an inertial frame. Having found one such frame, a second frame that moves with constant velocity relative to the first, while preserving a fixed orientation (i.e. not rotating), will also be an inertial frame (see Figure 1.6). This must be so, because the three force-free particles would also have uniform velocities in that second frame. Since the relative velocity of the two inertial frames may have any constant value, and since the relative orientation of the two frames may be anything we choose, it is clear that having found one inertial frame we may indeed establish infinitely many others.

> An inertial frame is a frame of reference in which Newton's first law holds true. Any frame that moves with constant velocity relative to an inertial frame, while maintaining a fixed orientation, will also be an inertial frame.

In practice, a frame of reference fixed on the Earth's surface provides an adequate approximation to an inertial frame for most purposes, and this is the frame we shall adopt for most of our applications of Newton's laws. However, it is worth noting that the Earth's daily rotation about its axis, and its annual revolution around the Sun mean that a reference frame fixed to the Earth's surface is only *approximately* inertial.

Although the non-inertial effects associated with the rotation of the Earth are small, the fact that a frame fixed to the Earth is, strictly speaking, non-inertial is significant in some contexts. For instance, non-inertial effects arising from the rotation of the Earth are of great importance to meteorologists studying the weather and the associated movement of air masses relative to the Earth's surface. They find that in the Northern Hemisphere moving air masses act as though they are subject to a force that tends to deflect them to the right. This apparent force is called the **Coriolis force** and is partly responsible for the tendency of air to circulate in an anticlockwise direction around centres of low pressure in the Northern Hemisphere, giving rise to the meteorological phenomenon known as a *cyclonic depression* (Figure 1.7). In fact, there is really no deflecting force acting on moving air masses. Their very real tendency to be deflected to the right in the Northern Hemisphere is not the result of an unbalanced force, but a consequence of the fact that their motion is being observed from a (strictly speaking) non-inertial frame attached to the turning Earth.

Similarly, a passenger in a car or bus turning a corner often has the feeling of being pushed outwards, away from the centre of the turn, by some kind of force, and may actually be flung outwards unless restrained in some way. Objects that are free to move around inside a turning vehicle will also be flung outwards in the same way, especially if the cornering is sufficiently sharp. The cause of these outward directed motions is usually referred to as **centrifugal force**, but once again the 'force' is illusory even though the effect is real. Viewed from an inertial frame, what is really happening as a car or bus turns a corner is that its passengers, along with any other objects that it contains, have a natural tendency to keep moving in a straight line. This soon brings them into contact with the outer wall of the vehicle or some other part of its structure that causes them to participate in the cornering by pushing them inward towards the centre of the turn. This inward force is real enough, but to the passenger it feels as though it is merely a reaction to an outward directed 'centrifugal force' that is actually not real at all.

Coriolis force and centrifugal force are both examples of **fictitious forces**. They are not real forces, but they are associated with real phenomena that arise whenever we use a non-inertial frame of reference. By introducing fictitious forces in such a frame, and by treating them as though they are real forces in that frame, it is possible to carry on using Newton's laws even though those laws do not, strictly speaking, hold true in a non-inertial frame.

Question 1.2   In two or three sentences, summarize the main aspects of Newton's first law that have been introduced so far.   ■

2.3 Newton's second law: force, mass and acceleration

Newton's second law of motion lies at the heart of dynamics. The first law allows us to recognize the action of an unbalanced force by means of the acceleration it produces, but it is the second law that allows us to quantify that force. The second law achieves this by providing a precise mathematical relationship between the force that acts on a given body and the amount of acceleration that the force produces.

Before stating Newton's second law let us note a few important features of its three ingredients: force, acceleration and mass.

- *Force*, as has already been noted, is a vector quantity that is characterized by both a magnitude (the strength of the force), and a direction. Roughly speaking, it is a directed push or pull.

- *Acceleration* is also a vector quantity, so it too has a magnitude and a direction. The acceleration of a body (a particle, say) is defined as the rate of change of its velocity. Note that this definition means that a body may accelerate by changing its direction of motion, without necessarily altering its speed. It also means that a body that is slowing down is undergoing acceleration just as much as a body that is speeding up. This technical meaning of acceleration is clearly at odds with the more restricted everyday use of the term to mean 'speeding up'.

- *Mass* is a positive scalar quantity, i.e. one that can be specified by the product of a number and a suitable unit of measurement, as in 2 kilograms. It is often described as 'a measure of the amount of matter in a body', but this really isn't very helpful. One important aspect of Newton's second law is its ability to provide a much clearer insight into the significance of mass. We shall return to this shortly.

Pausing only to note that Newton's second law, like his first, is only strictly applicable in an inertial frame of reference, we can now state the second law:

Newton's second law of motion

An unbalanced force acting on a body of fixed mass will cause that body to accelerate in the direction of the unbalanced force. The magnitude of the force is equal to the product of the mass and the magnitude of the acceleration.

Although every physicist is able to quote this statement or some equivalent form of words, what usually pops into the mind of a physicist when you mention Newton's second law is the equation that encapsulates the second law:

$$\boldsymbol{F} = m\boldsymbol{a}. \tag{1.1}$$

Here, $\boldsymbol{F}$ represents the unbalanced force, m the mass and $\boldsymbol{a}$ the acceleration produced by the force. Note that bold italic symbols are used for force and acceleration to remind us that they are vector quantities, with a direction as well as a magnitude. (In handwritten work you should indicate vector quantities by drawing a wavy underline, as in $\underset{\sim}{F} = m\underset{\sim}{a}$.) The fact that Equation 1.1 is a vector equation is very significant; it means that if $\boldsymbol{F}$ and $\boldsymbol{a}$ are expressed in terms of their components, as $\boldsymbol{F} = (F_x, F_y, F_z)$ and $\boldsymbol{a} = (a_x, a_y, a_z)$, then the following three relationships must be individually true:

$$F_x = ma_x \tag{1.2a}$$

$$F_y = ma_y \tag{1.2b}$$

$$F_z = ma_z. \tag{1.2c}$$

On the other hand, if we only know the magnitudes of the force and the acceleration,

$$F = |\boldsymbol{F}| = \sqrt{F_x^2 + F_y^2 + F_z^2} \quad \text{and} \quad a = |\boldsymbol{a}| = \sqrt{a_x^2 + a_y^2 + a_z^2},$$

then taking magnitudes of both sides of Equation 1.1, we see that

$$F = ma. \tag{1.3}$$

Armed with these equations we can now make more sense of mass. You can see from Equation 1.1 that the acceleration of a body is proportional to the unbalanced force that acts on it, and the mass is the proportionality constant in that relationship. More specifically, it follows from Equation 1.3, that if a force of fixed magnitude F is applied in turn to two different bodies with respective masses m_1 and m_2, then those bodies will respond by accelerating with accelerations of magnitude a_1 and a_2, respectively, where

$$F = m_1 a_1 = m_2 a_2. \tag{1.4}$$

This implies that, if m_1 is greater than m_2 then a_1 must be less than a_2. In other words, the accelerative effect of a fixed force decreases as the mass of the body being accelerated increases. This finding supports the view that the mass of a body is a measure of its resistance to acceleration. But you already know that the phenomenon causing bodies to continue moving uniformly is called inertia. It follows that mass, as used in Newton's second law, is a measure of the inertia of a body. Indeed, the quantity m that appears in Equation 1.1 is sometimes explicitly referred to as the **inertial mass** of the body being accelerated.

The (inertial) mass of a body is a measure of its resistance to acceleration.

It follows from Equation 1.4 that the masses of two bodies can be compared by comparing the magnitudes of their respective accelerations when both bodies are subjected to the same unbalanced force. Specifically,

$$\frac{m_1}{m_2} = \frac{a_2}{a_1} \quad \text{(for a fixed force)}. \tag{1.5}$$

We can use this general relationship to assign a numerical value to the mass of a body by comparing it with some other body of known mass. In the SI system of units the unit of mass is the **kilogram** (denoted by the symbol kg), and this is defined as the mass of a certain cylinder of platinum–iridium alloy that was constructed in 1878, and which is kept under carefully controlled conditions at the International Bureau of Weights and Measures in Sèvres, near Paris (Figure 1.8). Ultimately, all masses measured in kilograms are determined by comparison with this international standard kilogram.

It is interesting to note that the kilogram is now the only SI unit that is defined in terms of a manufactured object. Most of the other base units (such as the second and the metre) are now based on some kind of natural standard, such as a specified atomic oscillation, or the speed of light. Much effort has gone into the search for a more fundamental standard of mass and it is widely expected that the definition of the kilogram will eventually change. For the present, however, the platinum–iridium cylinder remains the international standard. Incidentally, the SI unit of mass is also unusual in being the kilogram rather than the gram. In the SI system, the gram (g) is simply defined as one thousandth of a kilogram. Some other physically interesting masses are shown in Figure 1.9.

Figure 1.8 The SI unit of mass, the kilogram, is equal to the mass of the International Prototype Kilogram, a cylinder of platinum–iridium alloy which is kept at the International Bureau of Weights and Measures in Sèvres, France. The UK national standard of mass (shown here) is Copy No. 18 of the International Prototype and is kept at the National Physical Laboratory in a specially designed glass and stainless steel enclosure.

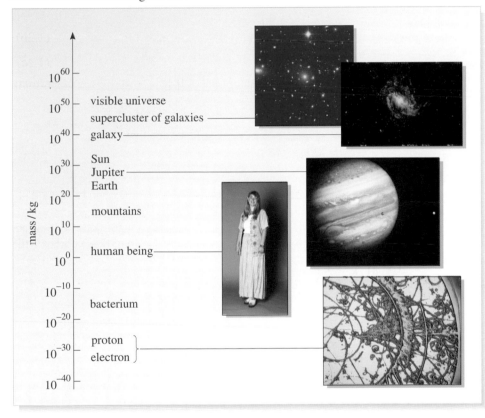

Figure 1.9 Some physically interesting masses.

There are two assumptions about mass which are widely used in Newtonian mechanics. The first is known as the **additivity of mass**, and asserts that if a body is made up of a number of different parts, the mass of the whole body is the simple arithmetical sum of the masses of its parts. The second is embodied in the law of **conservation of mass**. This states that if a body does not exchange matter with its surroundings, its mass remains constant. The development of physics in the twentieth century, particularly the development of Einstein's special theory of relativity, showed that both these assumptions are actually false. Nonetheless, they are very nearly valid over a wide range of conditions and we shall use them throughout most of this particular book, though we shall have to take their limitations into account later in the course.

The law of conservation of mass is sometimes called the principle of conservation of mass.

Now that you have learned how to quantify masses in terms of kilograms, you can also quantify force in terms of SI units.

Newton's second law implies that the unit of force must be the same as the product of the units of mass and acceleration. Thus, if a body with a mass of 1 kg undergoes an acceleration of magnitude $1 \, \text{m s}^{-2}$, then the force causing that acceleration must be of magnitude $1 \, \text{kg} \times 1 \, \text{m s}^{-2} = 1 \, \text{kg m s}^{-2}$.

This unit of force is called the **newton** and is represented by the symbol N.

$$1 \text{ newton} = 1 \, \text{N} = 1 \, \text{kg m s}^{-2}.$$

It's obviously important to know the definition of the newton, but it's also important to have some physical feeling for it. The following question does not have a uniquely correct answer, but it should help you to develop some feeling for the size of the newton, as well as giving you practice at the important skill of making physical estimates.

Question 1.3 Imagine you are holding an apple in your hand; the mass of an apple might typically be about 0.1 kg. Now suppose you hurl the apple as hard as you can. What speed do you think the apple would have when it left your hand? How long do you think it would have taken you to accelerate the apple to that speed from rest (i.e. what was the duration of your throwing action)? Assuming that you accelerated the apple uniformly, what would you estimate the magnitude of that acceleration to be? Using Newton's second law, what do you estimate to be the magnitude of the constant force that you would have had to apply to the apple to produce that uniform acceleration? ■

As you'll be aware, any object near the Earth's surface is pulled downwards by gravity. This downward force is called the *weight* of the object. When you hold a 0.1 kg apple in your hand you have to exert an upward force of about 1 N to balance the weight of the apple. We shall have more to say about weight and its distinction from mass in Section 3.1. Some other physically interesting force magnitudes are shown in Figure 1.10.

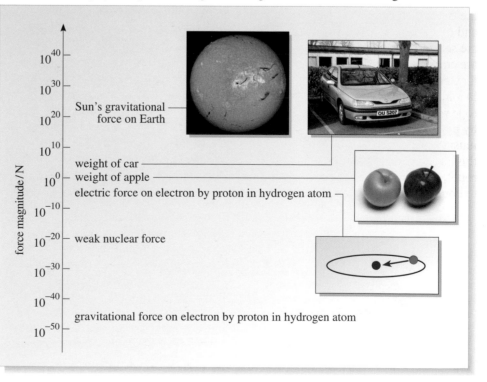

Figure 1.10 Some physically interesting force magnitudes.

2.4 Newton's third law: action and reaction

To complete the discussion of Newton's laws of motion, we now consider the third law which imposes an important condition that all forces must satisfy. Suppose you have to push a car to try to get the engine to start. When you push against the car you feel a force exerted by the car pushing back on you (Figure 1.11). The greater the force you apply to the car, the greater is the force that the car exerts on you. Indeed, the magnitudes of these two forces are the same, but the forces act in opposite directions. This applies even when the car moves, either at constant velocity or with some acceleration. This example illustrates Newton's third law of motion which may be stated as follows:

Newton's third law of motion

If body A exerts a force on body B, then body B exerts a force on body A. These two forces are equal in magnitude, but act in opposite directions.

Figure 1.11 When a car is pushed by a person, the car pushes back equally on the person.

Newton's third law implies that all forces occur in pairs. For any given force, there must be a second force, acting on some other body, that forms a 'third-law pair' with the first. Such pairs are sometimes said to consist of an **action** and a **reaction** to it, and are correspondingly referred to as 'action–reaction pairs'. Be careful when you are seeking to identify the second force of a third-law pair. Just because two forces are equal in magnitude and act in opposite directions does *not* necessarily mean that they constitute a third-law pair. For example, imagine that two people are indulging in a mini tug of war by pulling in opposite directions on a rope in such a way that the rope remains stationary, as shown in Figure 1.12. For this to occur, the force exerted by person A must be equal in magnitude to that exerted by person B, and the forces must be oppositely directed. However, they do *not* constitute a third-law pair. Why is this? The answer lies in the fact that both forces are exerted on the same body (the rope), whereas the two forces of a third-law pair must act on *different* bodies in order to comply with Newton's third law.

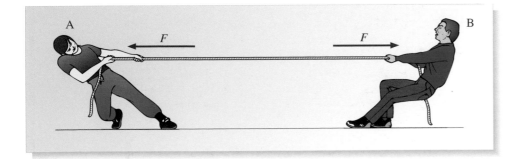

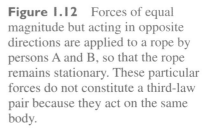

Figure 1.12 Forces of equal magnitude but acting in opposite directions are applied to a rope by persons A and B, so that the rope remains stationary. These particular forces do not constitute a third-law pair because they act on the same body.

● Identify the third-law pairs that are involved in the tug of war example (Figure 1.12).

○ If person A exerts a force on the rope, then the force the rope exerts on person A completes one third-law pair. Similarly, the force exerted by person B on the rope and the force the rope exerts on person B constitute a second third-law pair of forces. ■

If person A tries harder and manages to pull person B in the direction of person A, it is clear that the magnitude of the force exerted by A has become greater than that exerted by B. Since these two forces are no longer equal in magnitude, it is then even more obvious that they *cannot* constitute a third-law pair. However, even if the rope is accelerating towards person A, it will continue to be the case that each person will be subject to a force exerted by the rope that is equal in magnitude but opposite in direction to the particular force that each exerts on the rope.

One point that emerges very clearly from the third law is that force can be exerted by inanimate objects such as ropes or cars just as well as by living agencies such as people. The false belief that only living things can exert forces is a surprisingly widespread misconception. All the objects you see around you in the everyday world are subject to forces all the time, even when no living being is anywhere near them.

2.5 Newton's laws and vectors: resultant force, resolution and redirection

In stating Newton's three laws of motion it has already been pointed out that both force and acceleration are vector quantities that are characterized by direction as well as magnitude. Although this is a simple point, it deserves further emphasis since it is of the utmost importance and some of its implications are easy to overlook if you are not accustomed to working with vectors and vector notation. In this subsection we examine three implications of the vector nature of force under the headings of *resultant force*, *resolution* and *redirection*.

Figure 1.13 (a) Three horizontal forces, F_1, F_2 and F_3, acting on a ship. (b) Their resultant, $F = F_1 + F_2 + F_3$, may be determined by first using the triangle rule to determine $F_1 + F_2$, and then using the triangle rule again to determine $(F_1 + F_2) + F_3$.

Resultant force

It is important to remember that *any* of the forces involved in Newton's laws may be the *resultant* (i.e. the vector sum) of *several individual forces*. An example of this is shown in Figure 1.13a, where the horizontal force on a moving ship is due to the combination of three separate forces F_1, F_2 and F_3. Two of these, F_1 and F_2, are applied forces

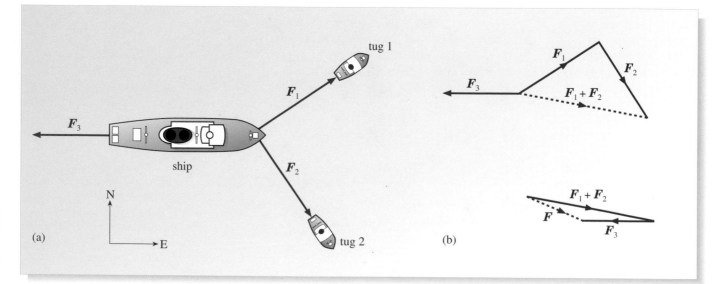

exerted by tugs; the third, F_3, is a force of resistance that opposes the motion of the ship and therefore points in the opposite direction to the ship's instantaneous velocity. The resultant horizontal force F is given by the vector sum $F = F_1 + F_2 + F_3$, and does not necessarily act in the direction of any of the individual forces. In the particular case shown in Figure 1.13, the resultant can be determined graphically by using the triangle rule (introduced in *Describing motion*) to combine F_1 and F_2, and then using the triangle rule again to combine that result with F_3, as shown in Figure 1.13b. However, if the components of the forces are known, it is usually easier to work out the resultant algebraically, as shown in Example 1.1.

Example 1.1

Suppose that in a coordinate system where the y-axis points north and the x-axis points east, the three horizontal forces shown in Figure 1.13 have the following components:

$$F_1 = (F_{1x}, F_{1y}) = (4.0 \times 10^5 \, \text{N}, \ 3.0 \times 10^5 \, \text{N})$$

$$F_2 = (F_{2x}, F_{2y}) = (3.0 \times 10^5 \, \text{N}, \ -4.0 \times 10^5 \, \text{N})$$

$$F_3 = (F_{3x}, F_{3y}) = (-4.0 \times 10^5 \, \text{N}, \ 0).$$

If the mass of the ship is 20 000 tonne (where 1 tonne = 1000 kg), what is the resultant horizontal force on the ship and what is the acceleration caused by that force? What is the magnitude of the acceleration?

Solution

According to the general rule for adding vectors (Equation 2.15 of *Describing motion*) the resultant of the three forces is

$$F = F_1 + F_2 + F_3 = (F_{1x} + F_{2x} + F_{3x}, \ F_{1y} + F_{2y} + F_{3y})$$

so $\quad F = (4.0 + 3.0 - 4.0, \ \ 3.0 - 4.0) \times 10^5 \, \text{N},$

i.e. $\quad F = (3.0 \times 10^5 \, \text{N}, \ \ -1.0 \times 10^5 \, \text{N}).$

It follows from Newton's second law that $a = F/m$, so the acceleration of the ship will be

$$a = \frac{1}{2 \times 10^7 \, \text{kg}} (3.0 \times 10^5 \, \text{N}, \ -1.0 \times 10^5 \, \text{N})$$

i.e. $\quad a = (1.5 \times 10^{-2} \, \text{m s}^{-2}, \ -0.5 \times 10^{-2} \, \text{m s}^{-2}).$

The magnitude of this acceleration is

$$a = |a| = \sqrt{a_x^2 + a_y^2} = \sqrt{(1.5 \times 10^{-2})^2 + (-0.5 \times 10^{-2})^2} \, \text{m s}^{-2}$$

$$= 1.6 \times 10^{-2} \, \text{m s}^{-2}.$$

In general, if a body is acted upon by N distinct forces, F_1, F_2, etc. up to F_N, then the resultant force F on that body will be given by the vector sum

$$F = F_1 + F_2 + \ldots + F_N.$$

This summation is lengthy to write and may be indicated more conveniently using the following standard mathematical shorthand:

$$\boldsymbol{F} = \sum_{i=1}^{N} \boldsymbol{F}_i \tag{1.6}$$

where the summation symbol $\displaystyle\sum_{i=1}^{N}$ tells us that terms of the form $\boldsymbol{F}_i$ should be added together for all the integer (whole number) values of i in the range $i = 1$ to $i = N$.

Using this compact notation we can now rewrite Newton's second law, $\boldsymbol{F} = m\boldsymbol{a}$, in the more explicit form

$$\sum_{i=1}^{N} \boldsymbol{F}_i = m\boldsymbol{a} \tag{1.7}$$

where the summation includes all the forces that act on the body. You will not often see Newton's second law written this way, but it is what the second law always implies.

Resolution

Resolution is the process of splitting a given vector into components. When a vector is expressed in terms of its Cartesian components, as in $\boldsymbol{F} = (F_x, F_y, F_z)$, we can say that it has been *resolved* along the directions of the x-, y- and z-axes, since we then know the component of the vector along each of those axes. However, a given vector may, in fact, be resolved along any chosen direction. In general, if the angle between a given vector $\boldsymbol{F}$ and a chosen direction is θ, then the component of the vector along the chosen direction is $F\cos\theta$, where F is the magnitude of $\boldsymbol{F}$. (See Figure 1.14.)

The important point about resolution in the context of Newton's laws is this: a vector equation such as Newton's second law, $\boldsymbol{F} = m\boldsymbol{a}$, implies a corresponding *equality between components resolved along any chosen direction*. We noted earlier (in Equation 1.2) that $\boldsymbol{F} = m\boldsymbol{a}$ implies that $F_x = ma_x$, $F_y = ma_y$ and $F_z = ma_z$, but it's worth remembering that similar resolutions can be carried out in *any* direction. It's also worth noting that if a body has no component of acceleration in a certain direction (i.e. if its component of acceleration in that direction is zero), then the net component of force in that direction must also be zero. Example 1.2 makes use of this.

θ is the Greek letter theta.

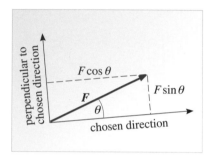

Figure 1.14 Resolving a given force $\boldsymbol{F}$ along a chosen direction. Note that the component in the chosen direction is $F\cos\theta$ and that the component in the perpendicular direction is $F\sin\theta$.

Example 1.2

Figure 1.15 shows a picture hanging from a nail. At the point of suspension the string that supports the picture is subject to three coplanar forces; an upward force $\boldsymbol{F}_1$ of magnitude $20\,\text{N}$, and two other forces $\boldsymbol{F}_2$ and $\boldsymbol{F}_3$, each directed at $40°$ to the vertical, along the two parts of the string. Given that the string is stationary, determine the magnitudes of $\boldsymbol{F}_2$ and $\boldsymbol{F}_3$.

Solution

Since the string is not accelerating, Newton's second law tells us that the total force on any part of the string must be zero. In particular, this must be true at the point of suspension. It further follows that, at the point of suspension, the sum of the force components in any chosen direction must be zero.

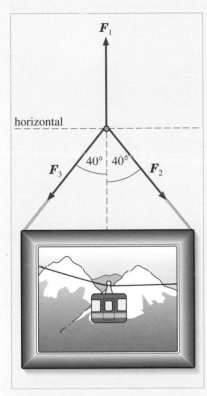

Figure 1.15 A picture hanging from a nail.

Resolving in the horizontal direction (noting that the string is at 50° to the horizontal) and treating components that point to the right as positive, we see that

$$F_2 \cos 50° - F_3 \cos 50° = 0. \qquad (1.8)$$

It follows from this that $F_2 = F_3$, so the two forces must have the same magnitude to ensure horizontal equilibrium. (This is intuitively obvious since only these two forces act in the horizontal direction, and there is no horizontal acceleration.)

In order to determine the unknown magnitude we can resolve the three forces in the vertical direction. Taking upwards as the positive direction this gives

$$20\,\text{N} - F_2 \cos 40° - F_3 \cos 40° = 0.$$

Using the fact that $F_2 = F_3$, to eliminate F_3 from the above equation we see that

$$20\,\text{N} - F_2 \cos 40° - F_2 \cos 40° = 0,$$

it follows that

$$2 \times F_2 \cos 40° = 20\,\text{N}$$

i.e. $F_2 = \dfrac{20\,\text{N}}{2 \cos 40°} = 13\,\text{N}.$

Thus F_2 and F_3 each have magnitude 13 N.

A point to note from Example 1.2 is that in order to determine two quantities, F_2 and F_3, we have had to use two separate conditions, and that these were obtained by resolving in two perpendicular directions.

Redirection

A final point about Newton's second law that deserves emphasis is this: *When an unbalanced force acts on a moving body, the direction of the resulting acceleration may well be different from the original direction of motion.* For example, if you kick a stationary ball it will accelerate in the direction of the kick and then move off in that direction. However, if you kick a moving ball at right angles to its direction of motion it will still accelerate in the direction of the kick, but its final velocity will be directed somewhere between the direction of the kick and the direction of motion prior to the kick.

Always remember that the effect of an unbalanced force is to cause acceleration, which results in a *change* of motion. There is no reason why that change should be in the direction of any motion that preceded it, and, in general, there is no guarantee that the final motion will be in the direction of the force causing the change. (This is illustrated in Example 1.3.)

Example 1.3

A small 20 kg spacecraft, that is free of external influences, is travelling with velocity $u = (2.25, 0, 0)\,\text{m s}^{-1}$ in a certain inertial frame. If an on-board thruster, that produces a constant force $F = (3.60, 3.60, 0)\,\text{N}$, is fired for 12 s, what will be the velocity of the spacecraft after the firing of the thruster?

Solution

According to Newton's second law, the uniform acceleration caused by the thruster will be

$$a = \frac{(3.60, 3.60, 0)\,\text{N}}{20\,\text{kg}} = (0.18, 0.18, 0)\,\text{m s}^{-2}.$$

It follows from the uniform motion equations (introduced in *Describing motion*) that the final velocity of the spacecraft will be

$$v = u + at$$

where t is the duration of the thruster firing. (Note that at represents the *change* in the spacecraft's velocity.) Since $t = 12\,\text{s}$ in this case, it follows that the final velocity will be

$$v = (2.25, 0, 0)\,\text{m s}^{-1} + 12 \times (0.18, 0.18, 0)\,\text{m s}^{-1},$$

i.e. $v = (4.41, 2.16, 0)\,\text{m s}^{-1}.$

Note that the initial velocity is in the x-direction, but the final velocity points in some other direction in the xy-plane. (The angle θ between v and the x-axis is given by $\tan\theta = 2.16/4.41$, from which it follows that $\theta = \arctan 0.490 = 26.1°$.)

Having just said that an object will not necessarily end up moving in the direction of the force that accelerates it, there *are* conditions under which this *will* happen.

● Under what conditions will an object end up moving in the direction of a constant resultant force that acts on it?

○ The object's final velocity is guaranteed to be parallel to that of the constant resultant force only if the object was initially at rest, or if it was initially travelling in the direction of that resultant force. ■

2.6 Newton's laws and rigid bodies: centre of mass

Although Newton's laws refer to the behaviour of 'bodies', and although we have occasionally considered the effect of forces on 'real' objects such as satellites, ships and people, we have not given any detailed consideration to the way extended objects differ in their behaviour from point-like **particles**. In this subsection we take a first look at the behaviour of finite (i.e. not point-like) sized bodies when they are subjected to forces.

In what follows we shall restrict our attention to what are known as **rigid bodies**. By definition, such bodies cannot be deformed; each of their parts always maintains the same separation from every other part. Restricting our attention in this way will force us to overlook the effect of applying a force to a flexible body such as a jelly, or a lump of clay, but we will gain the comfort of knowing that we don't have to worry about complications arising from the distortion of the body.

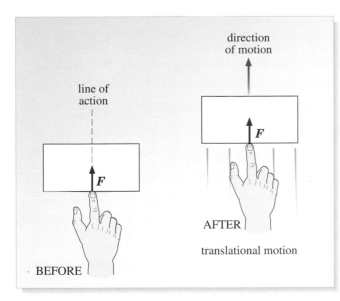

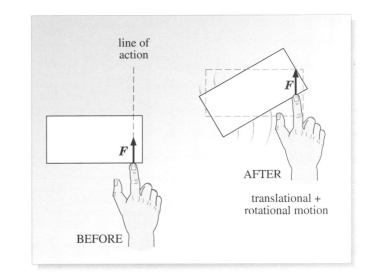

Figure 1.16 A horizontal force applied at the centre point of one of the longer edges of a rectangular body, at right angles to that long edge, will cause purely translational motion.

Figure 1.17 A horizontal force applied close to the end of one of the longer edges of a rectangular body, at right angles to that long edge, will cause a combination of translational and rotational motion.

If an unbalanced force is applied to a stationary particle, the particle will accelerate in the direction of the force in accordance with Newton's second law: $F = ma$. However, if the same force is applied to a stationary rigid body of finite size, the outcome generally depends on the point at which the force is applied. You can verify this for yourself by performing the following simple experiment. Place a reasonably heavy, rectangular object, such as this book, on a smooth horizontal surface. Then apply a horizontal force perpendicular to one of the longer edges by pushing the centre of that edge with your finger, as in Figure 1.16. You should find that the whole object moves in a straight line in the direction of the force, while the orientation of the object remains unchanged. Such motion, in which every part of the body moves parallel to every other part and there is no change of orientation, is described as **translational motion**. However, if you apply the same horizontal force at a point close to the end of the longer edge, taking care still to apply the force at right angles to the edge, as in Figure 1.17, you should now observe a very different outcome. This time the resulting motion should be a combination of translational and **rotational motion**. The object should turn about its centre point as well as moving as a whole. More specifically, the centre of the rectangular object should move in a straight line, as before, but all other parts of the object should rotate about that centre point.

These observations suggest that there is something special about the centre point of a rectangular body in relation to the action of a force on that body. This is indeed the case. The point at which the force is applied, together with the direction of the force, jointly determine a line called the **line of action** of the force. When the line of action of an unbalanced force passes through the centre point of a uniform rigid rectangular body, the effect of that force is to cause purely translational acceleration. When the line of action does not pass through the centre point, the effect of the unbalanced force is to cause a combination of translational *and* rotational acceleration. This idea can be generalized to any rigid body in the following way.

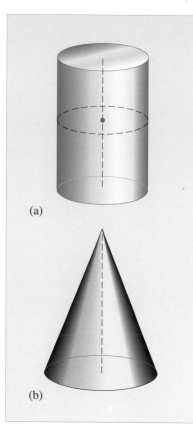

(a)

(b)

Figure 1.18 (a) The centre of mass of a uniform cylinder is half-way up the axis of symmetry. (b) The centre of mass of a uniform cone is also on the central axis, but its precise location is less obvious.

Associated with each rigid body there is a unique point, called the **centre of mass**, with the property that if any unbalanced force is applied to the body in such a way that its line of action passes through the centre of mass, then the only effect of that force will be to cause translational acceleration of the body.

It is not appropriate here to give a detailed mathematical prescription from which you could determine the position of the centre of mass of any given body (though such a prescription does exist). However, it can be said, somewhat imprecisely, that the centre of mass represents the 'average' position of the mass of a body. For bodies with a high degree of symmetry, the centre of mass can be easily guessed. For example, the centre of mass of a uniform flat disc is at the centre of the disc, and the centre of mass of a uniform sphere is at the centre of the sphere. Similarly, the centre of mass of the uniform solid cylinder shown in Figure 1.18a is half way up the central axis. However, it is by no means obvious where the centre of mass of the uniform solid cone in Figure 1.18b is located. Clearly, it will be somewhere on the axis of symmetry of the cone; but how far up? Common sense tells us that it will be less than half-way up because there is more mass in the lower half of the cone than in the upper half. However, that is about as far as common sense will take us. A detailed mathematical analysis shows that the centre of mass of the cone is actually one-quarter of the way up the axis from the base, but this is far from obvious.

The centre of mass plays an important role in the analysis of the motion of extended bodies. When external forces act on an extended body the motion of its centre of mass is identical to that of a particle with the same mass as the whole body subjected to the resultant of those external forces. This is illustrated schematically in Figure 1.19. Note that the lines of action of the individual forces have no influence on the motion of the centre of mass, they are only of significance in determining the rotation about the centre of mass.

So, as far as the motion of its centre of mass is concerned, an extended body subjected to external forces behaves as though its entire mass is concentrated at its

Figure 1.19 The centre of mass of a rigid body moves as though all the mass of the body is concentrated there, and all the external forces on the body act there. As far as the centre of mass of the brick is concerned, the forces F_1 and F_2 have the same effect as the single force $F_1 + F_2$.

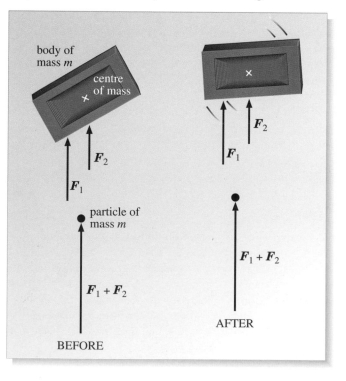

body of mass m

centre of mass

F_2

F_1

particle of mass m

$F_1 + F_2$

BEFORE

F_2

F_1

$F_1 + F_2$

AFTER

centre of mass (CM), with the resultant of those external forces acting on that concentrated mass. We can express this in terms of Newton's second law:

If a resultant external force F acts on a body of fixed mass m, then the acceleration of the body's centre of mass, a_{CM}, is given by:

$$F = ma_{CM}. \tag{1.9}$$

One consequence of this result is shown in Figure 1.20. It was demonstrated in Chapter 2 of *Describing motion* that a point-like projectile launched at an angle to the horizontal will follow a parabolic trajectory (neglecting air resistance). This is a result of the constant downward acceleration caused by gravity. Figure 1.20 shows that an extended object, such as a spanner, launched in a similar way, may rotate as it moves, but its centre of mass will still follow the same parabolic path because it is subject to the same downward acceleration.

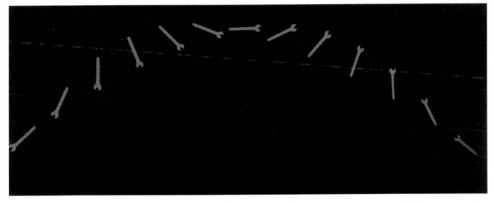

Figure 1.20   A stroboscopic picture of a spanner thrown across a room. The centre of mass of the spanner follows a parabolic trajectory.

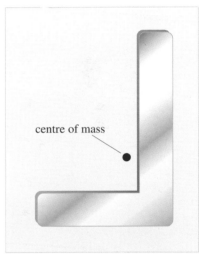

Figure 1.21   The location of the centre of mass of the L-shaped piece of metal lies outside the body.

Another, quite different, consequence of Equation 1.9 is this: *If the total external force on an object is zero, then the object's centre of mass cannot accelerate; it must move with constant velocity* (including the possibility that it remains at rest).

As a further example, consider the L-shaped piece of metal shown in Figure 1.21. What may surprise you in this case is that the centre of mass lies *outside* the body, as shown. When a body of this kind is launched in a vertical plane (e.g. thrown across a room), it is still the case that its centre of mass will follow a parabolic trajectory. This can be seen in Figure 1.22 thanks to a light marker fixed at the centre of mass by a fine thread. Note that generally a point in the body, such as an external corner, follows

Figure 1.22   A stroboscopic photograph shows that the centre of mass of the L-shaped body travels along a parabolic path when the body is thrown across a room. A general point in the body, such as the external corner, follows a more complicated path.

a more complicated path, as the body rotates about its centre of mass. This example reinforces the point that the motion of the centre of mass of a body may be analysed relatively simply by applying Newton's second law: $F = ma_{CM}$.

Although we have now discussed the motion of the centre of mass of a rigid body in some detail, we have still said very little about the rotation of the rigid body. This is a subject to which we shall return in Chapter 4.

3 Some familiar forces

By observing the instantaneous acceleration of a body, and using Newton's second law, we can easily work out the resultant force acting on that body at any time. However, in order to *predict* the acceleration of the body we need to know the force without measuring the acceleration that it causes. Fortunately, there are many situations in which this is possible thanks to our knowledge of certain **force laws** that determine the magnitude and direction of various forces in specified circumstances. The most famous of these force laws is undoubtedly Newton's law of universal gravitation. In this section we shall introduce this law along with some of the other force laws that play such a vital part in predicting motion.

3.1 Weight and terrestrial gravitation

All objects near the Earth's surface are subject to a downward pull due to the Earth's gravity. In response to this pull, any object that is free to do so will accelerate downwards with an acceleration g, called the **acceleration due to gravity**.

The magnitude of g varies slightly with altitude and with location. However, these variations are quite modest (between about 9.78 m s^{-2} at the Equator and 9.83 m s^{-2} at the North Pole), so g $(= |g|)$ can be taken to be 9.81 m s^{-2} across much of the Earth's surface and that is the value we shall generally adopt. Nonetheless, it is important to note that the value of g is the same for *all* objects at a given place; in particular, it *does not depend on the mass of the object*. This statement forms part of the **law of terrestrial gravitation**.

> At any given location on the Earth, all objects that are subject only to the effect of gravity have the same downward acceleration g, irrespective of their mass and composition.

It is often said that Galileo demonstrated the truth of this (apart from the effects of air resistance, etc.) by simultaneously dropping two bodies of different mass from the Leaning Tower of Pisa and showing that they reached the ground at the same time (Figure 1.23). Sadly, most scholars seem to agree that he is unlikely to have actually performed such an experiment, even if he spoke of doing so. Still, the legend lives on, and there is no doubt that Galileo could have conducted such a demonstration had he wished to.

Now, Newton's second law tells us that if a body has an acceleration g, then that body must be subject to a force, $F = mg$. This force, the gravitational pull on the body due to the Earth, is called the **weight** of the body. In the case of an extended body, its line of action passes through the body's centre of mass. The weight of a body is normally represented by the vector symbol W, so for any body of mass m

$$W = mg \tag{1.10}$$

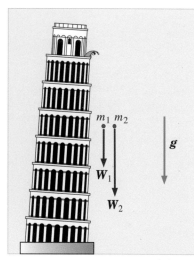

Figure 1.23 The experiment Galileo probably didn't perform. Two bodies with different mass, released simultaneously, reach the ground at the same time because each is accelerated downwards by a weight that is proportional to its mass. (Effects arising from air resistance and the rotation of the Earth are assumed to be equal for the two bodies.)

where g is the local acceleration due to gravity at the location of the body. This expression for the weight of a body is the first of the 'force laws' that were mentioned at the start of this section. It gives us a way of determining a particular force (the weight of a body) without having to measure the acceleration that particular force produces. Since Equation 1.10 is essentially a direct consequence of the law of terrestrial gravitation, it too is sometimes referred to by that name.

Note that the weight of an object is very different from its mass. Weight is a vector quantity; it is a *force*, that acts downwards near the Earth's surface and is measured in newtons. Mass is a scalar quantity; it quantifies the translational inertia of a body, and is measured in kilograms. Mass is an intrinsic property of a body. Weight arises from the gravitational interaction of the body and its external environment. In everyday language, the terms weight and mass are often used interchangeably, as though they mean the same thing. In science, however, the distinction between them is clear and it is important to use the terms correctly.

Despite the differences, the weight of an object *is* directly proportional to its mass. Weighty objects (those that are attracted strongly to the Earth) are also massive (they have a high inertia). This is why bathroom scales (which actually respond to your *weight*) are calibrated in kilograms and can be used to measure your mass. Provided you stay reasonably close to the Earth's surface (within a few kilometres), your weight is practically independent of altitude and a pair of bathroom scales will indicate the same value for your mass whether you are at the top of a mountain or at its base. However, if you went into space your mass would not be affected, but your weight certainly would.

Question 1.4 A bag of sand has a mass of 50 kg. What is its weight? ■

3.2 Universal gravitation

It was Isaac Newton, allegedly inspired by a falling apple, who realized that gravitation was a universal phenomenon, and not just a purely terrestrial one. By supposing that the force holding the Moon in its orbit is of the same nature as that which causes an apple to fall to the Earth, Newton was led to the conclusion that every particle of matter is attracted towards every other particle of matter by a gravitational force, wherever those particles might be located. Expressed in modern terminology, Newton's **law of universal gravitation** may be stated thus:

> Every particle of matter attracts every other particle of matter with a gravitational force, whose magnitude is directly proportional to the product of the masses of the particles, and inversely proportional to the square of the distance between them.

Two examples of this attraction are shown in Figure 1.24, where a particle of mass m_1 is separated by a distance r from a second particle of mass m_2. The gravitational force on the second particle due to the first is denoted $\boldsymbol{F}_{21}$ and is directed towards the first particle. Similarly, the gravitational force on the first particle due to the second is denoted $\boldsymbol{F}_{12}$ and is directed towards the second particle. The forces are oppositely directed, but they have the same magnitude, given by

$$F_{21} = F_{12} = \frac{Gm_1m_2}{r^2}, \tag{1.11}$$

where G is a fundamental constant called the **universal gravitational constant**, with the value $G = 6.67 \times 10^{-11} \text{ N m}^2 \text{ kg}^{-2}$. Because the magnitude of the gravitational

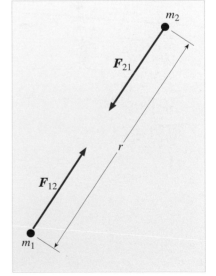

Figure 1.24 The gravitational forces $\boldsymbol{F}_{12}$ and $\boldsymbol{F}_{21}$ acting on particles of masses m_1 and m_2, separated by a distance r. The forces are oppositely directed, but they have the same magnitude, Gm_1m_2/r^2.

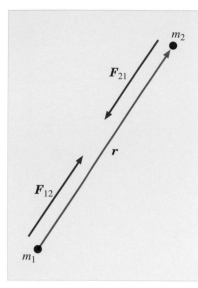

Figure 1.25 If mass m_1 is taken to be at the origin, then the position of mass m_2 can be specified by the position vector r. The force on m_2 due to m_1 is directed towards m_1, so is in the opposite direction to r.

force is inversely proportional to the square of the separation r, the law of universal gravitation is often referred to as the **inverse square law** of gravitation.

It is interesting to note that, in principle, Equation 1.11 provides a way of defining mass that does not depend on comparing accelerations. Mass defined in this way is called **gravitational mass**. The experimental fact that the gravitational mass of any body is equal to its inertial mass (Section 2.3) is very remarkable.

For the purposes of predicting motion, we would like to write down a precise force law corresponding to the law of universal gravitation. Equation 1.11 goes some way towards meeting this ambition, but it only describes the magnitude of the force, not its direction. How can the direction of the gravitational force be described? Perhaps the easiest way of taking direction into account is to choose the position of the first particle as the origin of a coordinate system in which the location of the second particle is specified by the position vector r (see Figure 1.25). The vector r then has magnitude r, equal to the separation of the two particles, but the quantity

$$\hat{r} = \frac{r}{r} \tag{1.12}$$

is a vector pointing in the direction of r that has magnitude 1 (since $|\hat{r}| = |r|/r = r/r = 1$). A vector of magnitude 1 is generally called a **unit vector** and is written with the caret character (^) placed above the vector; so in this case $\hat{r}$ is the unit vector in the direction of r, and it provides just the kind of directional information we need. (Pay particular attention to the fact that a unit vector has magnitude 1, not 1 metre, nor even 1 unit, just 1, i.e. it is dimensionless.) Remembering that the unit vector defined by Equation 1.12 points from the first particle to the second, we can now write the force on the second particle due to the first as

$$F_{21} = \frac{-Gm_1m_2}{r^2}\hat{r} \tag{1.13}$$

where the minus sign indicates that F_{21} points in the opposite direction to $\hat{r}$, that is, from the second particle towards the first. Equation 1.13 is exactly the force law we want.

Equation 1.13 gives the gravitational force between two *particles* of matter. However, it can also be used to determine the force between two *extended* bodies of finite size. This can be achieved by imagining each body to consist of a large number of particles, and then determining the vector sum of all the gravitational forces that the particles of one body exert on each of the particles of the other body. In general, this is easier said than done. However, in the special case of objects with **spherical symmetry**, that is spherical objects in which the density depends only on the distance from the centre, the following remarkably simple result applies.

The gravitational effect of any spherically symmetric body, outside its own surface, is identical to that of a single particle, with the same mass as the body, located at the centre of the body.

So, as far as gravity is concerned, spherically symmetric bodies such as billiard balls, or, to a good approximation, stars and planets, behave as though their entire mass is concentrated at their geometric centre. This very useful result is known as **Newton's theorem**. Apparently, even Newton took some time to find a proof of its correctness, and it is believed that this delayed some of his other work.

If we assume the Earth to have spherical symmetry, we can use Newton's theorem, together with Equation 1.13, to determine the gravitational force that the Earth exerts on a particle of mass m close to its surface. That force will be

$$\boldsymbol{F} = \frac{-Gmm_E}{r^2}\hat{\boldsymbol{r}} \tag{1.14}$$

where m_E is the mass of the Earth, $\hat{\boldsymbol{r}}$ is a unit vector pointing from the centre of the Earth to the location of the particle, and r is the distance of the particle from the centre of the Earth (which must be greater than or equal to the Earth's radius, R_E, for Newton's theorem to apply).

However, from Section 3.1, we know that the gravitational force on a particle close to the Earth's surface is nothing other than its weight, $W = m\boldsymbol{g}$. Equating the right-hand side of Equation 1.14 with the weight then gives:

$$m\boldsymbol{g} = \frac{-Gmm_E}{r^2}\hat{\boldsymbol{r}}.$$

Dividing both sides of this equation by m, we see that Newton's law of universal gravitation predicts that the acceleration due to gravity, $\boldsymbol{g}$, at any point above the Earth's surface is directed vertically downwards (in the $-\hat{\boldsymbol{r}}$ direction), and that its magnitude is given by

$$g = \frac{Gm_E}{r^2} \quad \text{where } r \geq R_E. \tag{1.15}$$

This explains several of the features of terrestrial gravitation. It predicts that at the surface of the Earth, where $r = R_E$, the magnitude of the acceleration due to gravity should have the constant value Gm_E/R_E^2 which is approximately correct. (For the prediction to be exactly right the Earth would have to be exactly spherically symmetric, which it isn't.) It also shows why the value of g will diminish slightly as altitude above the Earth increases; increasing r will have the effect of reducing g, but the effect won't be very great over a height of a few kilometres because that represents such a small change compared with the radius of the Earth. Equation 1.15 can even go some way towards explaining the variation in g between the Equator and the poles. The Earth bulges slightly at the Equator (another effect that Newton predicted), so the Earth's radius is effectively greater at the Equator and g is correspondingly less there, as we noted earlier. Even from these simple consequences, you can see the amazing power of universal gravitation to explain phenomena that might otherwise seem mysterious or arbitrary. The law of universal gravitation is clearly a much more profound result than the law of terrestrial gravitation. Its formulation was, perhaps, Newton's greatest single achievement.

Question 1.5 Determine the magnitude of the acceleration due to gravity on the surface of the planet Mars, given that the mass and radius of Mars are 6.42×10^{23} kg and 3.38×10^6 m, respectively.

Question 1.6 (a) The mass of the Earth is $m_E = 5.98 \times 10^{24}$ kg, and its average radius is $R_E = 6.37 \times 10^6$ m. Use these values together with Equation 1.15 to evaluate the acceleration due to gravity at the surface of the Earth. (b) Suggest a reason why the answer to part (a) is greater than the average measured value of 9.81 m s^{-2}. (*Hint*: Think about the effects discussed in Section 2.2.) ■

3.3 Contact forces and normal reactions

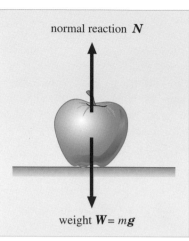

Figure 1.26 The forces acting on an apple, at rest on a flat rigid surface.

The word 'normal' here comes from Cartesian geometry and means perpendicular to, as in 'the normal to a surface'.

The electrical forces between atoms in a solid are generally attractive when the atoms are well separated, but become repulsive when the atoms get too close together. These forces are discussed in detail in later books.

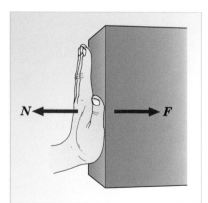

Figure 1.27 A hand exerts a horizontal contact force F on the side of a box. The box exerts a horizontal normal reaction $N = -F$ on the hand.

Figure 1.26 shows an apple at rest on a rigid horizontal surface. The apple is subject to terrestrial gravity, so the apple's own weight, W, acts downwards, through the apple's centre of mass. Even so, the apple is not accelerating, implying that the resultant force on the apple must, according to Newton's first law, be zero. Thus, the apple must also be subject to an upward force that exactly balances its weight. This is the force N shown in Figure 1.26. What is this force, and how does it arise?

The upward force N is exerted by the rigid surface on which the apple is resting. It is at right angles to that surface and is a reaction to the force that the apple applies to the surface. For these reasons it is said to be a **normal reaction force**. In principle, N is a result of the impenetrable nature of rigid surfaces. However, what happens in practice, on a real surface, is that the weight of the apple causes it to sink a little way into the surface, compressing the surface and forcing its atoms closer together, this leads to an increase in the repulsive electrical forces that keep the atoms in a solid apart. The compression of the surface stops when the electrical forces have increased sufficiently to compensate for the force applied by the apple. Thus, normal reaction forces are ultimately electrical in nature.

Note that in the case we have been discussing, the normal reaction force acting on the apple is *not* a reaction to the weight of the apple. The normal reaction force acts *on* the apple and is a reaction to the force that the apple exerts on the surface. The force exerted *by* the apple is an example of a **contact force**; like the normal reaction it is ultimately electrical in nature, arising from the forces between atoms. It is true that this contact force only arises because of the apple's weight, but that doesn't alter the fact that the weight itself acts on the apple, not on the surface. It is important to remember that the action and reaction that form a third-law pair always act on different bodies. It is also important to be clear about which forces act on which objects. In fact, when trying to predict how a body will move it is often useful to draw a diagram that shows *only* the body of interest and the forces that act upon it, including any reactions. Such diagrams are called **free-body diagrams**, and are especially useful in some of the engineering applications of physics.

● What is the force that forms a third-law pair with the weight of the apple?

○ The upward gravitational force that the apple exerts on the Earth. These are oppositely directed forces of the same magnitude that arise from a common cause but act on different bodies. ■

Normal reaction forces always act at right angles to the surfaces that exert them, but that doesn't mean that they have to act vertically. Figures 1.27 and 1.28 show two examples that illustrate this point. In Figure 1.27 a hand is exerting a horizontal contact force F at right angles to the side of a box. The reaction is the normal reaction N that the box exerts on the hand. This too is a horizontal force, but it acts in the opposite direction to F. In fact, $N = -F$ in this case. Note that this relationship between the contact force and the normal reaction holds true even if the box accelerates.

In Figure 1.28 the situation is rather different. A brick of mass m is sliding down a smooth inclined plane, free from friction or any other kind of resistance. In fact, the only forces on the brick are its weight $W = mg$ and the normal reaction N exerted by the plane. As usual, the weight acts vertically downwards, through the brick's centre of mass, and the normal reaction acts at right angles to the plane. But what will the normal reaction be in this case, where its direction is so different from that of the weight? In order to answer this we shall use the two-dimensional coordinate system shown in Figure 1.28, in which the x-axis points down the slope and the y-axis is perpendicular to the slope. (We could pick any two directions for our x- and y-axes, but as you will see, this choice makes the maths simple since there is no acceleration perpendicular to the slope.) Resolving the vectors W and N along these axes we see that

$$W = (W_x, W_y) = (mg \sin \theta, -mg \cos \theta), \tag{1.16}$$

(Note the minus sign arising from the direction of the weight relative to the y-axis.)

and

$$N = (N_x, N_y) = (0, N). \tag{1.17}$$

It follows that the total force on the brick is

$$W + N = (mg \sin \theta, N - mg \cos \theta). \tag{1.18}$$

Now, the motion of the brick is confined to the x-direction. Consequently, the brick is not accelerating in the y-direction, and, according to Newton's first law, the resultant force component in the y-direction must therefore be zero. It follows from Equation 1.18 that

$$N - mg \cos \theta = 0,$$

i.e. $\quad N = mg \cos \theta.$

It then follows from Equation 1.17 that in this case,

$$N = (0, mg \cos \theta). \tag{1.19}$$

This result is a special case of the following general rule:

> When a body exerts a contact force on a rigid surface, the surface exerts a normal reaction on the body. The normal reaction is directed at right angles to the surface and is equal in magnitude to the component of the contact force that is perpendicular to the surface.

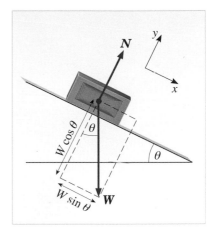

Figure 1.28 A brick of mass m slides down a frictionless plane inclined at an angle θ to the horizontal. Note that the geometry of the situation ensures that the angle between the weight W and the normal to the plane is also θ.

3.4 Friction, viscosity and air resistance

In Section 2.2 you were introduced to the phenomenon of friction and the associated frictional forces that oppose the relative sliding motion of two solid surfaces in contact. In many respects friction plays a vital part in our lives; it allows us to park cars on hills, to tie knots in ropes, and even to walk. However, it also causes problems since it impedes the motion of objects, causes moving machine parts to wear, and dissipates potentially useful mechanical energy. In this subsection we take a more detailed look at friction, and then consider two other causes of dissipative forces, *air resistance* and *viscosity*.

Imagine trying to slide a chest of drawers over a floor by pushing it horizontally (Figure 1.29). If you push very gently, the chest won't budge. Why? Because the horizontal force you apply is exactly balanced by a frictional force that arises between the bottom of the chest and the floor. This force is just sufficient to prevent motion, so its magnitude must be the same as that of the horizontal force you are applying, but it must act in the opposite direction to the applied force.

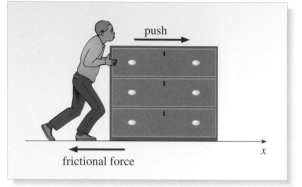

Figure 1.29 Forces on a static chest of drawers.

If you push harder, the frictional force increases, always remaining just large enough to cancel out your efforts. However, the frictional force has a maximum magnitude, F_{max}, beyond which it cannot increase. If you push harder than this, you can overcome friction and make the chest of drawers slide over the floor.

The maximum frictional force depends on several factors. First, it depends on the nature of the surfaces involved; it is easier to slide a wooden chest over marble than over rubber. It also depends on lubrication; an oily marble floor is more slippery than a dry one. One other important factor is the force with which the two surfaces are pressed together. If the drawers are empty, the chest rests quite lightly on the floor and the maximum frictional force is relatively small; if the drawers are loaded with heavy objects, its lower surface is pressed firmly against the floor and the maximum frictional force is large. This is not entirely a question of the chest's weight. If someone pushed down on the chest, its weight would remain the same, but it would be pressed more firmly against the floor and both the normal reaction force and the maximum frictional force would be greater.

Detailed experiments broadly confirm these everyday ideas about friction and lead to the following approximate value for the maximum magnitude of the frictional force acting on a stationary object:

$$F_{max} = \mu_{static}N, \tag{1.20}$$

where N is the magnitude of the normal reaction between the surface and the object, and μ_{static} is a proportionality constant called the **coefficient of static friction** (where μ is the Greek letter mu). The proportionality constant μ_{static} is dimensionless and so it can be represented by a simple number without any units. The value of μ_{static} depends on the nature of the two surfaces involved and their state of lubrication. Rather surprisingly, it does *not* depend on the apparent area of contact between the two surfaces.

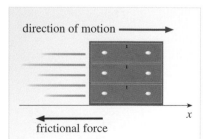

Figure 1.30 Forces on a sliding chest of drawers.

Once the chest has been set into motion, sliding across the floor, friction still acts on it. However, the direction of the frictional force is no longer determined by the applied force, but is opposite to the direction of motion. If the chest is sliding forwards along the x-axis, friction acts backwards, in the opposite direction (Figure 1.30).

The magnitude of the frictional force on a sliding object is given by a force law similar to Equation 1.20:

$$F = \mu_{slide}N, \tag{1.21}$$

where the proportionality constant μ_{slide} is called the **coefficient of sliding friction**. The value of μ_{slide} depends on the nature of the two surfaces and their state of lubrication, but is largely independent of other factors, including the apparent area of contact and the speed of motion. For two given surfaces, the value of μ_{slide} is usually slightly less than that of μ_{static}. For example, for rubber on wet concrete, $\mu_{slide} \approx 0.25$, while $\mu_{static} \approx 0.30$. This is why pushing heavy objects becomes a little easier once they are in motion.

Question 1.7 A car of mass m is travelling along a motorway. The brakes are suddenly applied, the wheels lock, and the tyres slide over the wet concrete surface of the road. Use the information given in the preceding paragraph and the value of g to estimate the magnitude of the car's acceleration. ■

Every cyclist who has ever pedalled against a wind, will be familiar with another phenomenon that impedes motion — **air resistance**. Any object that moves through a body of air is subject to a resistive force, called the **aerodynamic drag**, that arises from the flow of air around the moving object. This force opposes the relative motion of the object and the air, and has a magnitude that is determined by factors such as the size and shape of the object, and its speed relative to the air. Much effort goes into reducing air resistance (and therefore aerodynamic drag) by streamlining objects so that air is able to flow round them as easily as possible. This is responsible for the characteristic shape (Figure 1.31) of many modern vehicles and even some cyclists!

Figure 1.31 Some examples of streamlined design, intended to reduce air resistance.

As in the case of friction, it is difficult to give an exact description of air resistance, but we can describe the essence of the phenomenon. From your own experience, you probably know that air resistance has little effect at low speeds, but becomes more noticeable as the speed increases, making it difficult to cycle faster than about

$18\,\mathrm{m\,s^{-1}}$ (about 40 mph). The effect is complicated by its dependence on size and shape, which are difficult to quantify. Nonetheless, air resistance is such an important factor in modern life (in the design of cars and aircraft, for example) that detailed experiments have been carried out to discover its properties. The results quoted here are valid for smooth spheres moving through ordinary air, but broadly similar results apply to other shapes.

Suppose that a sphere of radius R is moving at speed v through still air. If the sphere is very small and slow moving (with the product of radius and speed, Rv, being less than $10^{-4}\,\mathrm{m^2\,s^{-1}}$), then the force due to air resistance has magnitude

$$F_{\mathrm{air}} = k_1 R v \qquad\qquad (1.22)$$

where $k_1 = 3.4 \times 10^{-4}\,\mathrm{N\,s\,m^{-2}}$. If the sphere has moderate size and speed (with the product Rv in the range from $10^{-4}\,\mathrm{m^2\,s^{-1}}$ to $1\,\mathrm{m^2\,s^{-1}}$), the force of air resistance has magnitude

$$F_{\mathrm{air}} = k_2 R^2 v^2 \qquad\qquad (1.23)$$

where $k_2 = 0.8\,\mathrm{N\,s^2\,m^{-4}}$.

Exceptionally large or fast spheres (those with $Rv > 1\,\mathrm{m^2\,s^{-1}}$) obey more complicated laws, but, in practice, most everyday objects (cricket balls, golf balls, etc.) experience the v^2 law of air resistance (Equation 1.23).

The fact that aerodynamic drag (the force due to air resistance) increases with speed has an important consequence for falling objects. In the absence of air resistance, an object falling under the influence of terrestrial gravity would undergo uniform acceleration, increasing its speed by about $9.8\,\mathrm{m\,s^{-1}}$ for every second of fall. However, in the presence of air resistance the acceleration will be somewhat reduced and, when the speed becomes high enough, the aerodynamic drag on the object will balance its weight causing it to cease accelerating altogether. Once this occurs the falling object will move with a constant speed known as its **terminal speed**. For a person falling feet first through the air, the terminal speed is about $70\,\mathrm{m\,s^{-1}}$ (160 mph). However, for a person wearing a fully opened parachute, the air resistance is increased and the terminal speed is more like 3 or $4\,\mathrm{m\,s^{-1}}$ (the precise value depends on the design of the parachute). Incidentally, this example shows that not all designers are concerned to minimize air resistance; parachutes are designed to maximize air resistance and thereby reduce the speed of an object relative to the air.

Question 1.8   A spherical raindrop has radius 0.50 mm and mass $5.00 \times 10^{-7}\,\mathrm{kg}$. Falling vertically downwards under the combined influence of gravity and air resistance it will eventually reach its terminal speed. What is that terminal speed? (Assume that the v^2 law of air resistance, Equation 1.23, applies, assume $k_2 = 0.80\,\mathrm{N\,s^2\,m^{-4}}$, and show that your final answer is consistent with this assumption.)   ∎

Air resistance is a particular example of a phenomenon that is common to all fluids (i.e. all liquids and gases). Whenever a solid body moves through a fluid, the body experiences a resistive force due to the fluid. The property of a fluid which gives rise to the resistive force is known as the **viscosity** of the fluid, and the resistive force is sometimes called the viscous force, or the viscous drag. Perhaps the best known force law concerning viscous forces is **Stokes' law**, which may be stated as follows:

When a sphere of radius R moves through a fluid at constant speed v, the magnitude of the viscous force that opposes the motion is

$$F = 6\pi\eta Rv, \tag{1.24}$$

where η is the **coefficient of viscosity** of the fluid.

η is the Greek letter eta.

By comparing Equation 1.24 with Equation 1.22 (which applies specifically to air), you should be able to convince yourself that the value of η for air is about $1.8 \times 10^{-5}\,\mathrm{N\,s\,m^{-2}}$. Typical values for more viscous fluids are $1.5 \times 10^{-3}\,\mathrm{N\,s\,m^{-2}}$ for water and $8.4 \times 10^{-2}\,\mathrm{N\,s\,m^{-2}}$ for olive oil.

3.5 Forces arising from tension and compression

If you fix one end of a rope and pull on the other end, the rope will become taut and pull back on you (see Figure 1.32). The phenomenon that enables the rope to exert force on you is called **tension**. It is not restricted to ropes, but may arise in strings, springs, solid rods or just about anything that might be stretched. Technically speaking, the tension in a body is a measure of its ability to resist being extended. Fundamentally, it is an electrical effect, since it arises from the electrical forces between atoms that oppose any increase in their separation. Tension is measured in the same units as force (i.e. newtons), though it is not itself a force, since it does not act in a particular direction. Nonetheless, tension certainly gives rise to forces, which are called **tension forces**. In the case of the stretched rope, for example, each part of the rope will be pulled towards the immediately adjacent parts of the rope, and whatever is attached to the ends of the rope will also be subject to a tension force, as Figure 1.32 indicates. If the tension in the rope is T, then the forces caused by that tension will have magnitude

$$F = T. \tag{1.25}$$

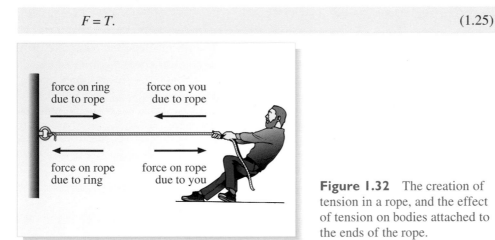

Figure 1.32 The creation of tension in a rope, and the effect of tension on bodies attached to the ends of the rope.

In practice, tension is often used to transmit forces from one place to another. For instance, if I wanted to tow your car I could attach your car to mine by means of a rope. I could then use my car to apply a force that would create tension in the rope. This would have the effect of causing the rope to apply a tension force to your car. In order for the force applied to the rope by my car to be equal to the force that the rope applies to your car, the rope would have to be straight, and it would also have to be light (i.e. effectively massless). These are idealizations, but they are not too far from reality, and are often assumed for the sake of simplicity.

tension
force
exerted by
rope

force
applied
to rope

Figure 1.33 Using the tension in a rope to redirect a force.

The Millennium Dome, London.

Of course, ropes don't have to be straight. If a rope passes over a pulley, as in Figure 1.33, then a downward force applied to one end of the rope will produce a tension in the rope that gives rise to an upward tension force at the other end. These two forces cannot be equal because they act in different directions, but they will have the same magnitude provided the rope is light and inextensible (i.e. incapable of being stretched), and provided the pulley is frictionless. Using the tension in a rope or string to redirect a force is a common practice with a host of applications. Structure's such as suspension bridges and the UK's Millennium Dome attest to the importance of tension forces in architecture and civil engineering.

Ideal ropes do not stretch when put under tension, but real ones do. When the ends of a real body are pulled in opposite directions, that body will generally respond by stretching to some extent. In such cases the tension forces tend to restore the body to its unstretched length and are referred to as **restoring forces**. For many bodies, provided they are not stretched too much, the magnitude of the restoring force is directly proportional to the increase in length. These bodies are said to obey Hooke's law which may be stated as follows:

> **Hooke's law**
>
> When a body is stretched, the magnitude of the restoring force is directly proportional to the increase in its length, provided the extension is not too great.

Note that Hooke's law is not a fundamental law that must be obeyed by all bodies. Rather it is an empirical statement that describes the behaviour of some bodies provided they are not stretched too far. (What constitutes too far will depend on the nature of the body.)

In order to obtain a force law representing Hooke's law, consider the situation shown in Figure 1.34. An extensible body such as a spring, with one end fixed and the other end free to move, is arranged along the x-axis of a coordinate system. When the body is unstretched, its free end is at $x = 0$, so any increase in its length, resulting from tension, will cause the free end to be displaced to some positive value of x. If the body obeys Hooke's law, then the restoring force F_x at the free end that arises in response to such an extension will be

$$F_x = -k_s x \qquad (1.26)$$

where k_s is a constant, called the **spring constant**, that characterizes the stiffness of the body, and the minus sign on the right-hand side indicates that the restoring force points in the opposite direction to that of increasing extension.

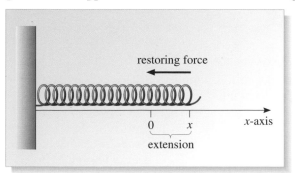

Figure 1.34 An extensible body, with one end fixed, is oriented along the x-axis. When the body is unstretched, its free end is at $x = 0$. So, when the body is stretched, the x-coordinate of its free end is a measure of the body's extension.

For a body such as a wire or an elastic string, the extension x must be positive for a restoring force to exist. However, if the body in Figure 1.34 really was a spring then a restoring force would exist even if the spring was compressed to something less than its unstretched length. In such a case, where x would be negative, the minus sign on the right of Equation 1.26 implies that the force F_x would point in the direction of increasing x, so it would still have the effect of restoring the spring to its unstretched length. Any spring that obeys Equation 1.26 for all values of x, irrespective of sign, is also said to obey Hooke's law, and is called an **ideal spring**. Many real springs behave in approximately this way, provided they are neither stretched, nor compressed, too much.

In the case of an ideal spring, a graph showing the magnitude of the restoring force against the extension, would be a straight line (see Figure 1.35). For this reason, the restoring force produced by such a spring is said to be a **linear restoring force**.

Question 1.9   A monkey of mass 5.0 kg is holding on to the end of a light ideal spring which is hanging from a hook in the ceiling. If the spring constant is $120\,\mathrm{N\,m^{-1}}$, determine the initial acceleration of the monkey when it is pulled down a distance of 0.10 m from its equilibrium position and then released from rest. ■

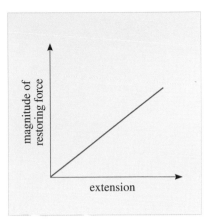

Figure 1.35 The straight-line graph that shows the magnitude of the restoring force in an ideal spring as a function of the spring's extension.

When dealing with fluids (gases or liquids), the concept of tension, or of a force due to tension, ceases to be of much use. On the whole, fluids are incapable of supporting tension. However, fluids can be compressed, and they exert forces as a result. These forces act on the walls of any vessel that contains a fluid, and on any object that is immersed (even partially) in the fluid; they are a consequence of the **pressure** in the fluid and can be very large. For example, at a depth of 1 km in the ocean, the magnitude of the force on one square metre of the outside of a submarine's hull would be about $10^7\,\mathrm{N}$, which is the equivalent of the combined weight of 16 050 people, each of mass 63.5 kg (10 stone).

Like the tension in a rope, the pressure in a fluid has no particular direction associated with it. Indeed, the pressure at any point in a fluid is usually described as acting equally in all directions. To better understand pressure, and its relation to force, imagine a small flat surface of area A at a point in a fluid where the pressure is P. This is illustrated in Figure 1.36. Whatever the orientation of the surface may be, the pressure on one side will create a force normal to the surface, which is directed towards the other side and which has magnitude

$$F = PA. \tag{1.27}$$

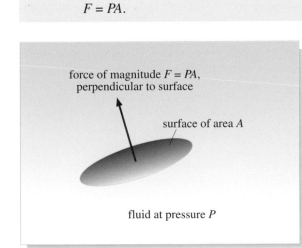

force of magnitude $F = PA$, perpendicular to surface

surface of area A

fluid at pressure P

Figure 1.36 A small flat surface of area A immersed in a fluid. Fluid at pressure P adjacent to one side of the surface will exert a force perpendicular to the surface that is directed towards the other side of the surface and which has magnitude $F = PA$.

Rearranging this relationship to give

$$P = \frac{F}{A} \tag{1.28}$$

shows that we can define the pressure at any point in a fluid as *the magnitude of the perpendicular force per unit area that the fluid would exert on a surface at that point*. Remember that this definition only concerns the force exerted by the fluid on one side of the surface. If there is fluid at pressure P on both sides of a surface, then there will be perpendicular forces of equal magnitude acting each way and their resultant will be zero.

Although mainly used in the context of fluids it should be noted that pressure can be a valuable concept in other settings, where force is applied over an area, so that we can still make sense of the formula $P = F/A$, and the phrase '*magnitude of the perpendicular force per unit area*'.

From Equation 1.28, it follows that the unit of pressure is that of force divided by area. In the SI system this unit will be the newton per square metre ($N\,m^{-2}$), which is usually called the **pascal** (Pa):

$$1 \text{ pascal} = 1 \text{ Pa} = 1\,N\,m^{-2}.$$

As you read this sentence you are immersed in the Earth's atmosphere, a body of fluid that extends over a hundred kilometres above your head. You may be surprised to discover how large a pressure you are subjected to by the atmosphere. At ground level, normal atmospheric pressure is about 1.013×10^5 Pa. This exerts a force on a square metre of surface that is equal in magnitude to the combined weight of 163 people, each of mass 63.5 kg. You might wonder why your chest, which probably has an area of about 0.1 m^2, is not crushed by the 16.3 people who are effectively standing on it. The reason is simple, the air inside your body is also roughly at atmospheric pressure. This means that the external forces tending to crush you are balanced by internal forces tending to make you explode! It's not so much a case of 'may the force be with you' as of 'may all the forces be with you!'

3.6 Forces in action

The forces that have been catalogued so far play an important role in our everyday lives. You are only able to sit in a chair, lie on a bed or stand on the floor because the downward force of gravity is counterbalanced by upward forces, ultimately electrical in nature, arising from contact with physical objects such as seats, mattresses or carpets. You are only able to walk or run because of the existence of friction (think of how hard it is to walk on ice or any other surface where the coefficient of friction is small). You are only able to breathe because of the close balance that exists between the forces arising from pressure inside and outside your body.

Example 1.4 is designed to show the magnitude and nature of the forces that might arise during a car accident.

Example 1.4

A car travelling at $20\,m\,s^{-1}$ collides with a wall and is brought to rest as a result. The sole occupant is the 50 kg driver, who is wearing a seat belt. (Seat belts are designed to allow their users to move forward by about 0.3 m during a crash.)

Assuming that the crumpling of the front of the car allows the whole passenger compartment to move forward by 1 m, during the crash, and assuming that the acceleration suffered by the driver is uniform, estimate the force that acts on the driver and the pressure that the seat belt exerts on the driver's chest. State any additional assumptions that you make in arriving at the required estimates. What is the fundamental nature of the force exerted by the seat belt?

Solution

The total distance moved by the driver during the crash will be about 1.3 m. Now, if the acceleration is uniform, and the impact is of duration t, the distance s travelled by an object with initial speed u and final speed v during the crash will be

$$s = \left(\frac{u+v}{2}\right)t$$

In this case, $s = 1.30\,\text{m}$, $u = 20\,\text{m s}^{-1}$ and $v = 0\,\text{m s}^{-1}$, so

$$t = \left(\frac{2s}{u+v}\right) = \frac{2 \times 1.30}{20+0}\,\text{s} = 0.13\,\text{s}.$$

We could have obtained the same result using $v^2 = u^2 + 2as$.

The magnitude of the uniform acceleration suffered by the driver is therefore

$$a = \left|\frac{v-u}{t}\right| = \frac{20\,\text{m s}^{-1}}{0.13\,\text{s}} = 154\,\text{m s}^{-2}.$$

It then follows from Newton's second law that the magnitude of the force on a 50 kg driver will be

$$F = ma = 50 \times 154\,\text{N} = 7.7\,\text{kN}.$$

A typical European seat belt consists of a chest strap and a lap strap, each about 50 mm wide and each making fairly flat contact with the driver's body over a length of about 0.3 m. Assuming that the force of about 8 kN is evenly spread across these *two* rectangular areas, and is perpendicular to them everywhere (clearly wrong, but not wildly wrong) we can estimate the pressure as

$$P = \frac{F}{A} = \frac{8\,\text{kN}}{2 \times 0.05 \times 0.3\,\text{m}^2} \approx 2.7 \times 10^5\,\text{Pa}.$$

The force exerted by the seat belt is essentially a normal reaction to the contact force exerted by the driver. It follows that the force exerted by the seat belt is fundamentally electrical. Like all forces and reactions arising from contact it is ultimately due to the electrical repulsion between atoms in the bodies concerned.

Comment   *The forces and pressures involved in this example are quite substantial, but that's hardly surprising — you don't have to be a physicist to know that a car crash can be lethal. The force is actually equal in magnitude to about fifteen times the driver's weight. So, it is clear that seat belts need to be carefully designed to spread the force as evenly as possible across the driver's torso. The belts themselves also need to be strong and fixed to strong parts of the car.*

Cars are a plentiful source of examples of forces in action. Some more are discussed in Box 1.1.

Box 1.1   Physics and car safety

Antilock braking

When you put your foot on the brake pedal of a car, brake pads press on the brake discs creating frictional forces that cause the wheels to slow their rotation and eventually stop. But stopping the wheels does not immediately stop the car. If you slam on the brakes you can very easily lock the wheels, stopping them from turning altogether, but the car will continue to move, skidding along the road, and leaving you with little or no directional control. This lack of control has been a contributing factor in lots of road accidents.

During braking, the motion of a car's wheels is a combination of rolling and slipping. 0% wheel slip corresponds to pure rolling, 100% slip occurs when the wheels are locked and there is no rolling at all. Studies show that on a flat road the horizontal force that the road's surface exerts on a car tyre depends on the percentage of wheel slip. Figure 1.37 shows how the magnitude of the force opposing motion varies with slip on a dry road. The maximum force occurs at 15% wheel slip.

Antilock braking systems (ABS) use microprocessors to compare signals from wheel sensors, looking for signs of wheel locking. When such signs are detected, the system reduces the pressure exerted by the appropriate brake pad, so that slippage can return to its optimal level. For the driver, this means that when the brake pedal is pressed down hard, the car is brought to a halt as quickly as possible without the risk of skidding.

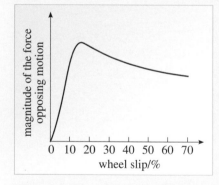

Figure 1.37   Magnitude of the force opposing motion plotted against wheel slip for braking on a dry horizontal surface.

Figure 1.38   The head of a properly belted driver can make contact with the steering wheel when a car stops suddenly.

Air bags

When a car stops suddenly the torso of the driver is restrained by a seat belt but the driver's head still has substantial freedom of movement (Figure 1.38). Even though steering columns are now designed to collapse on impact, the collision between a head and a steering wheel can still cause serious injury or even death. An air bag inflated between the head and the steering wheel can reduce the pressure on the head by 80%. For a front seat passenger, the dashboard can be just as lethal as the steering wheel. Due to the extra distance involved, passenger-side air bags are larger than those for drivers. Side and roof air bags are also increasingly available.

Cars equipped with air bags generally use at least two accelerometers — devices that measure the magnitude of the acceleration in a given direction — to provide the information that causes the air bags to be activated (Figure 1.39). A microprocessor monitoring the signals from the accelerometers ensures that only the massive acceleration of a head-on collision triggers the inflation of the air bag. (If this were not so, i.e. if minor collisions or bumps in the road could activate them, the air bags themselves could prove dangerous.) The accelerometers usually consist of small masses attached to springy strips of semiconductor (doped silicon). In a collision the mass continues to move forward as the rest of the accelerometer is stopped. This bends the semiconductor and alters its electrical resistance causing a signal to be registered on the microprocessor.

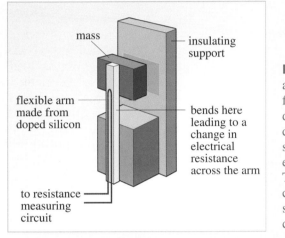

Figure 1.39 A piezoelectric accelerometer. Continued forward movement of the mass during a sudden slowing of the car bends the strip of semiconductor causing its electrical resistance to change. This is detected in a monitoring circuit and the information is sent to the microprocessor that controls the air bag.

Since a collision may last for as little as 100 ms, it is necessary that the air bags should be triggered and inflated in a shorter time than this. In practice, an air bag inflates in about 80 ms and deflates only 100 ms after firing. This should ensure that the driver regains a clear view of the road very quickly, in less time than it takes to blink.

Question 1.10 A person is sitting at rest on an office chair equipped with castors (small wheels). The person places both feet on the floor and pushes, propelling the chair (and themselves) backwards (Figure 1.40). Describe the forces involved in this process, starting with those that act while the person is sitting at rest. ■

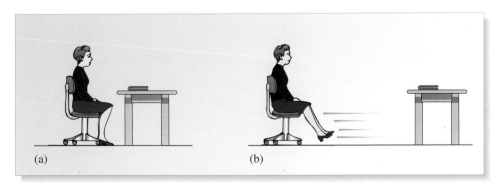

Figure 1.40 A person sitting in a chair equipped with castors. In (a) the person and chair are stationary. In (b) the person is propelled backwards by pushing on the floor.

4   Forces and motion

You have now been introduced to several different kinds of force and you have also learned about the way in which any force can cause a change of motion. But what sort of motion is predicted to arise as a consequence of a specific force, and which forces are responsible for some of the common forms of motion such as projectile motion and circular motion? These are the issues that are addressed in this section. We begin, however, by reviewing the terminology of functions and derivatives that was used to describe motion in *Describing motion*.

4.1   Rates of change: gradients and derivatives — a reminder

If a variable quantity x (such as the position coordinate of a particle) depends on an independent variable t (such as the time) then we say that x is a *function* of t and we indicate the dependence by writing x as $x(t)$. The fact that x is a function of t implies that there is an equation that relates the value of x to that of t; more specifically it implies that there will be a *unique* value of x corresponding to any given value of t, and consequently that we may plot a graph of x against t for a range of t values.

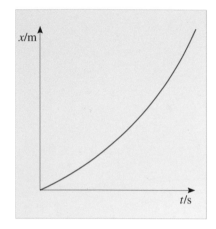

Figure 1.41   The graph of x against t for a function of the form $x(t) = At^2 + Bt + C$.

For example, suppose that

$$x(t) = At^2 + Bt + C, \tag{1.29}$$

where A, B and C are constants. The graph of x against t will then look something like Figure 1.41. (In drawing this particular graph we have actually used the values $A = 1 \, \mathrm{m \, s^{-2}}$, $B = 10 \, \mathrm{m \, s^{-1}}$ and $C = 0$.)

In such a case, at any particular value of t, the phrase 'the rate of change of x with respect to t' means the same as 'the gradient of the x against t graph' at the very same value of t.

In order to determine this gradient (or rate of change) you could draw the tangent to the graph at the relevant value of t (6 s, say) and then measure its gradient by determining the change measured along the vertical axis, Δx, that corresponded to a given change along the horizontal axis, Δt, as indicated in Figure 1.42. The gradient of the curve at the selected value of t is equal to the gradient of its tangent at that value, so

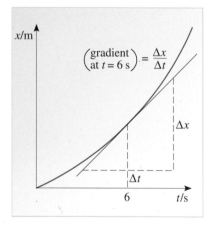

Figure 1.42   The rate of change of $x(t)$ at a chosen value of t is equal to the gradient $\Delta x / \Delta t$ of the tangent to the x against t graph at that value of t.

$$\text{gradient} = \frac{\text{change along vertical axis}}{\text{change along horizontal axis}} = \frac{\Delta x}{\Delta t}. \tag{1.30}$$

In the case of Figure 1.42 where $x(t) = At^2 + Bt + C$, if precise measurement of the tangent at $t = 6 \, \mathrm{s}$ had shown that when $\Delta t = 5 \, \mathrm{s}$ the corresponding change in x was $\Delta x = 110 \, \mathrm{m}$, then the gradient of the graph (or the rate of change of the function) at $t = 6 \, \mathrm{s}$, would have been $\Delta x / \Delta t = (110/5) \, \mathrm{m \, s^{-1}} = 22 \, \mathrm{m \, s^{-1}}$.

Why should we be interested in gradients or rates of change? Well, if x represents the position coordinate of a particle, and t represents the time, then the gradient of the x against t graph at $t = 6 \, \mathrm{s}$ represents the particle's instantaneous velocity in the x-direction at $t = 6 \, \mathrm{s}$, and this is clearly a quantity that may be of physical interest.

Given that we *are* interested in determining rates of change, we should also be interested in determining them as simply and as accurately as possible. That's where derivatives come in. Given a function such as $x(t)$, it is possible to define another function of t, called the *derivative* of $x(t)$, which is represented by the symbol $\dfrac{dx}{dt}$, or

$\dfrac{dx(t)}{dt}$, and which has the property that its value at any particular value of t is exactly equal to the gradient of the x against t graph at that value of t. Moreover, this derivative function can be determined from the algebraic expression for $x(t)$ (such as Equation 1.29) without having to go through the time consuming business of drawing graphs and measuring gradients.

There is a general mathematical procedure for determining the derivative of a function, but in practice derivatives are usually determined by applying certain standard rules, some of which were introduced in Table 1.6 of *Describing motion*. Table 1.1 lists some of those rules (together with examples) for your convenience.

Table 1.1 Some simple derivatives. The functions f, g and h depend on the variable t. The quantities A, n, ϕ and ω are constants that may be positive or negative (ϕ is the Greek letter phi). Note that n is not necessarily a whole number.

Function $f(t)$	Derivative $\dfrac{df(t)}{dt}$	Example
$f(t) = A$ (a constant function)	$\dfrac{df}{dt} = 0$	$f(t) = 6,\ \dfrac{df}{dt} = 0$
$f(t) = t^n$	$\dfrac{df}{dt} = nt^{n-1}$	$f(t) = t^3,\ \dfrac{df}{dt} = 3t^2$
$f(t) = At^n$	$\dfrac{df}{dt} = nAt^{n-1}$	$f(t) = 2t^3,\ \dfrac{df}{dt} = 6t^2$
$f(t) = A\sin(\omega t + \phi)$	$\dfrac{df}{dt} = A\omega\cos(\omega t + \phi)$	$f(t) = 2\sin(5t),\ \dfrac{df}{dt} = 10\cos(5t)$
$f(t) = A\cos(\omega t + \phi)$	$\dfrac{df}{dt} = -A\omega\sin(\omega t + \phi)$	$f(t) = \cos(5t + \pi),\ \dfrac{df}{dt} = -5\sin(5t + \pi)$
$f(t) = g(t) + h(t)$ (a sum of functions)	$\dfrac{df}{dt} = \dfrac{dg}{dt} + \dfrac{dh}{dt}$ (a sum of derivatives)	$f(t) = 2t^3 + 4t,\ \dfrac{df}{dt} = 6t^2 + 4$
$f(t) = Ag(t)$ (a constant $\times$ a function)	$\dfrac{df}{dt} = A\dfrac{dg}{dt}$ (a constant $\times$ a derivative)	$f(t) = 0.5 \times (2t^3 + 4t),$ $\dfrac{df}{dt} = 0.5 \times (6t^2 + 4) = 3t^2 + 2$

Applying these rules to a function of the kind given in Equation 1.29 shows that:

if $\qquad x(t) = At^2 + Bt + C,$ $\qquad\qquad\qquad\qquad$ (Eqn 1.29)

then $\qquad \dfrac{dx(t)}{dt} = 2At + B.$ $\qquad\qquad\qquad\qquad$ (1.31)

For the particular values of A and B that were used earlier ($A = 1\,\mathrm{m\,s^{-2}}$ and $B = 10\,\mathrm{m\,s^{-1}}$), Equation 1.31 implies that, at any value of t

$$\dfrac{dx(t)}{dt} = (2\,\mathrm{m\,s^{-2}})t + 10\,\mathrm{m\,s^{-1}} \qquad\qquad\qquad (1.32)$$

and hence, at the particular value $t = 6\,\text{s}$,

$$\frac{\mathrm{d}x(t)}{\mathrm{d}t} = 12\,\text{m}\,\text{s}^{-1} + 10\,\text{m}\,\text{s}^{-1} = 22\,\text{m}\,\text{s}^{-1}. \tag{1.33}$$

This is clearly a much easier way of determining the rate of change of $x(t)$ at $t = 6\,\text{s}$ than the graphical method. It is also much easier to generalize.

The mathematical process by which derivatives are determined is called *differentiation*, and the branch of mathematics concerned with differentiation is called *differential calculus*. Differential calculus is one of the physicist's main mathematical tools.

Putting all this into the context of motion, if $x(t)$ represents the position of a particle along the x-axis at time t, then the x-component of the velocity of the particle, v_x, is given by the rate of change of x with t. That is

$$v_x(t) = \frac{\mathrm{d}x}{\mathrm{d}t}, \tag{1.34}$$

and, since the rate of change of velocity with respect to time is acceleration, it follows that in calculus notation the x-component of the particle's acceleration is given by

$$a_x(t) = \frac{\mathrm{d}v_x}{\mathrm{d}t} = \frac{\mathrm{d}}{\mathrm{d}t}\left(\frac{\mathrm{d}x}{\mathrm{d}t}\right) = \frac{\mathrm{d}^2 x}{\mathrm{d}t^2}. \tag{1.35}$$

Here, the last expression on the right indicates that a_x is the *second derivative* of x with respect to t. Similar results apply to the y- and z-components of velocity and acceleration, with the consequence that if $\boldsymbol{r}$ is the position vector of the particle, so $\boldsymbol{r}(t) = (x(t), y(t), z(t))$ then the particle's velocity may be written

$$\boldsymbol{v}(t) = \frac{\mathrm{d}\boldsymbol{r}}{\mathrm{d}t} = \left(\frac{\mathrm{d}x}{\mathrm{d}t}, \frac{\mathrm{d}y}{\mathrm{d}t}, \frac{\mathrm{d}z}{\mathrm{d}t}\right) \tag{1.36}$$

and its acceleration may be written

$$\boldsymbol{a}(t) = \frac{\mathrm{d}\boldsymbol{v}}{\mathrm{d}t} = \frac{\mathrm{d}^2\boldsymbol{r}}{\mathrm{d}t^2} = \left(\frac{\mathrm{d}^2 x}{\mathrm{d}t^2}, \frac{\mathrm{d}^2 y}{\mathrm{d}t^2}, \frac{\mathrm{d}^2 z}{\mathrm{d}t^2}\right). \tag{1.37}$$

In practice, when writing instantaneous positions, velocities and accelerations, the (t) is often dropped to avoid clutter. Nonetheless, the position, $\boldsymbol{r}$, velocity, $\boldsymbol{v}$, and acceleration, $\boldsymbol{a}$, of a moving particle are all functions of time.

4.2 Some familiar forms of motion

In this subsection we review some of the well known forms of motion that were described in *Describing motion* and investigate the forces that might cause such motions to occur. In each case the moving object will be a particle of mass m, and its position, velocity and acceleration at time t will be denoted by the vectors $\boldsymbol{r}$, $\boldsymbol{v}$ and $\boldsymbol{a}$, respectively.

You will not find many surprises in this subsection. The descriptions of motion should already be familiar to you, and so should the forces that cause them. However, with the benefit of Newton's laws we can now link these ideas. It is important to take that small step, just to check that all of the ideas that we have discussed really do fit together and make sense. Also, some new ideas will emerge and the results that are listed here will be a valuable resource for future reference as you work through the rest of this book.

Uniform motion and the absence of force

A particle is said to move *uniformly* when it does not accelerate. For such a particle the position is given by

$$r(t) = r_0 + ut, \tag{1.38}$$

where the constant vectors r_0 and u, respectively, represent the particle's initial position and initial velocity. The velocity of the particle is constant

$$v(t) = \frac{dr}{dt} = u \tag{1.39}$$

and the acceleration, as already indicated, is zero

$$a(t) = \frac{d^2 r}{dt^2} = 0 . \tag{1.40}$$

It follows from Newton's second law that $ma = F = 0$. So there is no force in this case, as Newton's first law asserts.

Uniformly accelerated motion and constant force in two dimensions

A one-dimensional example of uniformly accelerated motion is that of a stone rising or falling vertically under the sole influence of a constant force mg due to terrestrial gravity. In two dimensions, uniform downward acceleration of magnitude g, combined with uniform horizontal motion accounts for projectile motion, which may be described by the following equations:

horizontal motion ($F_x = 0$):

$$x = x_0 + u_x t$$

$$v_x = u_x$$

vertical motion ($F_y = -mg$):

$$y = y_0 + u_y t - \frac{1}{2} g t^2$$

$$v_y = u_y - gt$$

where y is measured in the upward direction (so $a_y = -g$), the initial velocity of the particle is (u_x, u_y), and the position of the particle at time $t = 0$ is (x_0, y_0). Such a particle follows a parabolic trajectory in the xy-plane, as shown in Figure 1.43.

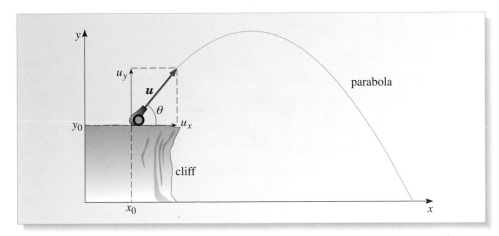

Figure 1.43 The parabolic motion of a projectile.

Uniformly accelerated motion and constant force in three dimensions

Generalizing to three dimensions, and writing the initial position as r_0, the initial velocity as u and the constant acceleration as a, we see that

$$r(t) = r_0 + ut + \tfrac{1}{2}at^2 \tag{1.41}$$

so,

$$v(t) = \frac{dr}{dt} = u + at \tag{1.42}$$

and

$$a(t) = \frac{d^2r}{dt^2} = a = \text{a constant vector}. \tag{1.43}$$

These equations imply that the particle will again follow a parabolic trajectory, but this time the parabolic path passes through the point r_0, and is in the plane that contains the vectors u and a.

Whether in one, two or three dimensions, if the acceleration of a particle is constant then the force causing that acceleration must also be constant. For a falling stone or a drag-free projectile, the constant acceleration is $a = g$, and the constant force causing that acceleration is the particle's weight, $W = mg$, which arises from terrestrial gravitation. However, weight is not the only source of constant force. A common problem in physics laboratories is to make an object accelerate uniformly in a horizontal direction. A simple solution is shown in Figure 1.44. The tension in a string provides the required force while the (constant) weight of an object is used to produce the necessary tension.

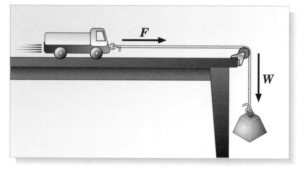

Figure 1.44 Using the tension in a string to exert a constant horizontal force. For a light inelastic string and a frictionless pulley, $|F| = |W|$.

Circular motion and centripetal force

Newton's first law tells us that a particle will travel with constant speed along a straight line unless it is acted on by an unbalanced force. But a particle travelling in a circle is continually deviating from a straight line, changing its direction of motion from one moment to the next. This means that such a particle must be acted on by an unbalanced force which is continually causing it to accelerate. This must be true even if the particle travels with constant speed v. More specifically, a particle moving in a circle of radius r, with uniform angular speed ω, will have the constant speed

$$v = r\omega \tag{1.44}$$

and must be subject to a *centripetal acceleration* that is always directed towards the centre of the circle, and which has magnitude

$$a = r\omega^2 = \frac{v^2}{r}. \tag{1.45}$$

It follows from Newton's second law that a uniformly circling particle, moving in this way, must be subject to an unbalanced force of magnitude

$$F = mr\omega^2 = \frac{mv^2}{r} \qquad (1.46)$$

that is always directed towards the centre of the circle. This force is called the **centripetal force**. Its magnitude is constant, but the force itself changes continually, as its direction alters from moment to moment. Since the force is always at right angles to the instantaneous velocity of the particle, it has the effect of changing the particle's velocity without altering its speed.

If we take the centre of the circle as the origin of coordinates, the instantaneous position vector $r(t)$ of a uniformly circling particle will always point directly away from the centre of the circle, and the centripetal force that keeps the particle moving in that circle will then be

$$\boldsymbol{F} = -mr\omega^2\hat{\boldsymbol{r}} = \frac{-mv^2}{r}\,\hat{\boldsymbol{r}} \qquad (1.47)$$

where $\hat{\boldsymbol{r}} = \boldsymbol{r}/r$ is a unit vector that points *away* from the centre towards the particle, and the minus sign indicates that the force on the particle acts in the opposite direction to $\hat{\boldsymbol{r}}$, i.e. *towards* the centre.

What might be the cause of such a force? Well, that depends on the situation. One of the simplest ways to make an object move in a circle is to tie it to a piece of string and whirl it around your head (Figure 1.45). If you do this, the horizontal force on the object at any moment must arise from the tension in the string (see Section 3.5). Incidentally, if the object moves in a horizontal circle, the string must also exert an upwards vertical force on the object, to balance its weight, as indicated in part (b) of the figure. This means that the string itself cannot be horizontal.

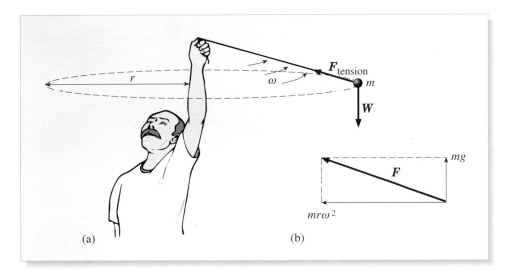

Figure 1.45 (a) A particle of mass m being whirled on the end of a string. The tension in the string exerts a force on the particle. (b) The horizontal component of this force provides the necessary centripetal force, while the vertical component balances the particle's weight.

In the case of a car turning a bend of radius r, the inward centripetal force must be supplied by the frictional force that the road exerts on the car's tyres (Figure 1.46). For an electron moving in a circle around the nucleus of an atom, as it was supposed to do in one of the earliest theoretical models of an atom, the required centripetal force is due to the electrical attraction between the negatively charged electron and the positively charged nucleus. For a geostationary satellite circling the Earth, or for the Moon moving in a circular orbit around the Earth, the centripetal force is provided by universal gravitation.

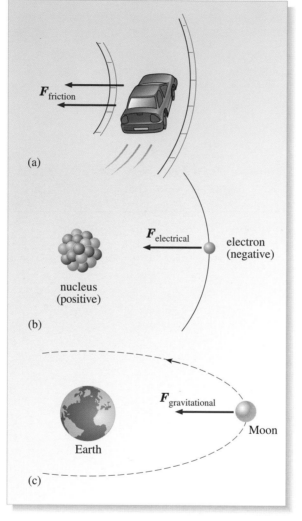

Figure 1.46 Sources of centripetal force in various situations. (a) Friction and a turning car. (b) Electrical attraction and an orbiting electron. (c) Gravitational attraction and the Moon.

The point of cataloguing all these different cases is to emphasize that centripetal force is not a particular 'kind' of force, like a contact force or a gravitational force, but rather a required 'quantity' of force, that may originate in a variety of ways.

Question 1.11 Write down a concise definition of the term 'centripetal force', and describe the major differences between centripetal force and centrifugal force. ∎

Simple harmonic motion and linear restoring force

Simple harmonic motion is the particular (and very common) kind of oscillatory motion that arises when a particle's acceleration is directed towards a fixed point (called the *equilibrium position*) and is proportional to the displacement from that point. This implies that if the particle moves along the x-axis, and its equilibrium position is $x = 0$, then

$$a_x = -\omega^2 x, \tag{1.48}$$

where ω is a positive constant, called the *angular frequency* of the oscillation, that is related to the period of oscillation by the equation

$$T = \frac{2\pi}{\omega}. \tag{1.49}$$

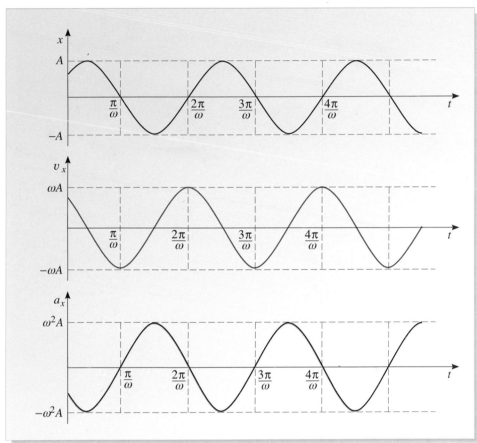

Figure 1.47 The displacement, velocity and acceleration of a simple harmonic oscillator, with equilibrium position $x = 0$.

For a particle in simple harmonic motion, the displacement, velocity and acceleration at time t may be written:

$$x(t) = A\sin(\omega t + \phi) \tag{1.50}$$

$$v_x(t) = \frac{\mathrm{d}x}{\mathrm{d}t} = A\omega\cos(\omega t + \phi) \tag{1.51}$$

$$a_x(t) = \frac{\mathrm{d}^2 x}{\mathrm{d}t^2} = -A\omega^2\sin(\omega t + \phi) = -\omega^2 x \tag{1.52}$$

where A and ϕ are constants called, respectively, the amplitude and the initial phase of the motion (Figure 1.47). The amplitude is equal to the oscillator's maximum displacement from its equilibrium position, and the initial phase determines the oscillator's position at $t = 0$, via the relation $x(0) = A\sin(\phi)$.

Using Newton's second law, it follows from Equation 1.48, that any force that causes simple harmonic motion ($F_x = -m\omega^2 x$) must be directed towards the equilibrium position and must be proportional to the displacement from that position. Hence, in the terminology of Section 3, the force must be a *linear restoring force*, and may be written, in general terms, as

$$F_x = -kx,\tag{1.53}$$

where the positive constant k is called the *force constant*, and is given by

$$k = m\omega^2.\tag{1.54}$$

It follows that the angular frequency of the motion is related to the force constant and the mass by $\omega = \sqrt{k/m}$. Substituting this into Equation 1.49 shows that the period of the oscillator is

$$T = \frac{2\pi}{\omega} = 2\pi\sqrt{\frac{m}{k}}.\tag{1.55}$$

Now, one obvious source of a linear restoring force is an ideal spring (Figure 1.48a). We saw in Section 3.5 that the force law describing an ideal spring (Hooke's law) is $F_x = -k_s x$, where k_s is the *spring constant*. So, a particle of mass m attached to such a spring will execute s.h.m. with $\omega = \sqrt{k_s/m}$, and hence with period

$$T = \frac{2\pi}{\omega} = 2\pi\sqrt{\frac{m}{k_s}}.\tag{1.56}$$

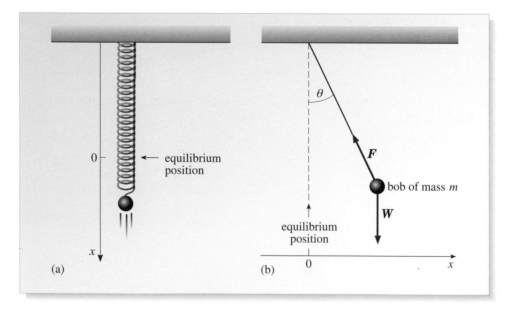

Figure 1.48 (a) A mass suspended from an ideal spring — a simple harmonic oscillator. (b) A simple pendulum — a simple harmonic oscillator for small θ.

However, there are many other possible sources for the required restoring force. For instance, a simple pendulum consisting of a bob of mass m at the end of a string of length l, swinging in a vertical plane, is subject to a horizontal force (see Figure 1.48b)

$$F_x = -F_T \sin\theta\tag{1.57}$$

where F_T is the tension in the string, and θ is the angle between the string and the vertical. For such a bob, its position along the x-axis is $x = l\sin\theta$, so

$$F_x = \frac{-F_T}{l}x.\tag{1.58}$$

We have used the symbol F_T here to represent tension rather than the more usual T (as was introduced in Section 3.5) so that no confusion is caused with the period T of oscillation in the discussion below.

Now, if the angle θ is sufficiently small the motion of the bob will be almost entirely horizontal. Under these circumstances we can ignore the vertical motion of the bob and say that the tension in the string must exert a vertical force on the bob that balances the weight of the bob

$$F_T \cos \theta = mg. \tag{1.59}$$

Using again the argument that θ is small, we can say that $\cos \theta$ is approximately 1 and that the restoring force on the bob is approximately

$$F_x = \frac{-mg}{l} x. \tag{1.60}$$

Comparing this with Equation 1.53, we see that F_x is a linear restoring force with force constant $k = mg/l$. It follows that provided θ is never too large, the pendulum will exhibit simple harmonic motion with angular frequency $\omega = \sqrt{g/l}$, and the pendulum's period will be

$$T = \frac{2\pi}{\omega} = 2\pi \sqrt{\frac{l}{g}}. \tag{1.61}$$

Note that the period is proportional to $\sqrt{l}$.

Orbital motion and inverse square forces

There's not much mystery about the kind of force that causes orbital motion. Orbits are almost synonymous with motion under (universal) gravitational forces. However, this is a good place to highlight a few of the consequences of universal gravitation, and to emphasize that there *are* other forces that can cause orbital motion.

According to Kepler's laws, for planets in the Solar System:

1 Each planet moves in an ellipse with the Sun at one focus. (See Figure 1.49a; Kepler's first law.)

2 A line from the Sun to any particular planet sweeps out equal areas in equal intervals of time. (See Figure 1.49b; Kepler's second law.)

3 The square of a planet's orbital period, T, is proportional to the cube of its semimajor axis, a. (See Figure 1.49c; Kepler's third law.) So

$$T^2 = Ka^3 \tag{1.62}$$

where K is constant.

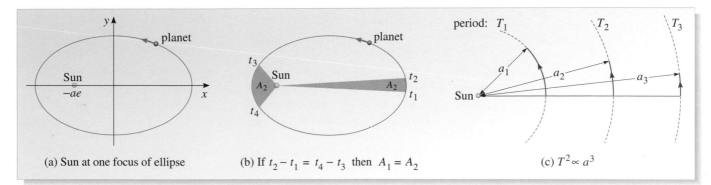

(a) Sun at one focus of ellipse   (b) If $t_2 - t_1 = t_4 - t_3$ then $A_1 = A_2$ (c) $T^2 \propto a^3$

Figure 1.49 Kepler's three laws of planetary motion.

By recognizing that the planets are held in their orbits by forces arising from universal gravitation, the somewhat mysterious constant K that appears in Equation 1.62 can be related to more familiar physical quantities such as Newton's gravitational constant and the mass of the Sun. The general proof of this for elliptical orbits is rather complicated, but it is easy to establish for the special case of circular orbits. This we shall now do.

A planet of mass m, moving in a circular orbit of radius r with uniform angular speed ω, must be subject to a centripetal force of magnitude $F = mr\omega^2$ (Equation 1.46). If this force is supplied by the gravitational attraction of the Sun (Equation 1.11) then,

$$mr\omega^2 = \frac{Gm_{Sun}m}{r^2}. \tag{1.63}$$

For a planet with angular speed ω, the orbital period is $T = 2\pi/\omega$, so

$$\omega = \frac{2\pi}{T}. \tag{1.64}$$

Replacing ω in Equation 1.63 by $2\pi/T$, and rearranging gives

$$T^2 = \frac{4\pi^2}{Gm_{Sun}}r^3. \tag{1.65}$$

Comparing Equation 1.65 with Kepler's third law (Equation 1.62), and remembering that in the case of a circular orbit the radius r is equal to the semimajor axis a, it is clear that the constant K in Kepler's third law is given by

$$K = \frac{4\pi^2}{Gm_{Sun}}. \tag{1.66}$$

This relationship is significant for at least two reasons:

1 Since K can be determined directly from observations of the planets, Equation 1.66 provides a way of determining the mass of the Sun from planetary observations. (It should be admitted that in deriving Equation 1.66 we have assumed that the Sun is so much more massive than the planet that the Sun remains at rest while the planet moves around it. In reality both the Sun and the planet move around their common centre of mass, but even when this is taken into account it has little effect on the result.)

2 Equation 1.66 also shows how we can generalize Kepler's laws so that they apply to any system in which particles of relatively small mass orbit a single body of very much greater mass. All we have to do is insert the appropriate value for the large mass in place of the value of m_{Sun} in Equation 1.66. (In this case, as in the Solar System, gravitational interactions between the orbiting bodies will lead to small deviations from Kepler's laws, but the laws will still provide a valuable first approximation to the true behaviour of the system.)

In deriving Equation 1.66 a crucial feature of the gravitational force on the planet was that its magnitude was inversely proportional to the square of the orbital radius ($F \propto 1/r^2$). In other words, the success of Kepler's third law depends on the fact that universal gravitation is described by an inverse square law. This suggests that orbital motion can also be expected to arise in other settings where bodies are attracted by a force that is described by an inverse square law. The most important case of this kind, apart from gravitation, is that of the electrical attraction between oppositely charged particles. We will consider such systems of orbiting charges later in the course, when we discuss the quantum physics of atoms.

Question 1.12 Determine the mass of the Sun, given that the orbital radius of the planet Mercury is 5.79×10^{10} m and that its planetary year lasts 88.0 Earth days. ∎

4.3 Derivatives and Newton's laws: equations of motion

The last subsection showed how certain well known kinds of motion can be accounted for in terms of specific forces, but nothing has yet been said about the general technique for predicting the motion that will arise when a particle, or a system of particles, is subjected to a specified force. Newtonian mechanics is certainly capable of providing such predictions. Its equations are *deterministic* in that, provided the positions and velocities of all the particles are known at one instant, and provided the relevant force laws are also known, then it is always possible (in principle) to predict how the particles will be moving at any time. But how is such a prediction to be made?

The key to making general predictions about motion lies in combining Newton's second law of motion with the idea of a derivative. When we first introduced Newton's second law we said that, for a body of fixed mass, it could be written as

$$\boldsymbol{F} = m\boldsymbol{a} \tag{Eqn 1.1}$$

Writing $\boldsymbol{a}$ as a second derivative of $\boldsymbol{r}$ and rearranging things slightly, we can now write this as

$$\frac{\mathrm{d}^2\boldsymbol{r}}{\mathrm{d}t^2} = \frac{\boldsymbol{F}}{m} \tag{1.67}$$

which, of course, is equivalent to the three scalar equations

$$\frac{\mathrm{d}^2 x}{\mathrm{d}t^2} = \frac{F_x}{m} \tag{1.68a}$$

$$\frac{\mathrm{d}^2 y}{\mathrm{d}t^2} = \frac{F_y}{m} \tag{1.68b}$$

$$\frac{\mathrm{d}^2 z}{\mathrm{d}t^2} = \frac{F_z}{m}. \tag{1.68c}$$

This is not of much help in itself, but it becomes significant when we try to predict the way in which a body will move in response to a force described by a known force law.

As an example, consider again the particle on an ideal spring that was shown in Figure 1.48a. We saw in the last subsection that such a particle will oscillate, exhibiting s.h.m. If we take the vertical direction to be the x-direction and let the equilibrium position of the particle be $x = 0$, then any vertical displacement of the particle from its equilibrium position will result in an unbalanced restoring force towards the equilibrium position that is described by Hooke's law:

$$F_x = -k_s x \tag{Eqn 1.26}$$

It then follows from Newton's second law (Equation 1.68a), that for this particle

$$\frac{\mathrm{d}^2 x}{\mathrm{d}t^2} = \frac{-k_s}{m} x. \tag{1.69}$$

This equation, the result of combining Newton's second law with the relevant force law, is said to be the **equation of motion** of the particle. Technically speaking, it is an example of a **differential equation**. Like all differential equations, it involves at least one derivative, in this case a second derivative ($\mathrm{d}^2x/\mathrm{d}t^2$), which makes it an example of a **second-order** differential equation.

Differential equations were introduced in Chapter 3 of *Describing motion*, using an example very similar to Equation 1.69. That earlier discussion involved the simple harmonic motion equation, a second-order differential equation of the form

$$\frac{\mathrm{d}^2 x}{\mathrm{d}t^2} = -\omega^2 x \qquad (1.70)$$

where ω is a positive constant representing the angular frequency of the oscillator. Equation 1.69 is really just a particular example of this kind of equation, with

$$\omega = \sqrt{k_\mathrm{s}/m}\,.$$

Now, the real point of all this is that a great deal is known about differential equations and there are standard techniques for solving most of those that are commonly encountered in physics. The solution of a differential equation is generally a function, rather than a number or value. In the case of the s.h.m. equation (Equation 1.70), for instance, the solution is

$$x(t) = A \sin(\omega t + \phi) \qquad \text{(Eqn 1.50)}$$

where A, ω and ϕ are constants. This, as we saw in Section 4.2, describes simple harmonic motion with amplitude A, angular frequency ω and initial phase ϕ. If the values of A, ω and ϕ are known, then Equation 1.50 can be used to make detailed predictions about the position x of the oscillating particle at any time t. So, once we have found the equation of motion of a particle, the problem of predicting its motion is reduced to the mathematical exercise of solving that (differential) equation and evaluating the constants that appear in the solution.

It is, of course, important for physicists to know how to solve the equations of motion they derive, but it is not a matter with which we shall be much concerned in this course. On the whole we shall simply quote solutions whenever we need them, just as we did for Equation 1.70. Nonetheless, you may wish to confirm for yourself that the quoted solutions are correct. If so, you can use the method informally referred to as 'plug it in and check it out'. In the case of Equation 1.50 this involves differentiating the given expression for $x(t)$ to obtain

$$v_x(t) = \frac{\mathrm{d}x}{\mathrm{d}t} = A\omega \cos(\omega t + \phi) \qquad \text{(Eqn 1.51)}$$

and then differentiating again to get

$$a_x(t) = \frac{\mathrm{d}^2 x}{\mathrm{d}t^2} = -A\omega^2 \sin(\omega t + \phi). \qquad \text{(Eqn 1.52)}$$

This expression for $\mathrm{d}^2x/\mathrm{d}t^2$, obtained from Equation 1.50, together with the original expression for x given in Equation 1.50, can then be substituted into Equation 1.70 to confirm that Equation 1.50 is indeed a solution, as claimed. The same technique may be used to check the validity of any given solution to a differential equation.

The function $x(t) = A \sin(\omega t + \phi)$ is certainly a solution to Equation 1.70, but it is actually more than that. It is in fact the **general solution** to Equation 1.70. What this means is that *any* solution to Equation 1.70 may be written in the form of Equation 1.50 simply by choosing the values of the constants appropriately. Clearly then, the constants that appear in a general solution (A, ω and ϕ, in the case of Equation 1.50) are of great significance, and deserve further consideration.

When analysing the general solution of a differential equation it is important to realize that the constants it contains fall into two groups:

1 The first group contains the constants that appear in the differential equation itself. In the case of Equation 1.50 the only constant of this kind is ω, since that was the only constant in Equation 1.70. Constants of this kind represent fixed properties of the system being studied; in the case of a particle on a spring, for instance, ω would be determined by the spring constant k_s and the particle's mass m: $\omega = \sqrt{k_s/m}$.

2 The second group contains the constants that appear in the solution but not in the differential equation. These are called **arbitrary constants**. In the case of Equation 1.50 they are represented by A and ϕ. One of the defining characteristics of a second-order differential equation is that its general solution always contains just *two* arbitrary constants.

Arbitrary constants are referred to by that name because they are determined by the way we choose to start the particle moving, i.e. by the **initial conditions** we arbitrarily impose on the motion. In the case of the harmonic oscillator for example, if we choose to start the oscillator from some point $x(0)$ at $t = 0$, with some initial velocity $v_x(0)$, then from Equations 1.50 and 1.51 we see that

$$x(0) = A \sin \phi \tag{1.71}$$

and $$v_x(0) = A\omega \cos \phi. \tag{1.72}$$

Which (remembering that ω is already determined by fixed properties of the system) shows that A and ϕ are determined by the initial position and the initial velocity we have chosen to give the oscillator. Example 1.5 shows how these ideas can be used.

Example 1.5

A particle of mass 2 kg is suspended from a fixed spring of spring constant $8\,\text{N}\,\text{m}^{-1}$. If the spring is pulled down a distance 0.5 m from its equilibrium position and then released from rest, the particle subsequently executes s.h.m. Write down a full mathematical description of this particular motion and use it to predict the position of the particle 10 s after release.

Solution

To start with we need a coordinate system. In this case let's choose the positive x-direction to be the downward vertical and let the equilibrium position of the oscillator be $x = 0$. With these choices the s.h.m. can be described by

$$x(t) = A \sin(\omega t + \phi). \tag{Eqn 1.50}$$

For a particle on a spring, the angular frequency is given by $\omega = \sqrt{k_s/m}$, so in this case, $\omega = \sqrt{4}\,\text{s}^{-1} = 2\,\text{s}^{-1}$. In addition, if we let the time at which the particle is released be $t = 0$, the initial conditions of the motion tell us that

$$x(0) = 0.5\,\text{m} = A \sin \phi$$

and $$v_x(0) = 0\,\text{m}\,\text{s}^{-1} = A \cos \phi.$$

Given that A is positive, these relations imply that $\phi = \pi/2$ and $A = 0.5\,\text{m}$. It therefore follows that this particular case of s.h.m. is described by

$$x(t) = (0.5\,\text{m}) \sin[(2\,\text{s}^{-1})t + \pi/2].$$

At $t = 10\,\text{s}$, therefore

$$x(10\,\text{s}) = (0.5\,\text{m}) \sin(20 + \pi/2) = (0.5\,\text{m}) \sin(21.57).$$

Using a calculator (and remembering to ensure that the 21.57 is treated as a number of radians, not degrees), it can be seen that

$$x(10\,\text{s}) = 0.2\,\text{m}.$$

This is a good point at which to summarize what has been said so far.

- The *equation of motion* of a system is a differential equation obtained by combining Newton's second law of motion with the force laws appropriate to the system.

- The *general solution* of the equation of motion is a function that predicts the general features of the motion. (That it is s.h.m., for example.)

- Detailed predictions concerning the motion require the evaluation of certain *arbitrary constants* that always appear in the general solution, but which are not present in the equation of motion. These are determined by the *initial conditions* of the motion itself.

This three-step programme — find the equation of motion, solve it, evaluate the arbitrary constants — is sometimes called the *Newtonian programme*. To a large extent it captures the essence of Newtonian mechanics. It shows how the forces that act on a system (which determine the equation of motion and hence the general solution), together with the state of motion of the system at some particular time (which determines the arbitrary constants in the general solution), fully determine the motion of the system at all other times. It therefore explains the determinism of classical mechanics and shows how it is that the past determines the future.

Formulating the equation of motion of a system is a crucial step in predicting the motion of the system. To end this section let's examine a few more examples of equations of motion and look at some of their consequences.

First, suppose that in place of a particle on a spring we have an oscillator that consists of a spherical body of radius R attached to an ideal spring, as shown in Figure 1.50. When the spring is in its equilibrium position ($x = 0$), the weight of the body is balanced by the force arising from tension in the spring. So, when the body is displaced from equilibrium, the only unbalanced forces on the body will be those arising from the additional extension of the spring and those due to air resistance. Newton's second law therefore implies

$$\frac{\mathrm{d}^2 x}{\mathrm{d}t^2} = \frac{F_x}{m} = \frac{F_x^{\text{spring}}}{m} + \frac{F_x^{\text{air}}}{m}. \tag{1.73}$$

(F_x^{air} is the force due to air resistance along the x-direction. This could equally have been written as $F_{\text{air},x}$.)

If, as discussed in Section 3.4, the force due to air resistance acts in the opposite direction to the velocity of the body, and has a magnitude that is proportional to the body's size and speed, then (as implied by Equation 1.22),

$$F_x^{\text{air}} = -k_1 R \frac{\mathrm{d}x}{\mathrm{d}t}, \tag{1.74}$$

where k_1 is a constant. Substituting this into Equation 1.73, and using Hooke's law to describe the spring force, we see that the equation of motion in this case is

$$\frac{\mathrm{d}^2 x}{\mathrm{d}t^2} = \frac{-k_s}{m} x - \frac{k_1 R}{m} \frac{\mathrm{d}x}{\mathrm{d}t}. \tag{1.75}$$

This is another second-order differential equation (the order is determined by the 'highest' derivative the equation contains). Its general solution is well known; it describes a *damped* harmonic motion in which the maximum displacement from

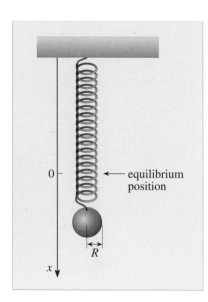

Figure 1.50 A spherical body of mass m and radius R suspended from an ideal spring of spring constant k_s. The downward vertical direction is taken as the x-direction, and the equilibrium position of the body is $x = 0$.

0 ⟵ equilibrium position

R

x

equilibrium decreases with time, as indicated in Figure 1.51. This decrease is, of course, a result of the resistance to the motion by the air. The solution is more complicated than that of Equation 1.70, so we will not write it out mathematically. Suffice it to say that the general solution again expresses x as a function of t, and involves the given constants k_s, m, k_1 and R, as well as two arbitrary constants that can only be determined by considering information about the motion itself, such as the initial position of the body and its initial velocity. It is these initial conditions that determine the precise value of x at $t = 0$ in Figure 1.51.

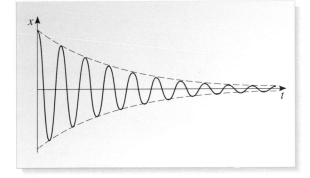

Figure 1.51 Damped harmonic motion in which the maximum displacement from equilibrium decreases with time.

As another example, consider a particle of mass m moving under the gravitational influence of a spherically symmetric body of mass M and radius R, located at the origin (Figure 1.52). In this case, provided the particle is outside the body, the force on the particle will be given by Newton's law of universal gravitation

$$\boldsymbol{F} = \frac{-GMm}{r^2}\hat{\boldsymbol{r}} \quad \text{provided } r \geq R, \tag{1.76}$$

where $\hat{\boldsymbol{r}}$ is a unit vector directed from the centre of the body towards the position of the particle. Substituting this particular force into Newton's second law we find

$$\frac{\mathrm{d}^2\boldsymbol{r}}{\mathrm{d}t^2} = \frac{-GM}{r^2}\hat{\boldsymbol{r}} \quad \text{provided } r \geq R, \tag{1.77}$$

and once again we have a second-order differential equation that is the equation of motion of the particle. (In the case of Equation 1.77, it's important to remember that $r = |\boldsymbol{r}|$, and that $\hat{\boldsymbol{r}} = \boldsymbol{r}/|\boldsymbol{r}|$, since this makes it clear that the only variables in the equation are $\boldsymbol{r}$ and t, and that the solution will be an expression for $\boldsymbol{r}$ as a function of t.) As you would expect, the solutions to Equation 1.77 include the functions that describe elliptical and circular orbits about the spherical body. However, these are only *amongst* the solutions; closed orbits do not constitute the general solution. The general solution allows motion along other curves belonging to the family of conic

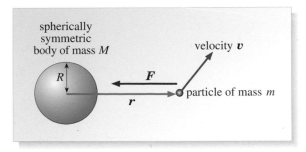

Figure 1.52 A particle of mass m subject to the gravitational attraction of a spherically symmetric body of mass M and radius R, located at the origin. Note that the velocity of the particle has no influence on the gravitational force.

sections, namely parabolas and hyperbolas, as well as ellipses and circles. Figure 1.53 shows the family of conic sections, while Figure 1.54 indicates the way in which the initial condition of the motion can determine the details of the motion.

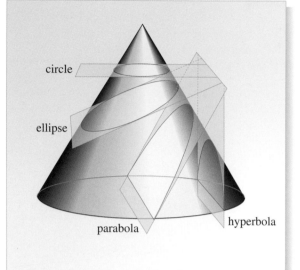

Figure 1.53 The family of conic sections. A circle is obtained by cutting the cone parallel to its base. More steeply inclined cuts result in ellipses, until the cut is parallel to the edge of the cone and a parabola is produced. Even steeper cuts result in hyperbolas.

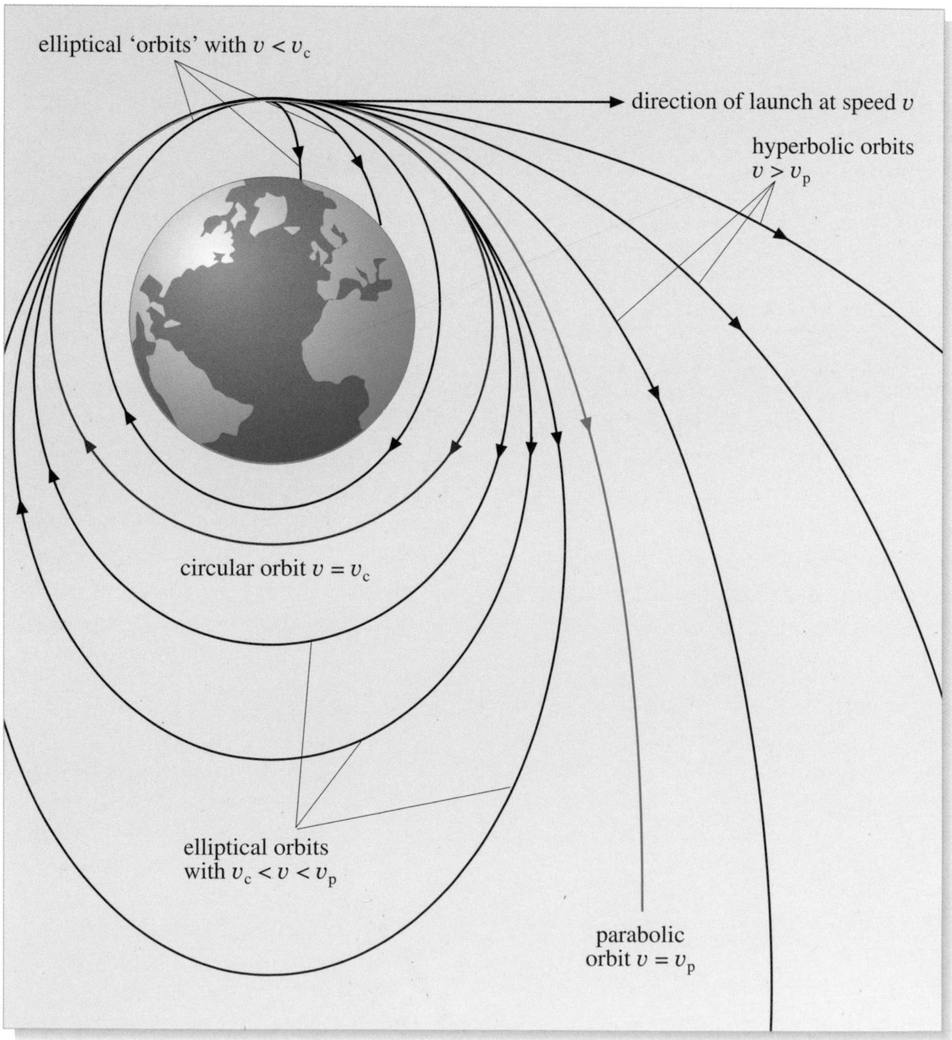

elliptical 'orbits' with $v < v_c$

direction of launch at speed v

hyperbolic orbits $v > v_p$

circular orbit $v = v_c$

elliptical orbits with $v_c < v < v_p$

parabolic orbit $v = v_p$

Figure 1.54 The family of orbits that result from launching a satellite in a fixed direction from a fixed point with a variety of speeds. Very specific launch speeds, denoted v_c and v_p, are required to achieve circular or parabolic motion. Other speeds result in ellipses or hyperbolas.

Planets move in elliptical orbits around the Sun, but a number of comets have been observed in parabolic trajectories, and the *Voyager 2* space probe, discussed at the beginning of this chapter, is now following a hyperbolic trajectory as it heads out of the Solar System.

4.4 Stepping through Newton's laws

The equation of motion of a system is obtained by combining Newton's second law of motion with the force laws that are relevant to the system. This subsection concerns computer based approaches to solving such equations and the importance of these methods in modelling dynamical processes such as collisions between galaxies and the gravitational capture of stars.

Open University students should leave the text at this point and use the multimedia package *Stepping through Newton's laws*, that accompanies this book. The package presents a more detailed treatment of the topics covered in this subsection, with many additional illustrations and examples. When you have completed this activity (which should occupy about 1 hour) you should return to this text.

When a particle or a system of particles moves under the influence of forces that are described by known force laws (such as Hooke's law or Newton's law of universal gravitation) then, provided the initial state of motion is known, it is generally possible to use a computer to follow the subsequent motion of the system in a step-by-step fashion. For many systems this is, in practice, the only way of predicting their motion. To illustrate how such predictions can be made, consider a particle of mass m moving in the x-direction under the influence of a single unbalanced restoring force $F_x = -k_s x$. It follows from Newton's second law that at time t, when the particle's position is $x(t)$, and its velocity is $v_x(t)$, the acceleration of the particle will be

$$a_x(t) = \frac{\mathrm{d}v_x}{\mathrm{d}t} = \frac{F_x}{m} = \frac{-k_s}{m} x. \tag{1.78}$$

Now suppose that we want to work out the position and velocity of the particle a very short time later. Let's denote this short time by Δt and suppose it is so short that the particle barely changes its position at all between t and $(t + \Delta t)$. Since the change in position is so small, both the force on the particle, F_x, and the acceleration it produces, F_x/m, can be considered to be constant throughout the interval Δt. It then follows from the equations of uniformly accelerated motion (see Equation 1.41), that the small changes in the position and velocity of the particle that will have occurred by the end of the time interval Δt will be given approximately by

$$\Delta x = v_x(t)\Delta t + \frac{1}{2}\left(\frac{F_x}{m}\right)(\Delta t)^2 \tag{1.79}$$

and $\qquad \Delta v_x = \left(\frac{F_x}{m}\right)\Delta t. \tag{1.80}$

If Δt is sufficiently small the term involving $(\Delta t)^2$ will be very small and can be neglected. If we then use Hooke's law to eliminate F_x we can write

$$\Delta x = v_x(t)\Delta t \tag{1.81}$$

and $\qquad \Delta v_x = \left(\frac{-k_s x(t)}{m}\right)\Delta t. \tag{1.82}$

Using these equations we can relate the position and velocity of the particle at time $(t + \Delta t)$ to the position and velocity at time t

$$x(t + \Delta t) = x(t) + \Delta x = x(t) + v_x(t)\Delta t \tag{1.83}$$

and $\qquad v_x(t + \Delta t) = v_x(t) + \Delta v_x = v_x(t) + \left(\frac{-k_s x(t)}{m}\right)\Delta t. \tag{1.84}$

So, given the position and velocity at any particular time, it is possible to work out the approximate position and velocity at some slightly later time. By repeating this process over and over again, using the results of one step as the starting values of the next, it is possible to follow the evolution of the system with time. In this particular case we already know, from Section 4.2, that the motion is actually a simple harmonic oscillation, so we can compare the results of our step-by-step approximation with the exact result, $x(t) = A \sin(\omega t + \phi)$. The result of such a comparison is shown in Figure 1.55. Comparisons of this kind indicate that reducing the size of the time step Δt tends to improve the accuracy of the results, but it also has the undesirable effect of increasing the length of the calculation. In practice, computer-based numerical methods of the kind we have just outlined almost always involve making compromises between accuracy and length or complexity.

Figure 1.55 A stepwise approximation to the simple harmonic motion of a particle on a spring (dashed line), compared with the exact description of the motion $x(t) = A \sin(\omega t + \phi)$ given by the solid line. In this case the period of the motion is 4.8 s.

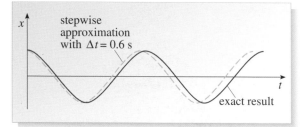

Numerical techniques, many of them far more sophisticated than the example we have discussed, are of great value in predicting motion. They permit the investigation of systems that are too complicated to be treated by purely algebraic methods. A well-known example of this is provided by the so-called **three-body problem**. The problem is very simple; it is to predict the motion of three bodies, each of which is free to move subject only to the gravitational attraction of the other two. Entire books have been written about this problem, and a few exact solutions, corresponding to very special initial conditions (see Figure 1.56) have been known since the eighteenth century. These special solutions are of interest because they indicate, for example, the stable locations of satellites moving under the combined gravitational influence of the Earth

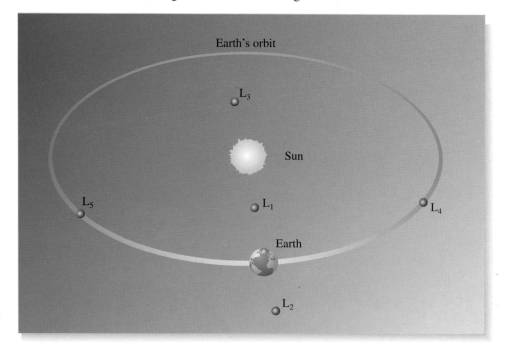

Figure 1.56 Some special solutions to the three-body problem. A satellite placed at any of the positions marked L_1 to L_5 (the so-called *Lagrange points*) will remain at that position relative to the other two bodies, in this case the Sun and Earth.

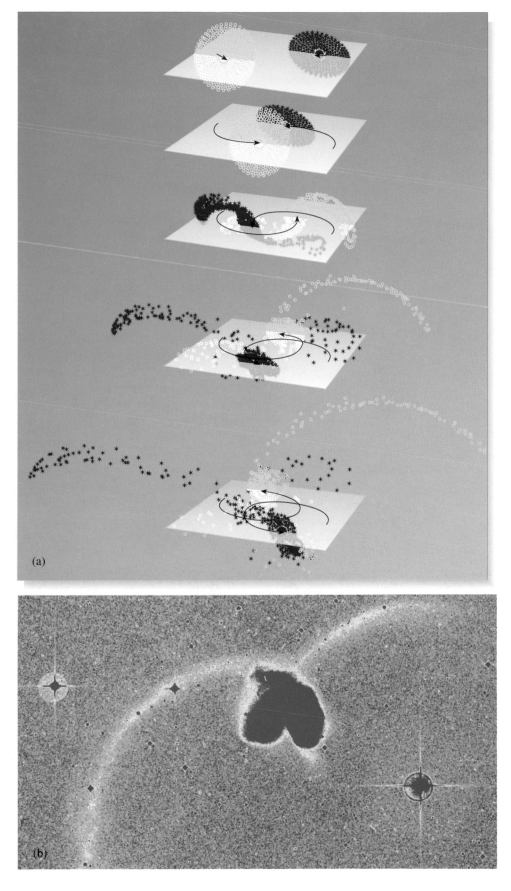

(a)

(b)

Figure 1.57 (a) A computer simulation of a collision between two galaxies, obtained using a stepwise approach to Newton's laws. (b) The Antenna Galaxy in Corvus, the result of a collision between real galaxies.

61

and the Sun. However, with the aid of modern computers it is now relatively easy to investigate three-body problems with specified masses and initial conditions, and by doing so to obtain insight into astronomical conundrums such as the likelihood that a system of two mutually orbiting stars will be broken up by an encounter with a third star. (Such encounters are thought to be quite common in dense clusters of stars and may explain some of the peculiar features seen there.)

Another astronomical application of the stepwise approach to the prediction of motion concerns the collision of galaxies. Galaxies are the building blocks of the Universe. Our own galaxy, the Milky Way, which is not atypical, is a large spiral composed of more than 10 000 000 000 stars. Simulating the collision of two such galaxies, in which every star interacts gravitationally with every other star (a 2×10^{10} body problem!) is well beyond even the most powerful of modern computers, but simulations involving several thousand particles can certainly be carried out. The results indicate that the collision may produce long 'tails' of the kind shown in Figure 1.57a, and can explain such peculiar sights as the Antenna Galaxy, seen in the constellation of Corvus (Figure 1.57b). Numerical studies of galactic dynamics constitute one of the most rapidly advancing areas of modern astrophysics.

5   Closing items

5.1   Chapter summary

1   The description of a body's motion depends on the frame of reference from which the motion is observed. A frame of reference is a system for assigning position coordinates and times to events.

2   An inertial frame is a frame of reference in which Newton's first law of motion holds true. Any frame that moves with constant velocity relative to an inertial frame, while maintaining a fixed orientation, will also be an inertial frame.

3   According to Newton's first law of motion: a body remains at rest or in a state of uniform motion unless it is acted on by an unbalanced force. The law therefore introduces force as a quantity that changes the motion of a body by causing it to accelerate.

4   According to Newton's second law of motion: an unbalanced force acting on a body of fixed mass will cause that body to accelerate in the direction of the unbalanced force. The magnitude of the force is equal to the product of the mass and the magnitude of the instantaneous acceleration. The law therefore enables force to be quantified and is usually expressed by the vector equation $\boldsymbol{F} = m\boldsymbol{a}$.

5   According to Newton's third law of motion: if body A exerts a force on body B, then body B exerts a force on body A. These two forces are equal in magnitude, but act in opposite directions and are said to constitute a third-law pair. The law therefore indicates that to every action there is an oppositely directed reaction of equal magnitude.

6   Observers who attempt to apply Newton's laws in a non-inertial frame will observe phenomena that indicate the existence of fictitious forces, such as centrifugal force and Coriolis force. These phenomena are real but the fictitious forces are not; they appear because of the acceleration of the observer's frame relative to an inertial frame.

7   The inertial mass of a body (measured in kilograms) is a measure of its resistance to acceleration when subjected to a given force (measured in newtons, where $1\,\mathrm{N} = 1\,\mathrm{kg\,m\,s^{-2}}$).

8 The centre of mass of a rigid body is the unique point which has the property that if any unbalanced force is applied to the body in such a way that its line of action passes through the centre of mass, then the only effect of that force will be to cause translational acceleration of the body. If a force is applied to the body in such a way that its line of action does not pass through the centre of mass, the body will undergo rotation as well as translation. When a resultant external force F acts on a body of fixed mass m, the acceleration of the body's centre of mass, a_{CM}, is given by $F = ma_{CM}$.

9 The weight of a body of mass m near the surface of the Earth is the gravitational force which the Earth exerts on the body. It acts vertically downwards and is given by $W = mg$, where g is the acceleration due to gravity.

10 According to Newton's law of universal gravitation: every particle of matter attracts every other particle of matter with a gravitational force, whose magnitude is directly proportional to the product of the masses of the particles, and inversely proportional to the square of the distance between them. The law therefore implies that the force on a particle of mass m_2 with position vector r, due to a particle of mass m_1 at the origin will be

$$F_{21} = \frac{-Gm_1m_2}{r^2} \hat{r},$$

where G is Newton's universal gravitational constant, and $\hat{r} = r/r$ is a unit vector pointing from the origin towards the particle of mass m_2.

11 According to Newton's theorem: the gravitational effect of any spherically symmetric body, outside its own surface, is identical to that of a single particle, with the same mass as the body, located at the centre of the body.

12 When a body exerts a contact force on a rigid surface, the surface exerts a normal reaction on the body. The normal reaction is directed at right angles to the surface and is equal in magnitude to the component of the contact force that is perpendicular to the surface.

13 The maximum magnitude F_{max} of the frictional force acting on a stationary object is given by $F_{max} = \mu_{static}N$, where μ_{static} is the coefficient of static friction, and N is the magnitude of the normal reaction acting on the object. When sliding occurs the value of the frictional force drops slightly, and is given by $F = \mu_{slide}N$, where μ_{slide} is the coefficient of sliding friction.

14 The viscous drag force exerted on a body moving through a fluid is a function of the speed of the body relative to the fluid. When the magnitude of this force equals that of the resultant accelerating force, a terminal speed is reached.

15 According to Hooke's law: when a body (such as a spring) is stretched, the magnitude of the restoring force arising from tension is directly proportional to the increase in the body's length. This is not a fundamental law, but an empirical relation, often written $F_x = -k_s x$, that describes some bodies, provided the extension is not too great.

16 At any point in a fluid, the pressure P is the magnitude of the perpendicular force per unit area that the fluid would exert on a surface at that point, so $F = PA$. The SI unit of pressure is the pascal, where $1\,Pa = 1\,N\,m^{-2}$.

17 The equation of motion of a system is a differential equation obtained by combining Newton's second law of motion with the force laws appropriate to the system. The general solution of the equation of motion predicts the general features of the motion. Detailed predictions concerning the motion require the evaluation of certain arbitrary constants that always appear in the general solution, but which are not present in the equation of motion. These are determined by the initial conditions of the motion itself.

18 In the absence of any force a particle moves with uniform motion. In two or more dimensions, a particle subjected to a constant force accelerates uniformly in the direction of the force, but moves uniformly at right angles to the force. In order to keep a particle in uniform circular motion, a centripetal force of magnitude $F = mv^2/r$ must be applied at all times. A linear restoring force ($F_x = -k_s x$) produces simple harmonic motion. An attractive force described by an inverse square law, such as universal gravitation, will cause a particle to move along a curve belonging to the family of conic sections, such as an ellipse, a parabola or a hyperbola.

19 Systems that are too complicated to be studied by purely algebraic methods can be investigated numerically by using a computer to follow the step-by-step evolution of the system through time. Such methods are of great importance in current research.

5.2 Achievements

Now that you have completed this chapter, you should be able to:

A1 Understand the meaning of all the newly defined (emboldened) terms introduced in the chapter.

A2 Recall Newton's first law of motion, explain how it leads to the concept of force and how an inertial frame of reference may be defined using the law.

A3 Recall Newton's second law of motion for a body of constant mass and show how it leads to the quantification of force.

A4 Recall Newton's third law of motion and identify third-law pairs of forces (actions and reactions) for various set of interacting bodies.

A5 Express and use Newton's laws of motion in terms of vectors.

A6 Appreciate the importance of the concept of the centre of mass of a rigid body, and its relevance to Newton's second law of motion.

A7 Explain the relation between weight and mass.

A8 Recall Newton's law of universal gravitation and describe how it leads to an explanation of terrestrial gravitation.

A9 Describe the behaviour of frictional forces between surfaces in contact and explain the distinction between static and sliding friction.

A10 Describe the dependence of air resistance on the speed of a body relative to the air through which it is moving. Explain how the speed dependence of viscous forces in general leads to a body reaching a terminal speed.

A11 Recall and use Hooke's law for an elastic string or spring.

A12 Describe the relation between tension and tension forces, and between pressure and forces arising from pressure.

A13 Write down the equation of motion of a system given the necessary information about the system and the relevant force laws. Also, explain the significance of the equation of motion and the initial conditions in determining the motion of the system.

A14 Relate forms of motion to the forces that cause them in a number of well known cases, including uniform motion, uniformly accelerated motion, uniform circular motion, simple harmonic motion and orbital motion.

A15 Describe some of the principles and results involved in computer-based numerical investigations of motion.

5.3 End-of-chapter questions

Question 1.13 A given force is applied in turn to three bodies with masses in the ratio $4:2:3$. What is the ratio of the magnitudes of the accelerations of the three bodies?

Question 1.14 A physics book lies on top of a chemistry book which lies on top of a table. Identify the third-law force pairs associated with the two books.

Question 1.15 A horizontal force of gradually increasing magnitude is applied to a wooden block which is resting on a horizontal surface. The block starts to slide when the force reaches a certain value. If the force is then kept constant at this value, find the magnitude of the acceleration of the block, given that $\mu_{static} = 0.50$ and $\mu_{slide} = 0.40$.

Question 1.16 A person falling from an aircraft is subject to a constant downward force due to their own weight and an upward force due to air resistance, the magnitude of which is proportional to their speed. (a) Taking the downward vertical as the x-direction, and assuming that the fall starts from $x = 0$ at time $t = 0$, write down the equation of motion describing the person's vertical motion, taking care to explain the significance of any new quantities you introduce. (b) You are not expected to solve the equation of motion, but you should use your knowledge of the behaviour of falling bodies to draw a velocity–time graph for the falling person indicating the general features of their motion.

Question 1.17 Determine the magnitude F_{ME} of the gravitational force exerted on the Moon by the Earth, given that the masses of the Moon and Earth are 7.35×10^{22} kg and 5.98×10^{24} kg, respectively, and that the radius of the Moon's orbit is 0.38×10^{9} m. Determine also the distance required between two particles of mass 1.0 kg each, to produce on each other a gravitational force of the same magnitude as F_{ME}. ■

Chapter 2 Work, energy and power

1 Escaping the Earth — an application of energy

The thought of escaping from the Earth has fascinated humankind for centuries. Early stories about travel to the Moon or the planets, often written with comic or satirical intent, suggested riding on a waterspout, being carried aloft by geese or simply rising with the morning dew (Figure 2.1). However, the real barrier to interplanetary flight is not the difficulty of trapping birds or bottling dew, but that of providing enough energy.

Another author, Jules Verne, had a more realistic idea. In his novel *From the Earth to the Moon* he imagined that a giant cannon might be used to launch a bullet-shaped spacecraft with a sufficiently high initial speed that it could reach the Moon (Figure 2.2). As a means of getting away from the Earth, this idea has merit, even though the initial acceleration would have killed the craft's occupants. Jules Verne had largely taught himself science, but he had learned enough to know that the launch speed of his Moon-bound projectile would have to be about $11 \, \text{km s}^{-1}$ if it was to escape from the Earth. This figure is quoted in Verne's book, by the proposer of the mission, who describes it as a result of 'incontestable calculations'. As you will see later in this chapter, that figure actually represents the *escape speed* from the Earth. In the absence of air resistance, it is the minimum launch speed that ensures a projectile has enough energy to completely escape the Earth's gravitational pull. You will evaluate this quantity for yourself a little later and you will see then that, by using arguments based on energy requirements, its 'incontestable calculation' need only occupy a few lines.

Figure 2.1 Cyrano de Bergerac's proposal for escaping the Earth involved fixing bottles of dew to his belt.

Figure 2.2 A bullet-shaped spacecraft is loaded into a space cannon. The still is from the 1936 film *Things to Come*, based on a novel by H. G. Wells, but the idea is from Jules Verne.

The term 'energy' was first used in its modern sense by Thomas Young, in 1807, and comes from the Greek words *en ergon*, meaning 'in work'. Following Young we can say:

> The **energy** of a system is a measure of its capacity for doing work.

It is by utilizing the energy stored in petrol that a car is able to move along a road, overcoming the friction and air resistance it inevitably encounters, and it is by using the energy you obtain from the digestion of food that you are able to work, lifting weights, pushing pens and so on. In a similar way, it is the energy provided at launch that enables Verne's bullet-like spacecraft to do the work required to overcome gravity as it moves away from the Earth.

In this chapter, you will learn both the precise definition of work and how to calculate the energy of a system in various circumstances. More importantly, you will also learn the relationship between these quantities, and the way in which they both relate to the *law of conservation of energy* that was mentioned in *The restless Universe*. We shall have quite a lot to say about energy conservation, since it was the discovery of this principle, around 1850, that persuaded physicists of the fundamental importance of energy. For the moment, however, let's end this section by noting that the take-off of a modern rocket (Figure 2.3) makes it readily apparent that a very great deal of energy is required to escape from the Earth.

Figure 2.3 An Ariane rocket being launched from the European Space Agency's launch facility at Kourou in French Guiana.

2 Work and kinetic energy

The concepts of work and energy fit together naturally. Starting from a quantitative definition of either quantity quickly leads to a similar definition of the other. We shall begin with energy.

2.1 Kinetic energy — energy due to motion

You will often hear people speak of different 'forms' of energy. Despite this, there are actually very few fundamentally different forms of energy. One of them, however, is undoubtedly kinetic energy, and this is where we shall begin our detailed discussion of energy.

> The **kinetic energy** of a body is the energy that the body has by virtue of its *motion*.

In this chapter we shall consider only translational motion, so the associated kinetic energy may be termed *translational kinetic energy*. In practice, the term 'translational' will sometimes be dropped for convenience, but you shouldn't forget it entirely since in later chapters you will encounter the rotational kinetic energy of a turning body. In general, the kinetic energy of a body is composed of both a translational and a rotational contribution. Having issued that warning, let's now concentrate on translational kinetic energy.

What factors determine the translational kinetic energy of a moving body and hence its capacity to do work? Well, if you take the ability to cause destruction as an indication of the capacity to do work, it's quite clear that mass is involved. If a lorry and a car crash into a wall at the same speed, the lorry, because of its greater mass, will generally do more damage than the car. It's also clear that speed is significant. Doubling the speed of impact of a vehicle will quadruple the damage that it causes. A car travelling at $80 \, \text{km h}^{-1}$ is four times as destructive as the same car travelling at $40 \, \text{km h}^{-1}$. These, and more precise considerations along similar lines, support the following definition of translational kinetic energy:

Figure 2.4 James Prescott Joule (1818–1889) was a British physicist whose work on the relationship between electrical, mechanical and chemical energy enabled him to contribute to the development of the law of conservation of energy. He was a brewery owner, who pursued his scientific studies as a hobby, yet his skill in designing experiments allowed him to obtain results of unrivalled accuracy.

The **translational kinetic energy** of a body of mass m, whose centre of mass is moving at a speed v is

$$E_{\text{trans}} = \tfrac{1}{2}mv^2. \tag{2.1}$$

Notice that translational kinetic energy is defined in terms of speed and not velocity; this means that it does not depend on the direction in which a body is moving. Kinetic energy is therefore a *scalar* quantity. Furthermore, it is a *positive* scalar, because m and v^2 must both be positive. A specified amount of translational kinetic energy can therefore always be represented by the product of a positive number and an appropriate unit of energy.

Using Equation 2.1, it is easy to work out what the unit of energy must be. Since the SI unit of mass is the kilogram (kg) and that of speed is the metre per second (m s^{-1}), translational kinetic energy can be measured in units of kg m^2 s^{-2}. However, because energy is such an important concept, this unit is given its own name. It is called the **joule**, in honour of James Joule (Figure 2.4), one of the contributors to the discovery and exploitation of the conservation of energy. The joule is represented by the symbol J, so we can write

$$1 \text{ joule} = 1 \text{ J} = 1 \text{ kg m}^2 \text{ s}^{-2}.$$

The joule is a moderate sized unit of energy. If a child runs into you in the street your body will probably have to absorb about half of their kinetic energy, which is likely to be about 60 J. Some other translational kinetic energies are shown in Figure 2.5. They range from the translational kinetic energy of the Earth as it moves around the Sun (2.5×10^{33} J) to that of a typical molecule in the air you are about to breathe in (6×10^{-21} J).

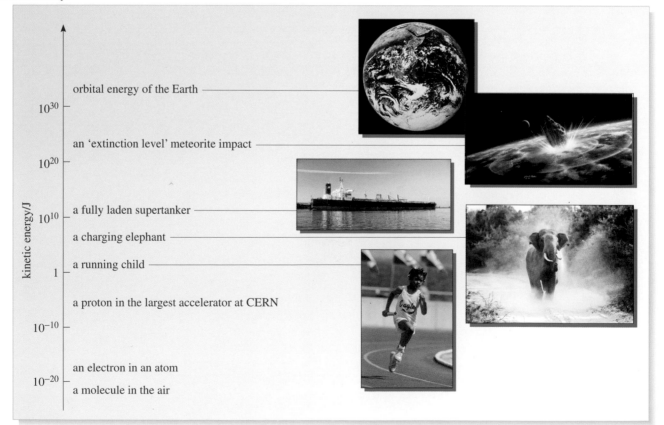

Figure 2.5 Some physically interesting kinetic energies.

● What is the translational kinetic energy of a beetle of mass 1 g moving at $2\,\mathrm{cm\,s^{-1}}$?

○ Using Equation 2.1,

$$E_{\mathrm{trans}} = \tfrac{1}{2}\,mv^2 = \tfrac{1}{2} \times (10^{-3}\,\mathrm{kg}) \times (2 \times 10^{-2}\,\mathrm{m\,s^{-1}})^2$$

i.e. $\quad E_{\mathrm{trans}} = 2 \times 10^{-7}\,\mathrm{kg\,m^2\,s^{-2}} = 2 \times 10^{-7}\,\mathrm{J}.$ ■

Before moving on to consider work and its relation to kinetic energy, it is worth saying that since the conservation of energy implies the possibility of 'exchanging' one form of energy for another, it must be the case that all forms of energy are scalar quantities and that they can all be measured in joules. In short, we can say quite generally that energy is a scalar quantity and that its SI unit is the joule.

2.2 Work and changes in kinetic energy

The relationship between speed and translational kinetic energy implies that any *change* in a body's speed must be accompanied by a corresponding change in its translational kinetic energy. However, we already know that a change of speed involves a change of velocity and hence (by Newton's first law) the action of an unbalanced force. So the application of a force to a body can bring about a change in its translational kinetic energy.

Since the energy of a body is a measure of its capacity to do work, any change in the energy must represent a change in the work that can be done. As these changes in its capacity to do work have been caused by the action of a force, it makes sense to say that they represent the work done on the body by the force. We are thus led to the following general statement:

> The **work done** on a body by any force is the energy transferred to or from that body by the action of the force.

In the case of a particle, we have the **work–energy theorem**, which asserts that:

> The change in the translational kinetic energy of a particle is equal to the work done by the total force that acts upon it.

So, if the total force on a particle causes it to change its speed from an initial value u to a final value v, then the work–energy theorem tells us that the amount of work done by that total force is

$$W = \Delta E_{\mathrm{trans}} = \tfrac{1}{2}\,mv^2 - \tfrac{1}{2}\,mu^2 \qquad (2.2)$$

where the symbol $\Delta E_{\mathrm{trans}}$ represents the change in translational kinetic energy, the result of subtracting the initial value from the final value. Note that although the initial and final kinetic energies must both be positive, their difference may be positive or negative, depending on whether the force causes the particle to speed up or to slow down. In other words, the work done by the total force may be positive or negative. Also note that this relation applies to the work done by the total force. A typical body is usually subject to several forces simultaneously, some of which do positive work while others do negative work. Equation 2.2 only applies to the work done by the *total* force and strictly only applies to a particle, though the result may be extended to a non-rotating rigid body.

Using the relationship between work and translational kinetic energy given by Equation 2.2, we can now determine precise algebraic expressions for the work done by a force in various situations. The next two subsections provide examples of this process. The results obtained should help to further clarify the nature of work.

2.3 The work done by a constant aligned force

Suppose that a body of mass m, moving in the x-direction with initial velocity u_x, is subject to a constant unbalanced force, F_x, that also points in the x-direction. As a particular example, consider the frictionless bobsleigh shown in Figure 2.6, where the constant force F_x is aligned with the direction of motion. Since the force is constant the bobsleigh will accelerate uniformly in the direction of the force, and its (constant) acceleration will be given by

$$a_x = \frac{F_x}{m}.$$

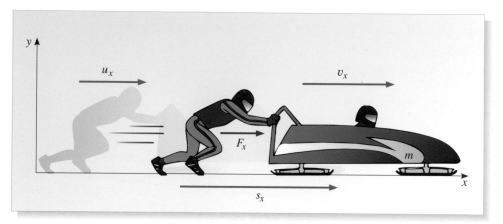

Figure 2.6   A constant force F_x is applied to a bobsleigh of mass m. As the bobsleigh travels through a displacement s_x, its velocity changes from u_x to v_x. Although F_x, u_x, v_x and s_x are all shown pointing the same way, any of them might have a negative value, in which case they would actually act in the opposite direction to that shown.

What effect will this acceleration have on the speed and hence the kinetic energy of the bobsleigh? Well, suppose that the force acts for some interval of time t and that at the end of that interval the bobsleigh has moved through a displacement s_x and has attained a final velocity v_x. The equations of uniformly accelerated motion tell us that

$$v_x^2 = u_x^2 + 2a_x s_x.$$

From which it follows that

$$v_x^2 - u_x^2 = 2\frac{F_x}{m}s_x.$$

Multiplying both sides by m and dividing by 2, we get

$$\tfrac{1}{2}mv_x^2 - \tfrac{1}{2}mu_x^2 = F_x s_x.$$

But the left-hand side of this equation (from Equation 2.2) is the change in the translational kinetic energy of the bobsleigh, ΔE_{trans}, and this, as we know from the work–energy theorem, is equal to the work done on the bobsleigh by the unbalanced force. Thus, we can say that, in the case of a constant unbalanced force F_x, acting along the direction of motion over a displacement s_x, the work done is

$$W = F_x s_x. \tag{2.3}$$

So the work done by a constant force that is aligned with a given displacement, is equal to the product of the force and the displacement.

Notice that the work done by a constant aligned force depends only on the *product* $F_x s_x$. This means that a large force acting over a small displacement can change the energy of a body just as much as a small force acting over a large displacement, provided the product $F_x s_x$ is the same in both cases. Also notice that the formula

$W = F_x s_x$ applies whether s_x and F_x are positive or negative. If the force and displacement are in opposite directions, then the work done by the force is negative, and it will cause a reduction in the energy.

Another point to note about work that follows directly from Equation 2.3, is that an SI unit of work is the newton metre (N m). However, since the work done on a body is equal to the energy transferred to that body (Equation 2.2), it is also the case that work can be measured in joules (J). The joule is the more conventional unit to use, but it is the case that $1\,J = 1\,N\,m$ since $1\,N = 1\,kg\,m\,s^{-2}$, while $1\,J = 1\,kg\,m^2\,s^{-2}$.

You will be aware, from everyday experience, that simply supporting a weight without displacing it can be quite tiring. You may therefore find it odd that Equation 2.3 implies that no work is done in such a situation, since $s_x = 0$. The reason for this apparent mismatch between physics and experience is that when you support a weight you are continually tensing and relaxing opposed sets of muscles, and doing work in this operation, rather than on the weight. This is why you become tired. A table, exerting a normal reaction force to support the same weight, would do no work, and it certainly wouldn't become tired! Simply exerting a force does not involve doing work; only when the point of application of the force moves in the direction of the force is work done.

Now attempt the following question by using Equations 2.2 and 2.3.

Question 2.1   (a) A 5 kg trolley stands at rest. If it is pushed with a constant unbalanced force of magnitude 40 N over a distance of 1 m, what is its final speed?

(b) If the same force is applied over the same displacement to a 10 kg trolley, what would be the final speed of this more massive trolley?

(c) The same force is applied to a 5 kg trolley that is already travelling at $2\,m\,s^{-1}$ in the direction of the force. If the force is again applied over a distance of 1 m, what is the final speed?

(d) In each case, find the change in kinetic energy of the trolley. ∎

2.4 The work done by a constant non-aligned force

Now, consider what happens in two or three dimensions, when a constant force, represented by the vector F, acts at a fixed angle θ to the displacement of a body, as indicated in Figure 2.7.

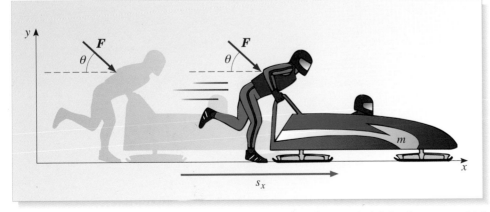

Figure 2.7   A constant force F is applied at an angle θ to a bobsleigh of mass m which moves through a displacement s_x.

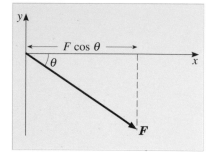

Figure 2.8 The x-component of the force F is $F_x = F \cos \theta$.

In this case, the vertical component of F will cause an increase in the normal reaction that is exerted on the bobsleigh, but the acceleration of the bobsleigh will be caused only by the unbalanced horizontal component of force, $F_x = F \cos \theta$, where F represents the magnitude of F (see Figure 2.8).

Arguments similar to those used in the last subsection show that in this case the work done by the force is entirely due to this horizontal component of the force, so

$$W = F_x s_x = (F \cos \theta)s_x. \tag{2.4}$$

However, our choice that the direction of motion was labelled the x-direction was totally arbitrary. The important point in deriving Equation 2.4 is that the effective part of the force is its component in the direction of motion, whatever that direction happens to be. We can thus generalize Equation 2.4 by writing

$$W = Fs \cos \theta. \tag{2.5}$$

This tells us that the work done by a constant force with a fixed orientation relative to some displacement, can always be calculated by multiplying together the magnitudes of the force and displacement vectors and the cosine of the angle between them. (The result we obtained in the last subsection was a special case of this corresponding to $\theta = 0°$ or $\theta = 180°$.)

Although the formula for the work done by a constant force has been determined by using arguments based on the change of translational kinetic energy, it's important to realize that Equations 2.3 and 2.5 remain true even in situations where the energy transferred does *not* result in a change of translational kinetic energy. For instance, if you apply a steady force to a book to make it slide across a table, then Equation 2.5 will enable you to calculate the work done on the book and hence the energy transferred to it, but at the end of the process, far from gaining kinetic energy, the book will come to rest. Work will have been done; and energy will have been transferred to the book, but there will be no increase in the final kinetic energy. There is nothing wrong with the formula for calculating the work done, it's simply that in this case the book is also subject to friction and this causes the energy transferred to the book to be dissipated as quickly as it is supplied. Similar dissipative forces are also present in the situation described in Example 2.1.

Example 2.1

A ship is pulled into a harbour by two horizontal ropes attached to tugs as shown in Figure 2.9. (a) How much work is done by each of the forces shown when the ship is moved 70 m in the x-direction? (b) What is the total energy transferred to the ship by the two tugs?

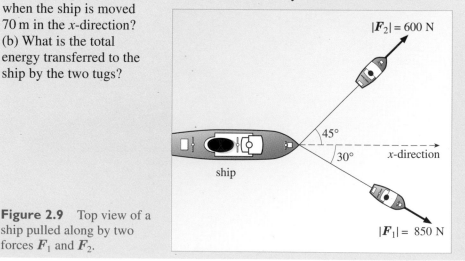

Figure 2.9 Top view of a ship pulled along by two forces F_1 and F_2.

Solution

(a) Using $W = Fs \cos \theta$ for each of the forces:

$$W_1 = F_1 s \cos 30° = 850\,\text{N} \times 70\,\text{m} \times \cos 30° = 5.2 \times 10^4\,\text{J}$$

$$W_2 = F_2 s \cos 45° = 600\,\text{N} \times 70\,\text{m} \times \cos 45° = 3.0 \times 10^4\,\text{J}.$$

(b) The work done by each force represents the energy transferred to the ship by that force and hence by the tug that is responsible for applying the force. Since energy is a scalar quantity, it follows that the total energy transferred to the ship by the two tugs will be the sum of the work done individually by each of the applied forces. In this case, the total energy transferred will therefore be $8.2 \times 10^4\,\text{J}$.

Question 2.2 (a) Calculate the work done by the force of terrestrial gravity when a skier, of mass 60 kg, travels 1 km, without friction, down a uniform slope at 30° to the horizontal. (b) Does the work done depend on the skier's initial speed? (c) How much work is done by the normal reaction force of the snow? (Assume the skier does not push with the ski poles.) ■

2.5 Work and the scalar product of vectors

There is another way of writing Equation 2.5 that involves an important mathematical concept called the *scalar product* of vectors. In this subsection we will introduce this concept in general terms (for later reference) and then return to its significance for calculations of the work done by a force.

The **scalar product** of two vectors is a scalar quantity obtained by multiplying the vectors together in a certain way. Any two vectors a and b, such as those shown in either part of Figure 2.10, have a scalar product. It is defined as follows:

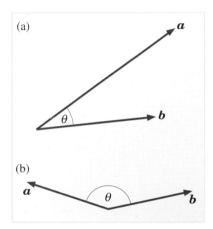

Figure 2.10 Any two vectors a and b have a scalar product $a \cdot b$ that depends on their magnitudes (a and b) and the cosine of the angle θ between them. Here are two cases. In (a) the angle is 30° and the magnitudes are $a = 5$ and $b = 3.5$, so $a \cdot b = 5 \times 3.5 \times \cos 30° = 15.16$. In (b) the angle is 150° and $a = b = 2.5$, so $a \cdot b = 2.5 \times 2.5 \times \cos 150° = -5.41$.

> Given two vectors a and b, which are at an angle θ to each other, where $0° \leq \theta \leq 180°$, their scalar product is given by
> $$a \cdot b = ab \cos \theta. \tag{2.6}$$

In situations where you know the magnitudes a and b of the vectors a and b, and the angle between them you can easily use Equation 2.6 to determine their scalar product $a \cdot b$. However, in practice it is often the case that vectors are specified in terms of their Cartesian components. In such cases another, entirely equivalent, way of expressing the scalar product is often of greater use.

> If $a = (a_x, a_y, a_z)$ and $b = (b_x, b_y, b_z)$ then the scalar product of a and b is given by
> $$a \cdot b = a_x b_x + a_y b_y + a_z b_z. \tag{2.7}$$

We shall not prove the equivalence of the two expressions we now have for $a \cdot b$, but it is worth noting that if we know the components of the two vectors then we can determine the angle between them by combining Equations 2.6 and 2.7 to write

$$\cos \theta = \frac{a_x b_x + a_y b_y + a_z b_z}{ab}.$$

It is very important to distinguish the scalar product $a \cdot b$ from the simple product of magnitudes ab, since the scalar product involves an additional factor of $\cos \theta$. Making the distinction is easy in print, but it requires care and attention in handwritten work. When *writing* a scalar product, you should employ the usual curly underline notation to denote the vectors, and take care to place a clear dot between them, as in $F \cdot s$. This may seem a bit tedious at first, but it will make you conscious of the fact that you are using vectors, and it will help you to avoid errors later. The scalar product is sometimes informally referred to as the 'dot product' for reasons that will become obvious once you have written a few.

Question 2.3   (a) What is $a \cdot b$ if a and b are at right angles? Express the answer in term of magnitudes a and b. (b) What is $a \cdot b$ if a and b are in exactly opposite directions? (c) What is $a \cdot a$, i.e. the scalar product of a vector with itself?

Question 2.4   If $F = (1, 2, 3)\,\text{N}$, and $s = (2, 3, 1)\,\text{m}$, what is the scalar product of F and s, and what is the angle between them?   ■

Since the work done by a constant force F, when its point of application is displaced by an amount s, is $Fs \cos\theta$, it follows from Equation 2.6 that we can write the work done by the force as

$$W = F \cdot s \tag{2.8}$$

where $F \cdot s$ represents the scalar product of the force and the displacement.

The scalar product is more than just a shorthand notation. It can be used to simplify a lot of complicated calculations, especially when the forces and displacements are specified in terms of their components, rather than their magnitudes and directions. It is used throughout mathematics and physics, and plays an important part in helping to develop expressions for the work done by a force in more general situations than those we have considered so far, including cases where the force changes in magnitude or direction during the motion.

One final point deserves particular emphasis. Since the scalar product of two vectors that are mutually at right angles is zero (see Question 2.3a), a force that is always directed at right angles to the direction of motion can do no work. This means that the centripetal force that maintains a body in uniform circular motion never does any work on that body because it always acts at right angles to the instantaneous velocity. This can be regarded as the reason why the kinetic energy of such a body remains constant.

3   Work and potential energy

Another 'form' of energy, quite different from kinetic energy, is *potential energy*. Informally, we can say

> The **potential energy** of a body is the energy that the body has by virtue of its *position*.

This kind of energy arises from interactions between bodies, typically electrical or gravitational interactions, and usually depends on the relative locations of bodies. For instance, a stone in the neighbourhood of the Earth has gravitational potential energy because of its gravitational attraction to the Earth. The higher the stone is above the Earth, the greater its gravitational energy due to the Earth. Similarly, an electron orbiting an atomic nucleus also has potential energy, in this case electrical potential energy due to the electron's electrical attraction towards the nucleus, and it too increases as the separation of the electron and the nucleus increases.

In general, a particle or a system of particles may have both kinetic energy and potential energy simultaneously, and when a force does work on a system, thereby transferring energy to or from the system, that energy transfer may change either the kinetic energy of the system or its potential energy, or both. In this section we shall be mainly concerned with potential energy and its relation to work. First, however, we need to consider the circumstances in which it is meaningful to talk about potential energy at all.

3.1 Conservative and non-conservative forces

Forces can be classified as either *conservative* or *non-conservative*, according to how they behave when doing work — basically, on whether the work they do is fully 'recoverable' or not. In the last section we discussed the work done on a book sliding across a table and we saw that in that case, due to the action of friction, the energy transferred to the book was dissipated; it was definitely not recoverable. A frictional force is a classic example of a non-conservative force. An equally classic example of a conservative force is the gravitational force exerted by the Earth. In this subsection we shall give precise definitions of the terms conservative and non-conservative, and in the next subsection we shall argue that it is the existence of conservative forces, such as that due to the Earth's gravity, that justifies the introduction of a new form of potential energy.

The gravitational force that the Earth exerts on a body beyond its surface is well described by Newton's law of universal gravitation. We shall consider this force later. For the moment however, let's restrict our attention to bodies lying close to the Earth's surface so that we can use the approximation of terrestrial gravitation. In such circumstances we can say that the gravitational force acting on a body of mass m is constant and given by the body's weight, $m\boldsymbol{g}$, where $\boldsymbol{g}$ is a constant vector representing the acceleration due to gravity. As usual, we shall assume that $\boldsymbol{g}$ is directed vertically downwards and that its magnitude is $g = |\boldsymbol{g}| = 9.81\ \mathrm{m\ s^{-2}}$.

To understand what a conservative force is, consider the process of lifting a stone of mass m slowly and vertically from the ground to a given height h, and then lowering it slowly and vertically back to the ground again (as shown in Figure 2.11). Of course, forces must be applied to cause this motion but for the moment we shall ignore all forces *except* for the constant gravitational force (i.e. the weight of the stone) which is directed downwards and has magnitude $m\boldsymbol{g}$ throughout the motion.

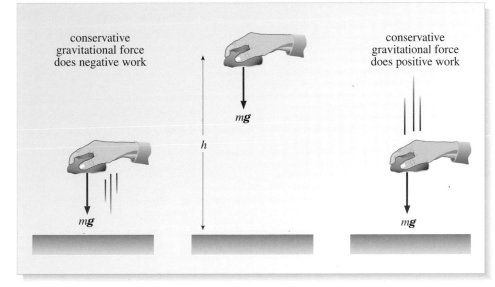

Figure 2.11 Raising and lowering a stone. The total work done by the gravitational force during this process is zero.

From Equation 2.5, the work done by this constant downward force, as a result of the stone's upward displacement, will be negative and will be given by $-mgh$. Similarly, the work done during the downward displacement, as the stone is returned to the ground, will be positive and will be given by, $+mgh$. Thus the total work done by gravity over the complete (*closed*) path is zero. (The positive work done in lowering recovers the negative work done in lifting the stone.)

We shall now show that this conclusion, that zero work is done by the gravitational force when the stone moves around a closed path, is true, irrespective of the precise shape of the path. The path does not have to go straight up and straight down; it can have any shape, as long as it's closed, and the total work done by the gravitational force, when the stone returns to its starting point, will still be zero. (Note that we are *only* talking about the work done by the *gravitational force*. The work that you would do in moving the stone around a closed path is certainly not zero, but that's not our concern at present.)

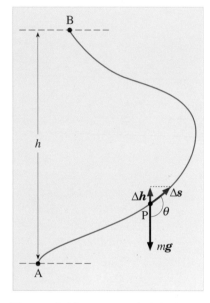

Figure 2.12 A stone is lifted from A to B via an arbitrary path through P.

Figure 2.12 shows a stone of mass m being lifted from point A to point B, through a height h, via an arbitrary path that passes through a point P. If the stone undergoes a very small displacement $\Delta \boldsymbol{s}$ from P then the small amount of work, ΔW, done by the gravitational force $\boldsymbol{F}_{\text{grav}} = m\boldsymbol{g}$ over that displacement will be:

$$\Delta W = \boldsymbol{F}_{\text{grav}} \cdot \Delta \boldsymbol{s}. \tag{2.9}$$

By choosing $\Delta \boldsymbol{s}$ to be sufficiently small, we can treat it as a small displacement along the (curved) path of the stone, and we can say that the work done by the gravitational force as the stone moves along that small part of its path is

$$\Delta W = mg\Delta s \cos\theta$$

where Δs is the magnitude of $\Delta \boldsymbol{s}$. However from Figure 2.12 it can be seen that

$$\Delta s \cos\theta = -\Delta h$$

where the minus sign arises from the fact that h increases in the *upward* direction, while $\Delta s \cos\theta$ is the *downward* component of $\Delta \boldsymbol{s}$. So

$$\Delta W = -mg\Delta h. \tag{2.10}$$

This tells us that the work done over the displacement $\Delta \boldsymbol{s}$ depends only on the height change Δh and not directly on Δs or on θ. It is then clear that the work done by gravity between A and B, which is just a succession of such small steps, also depends only on the total height difference and not on the details of the path chosen. We can indicate this mathematically by using the summation symbol (Σ). This symbol was introduced in Chapter 1, but on this occasion we will omit the limits and assume that the sum extends over all relevant values, so

$$W_{\text{AB}} = \sum \Delta W = -mg\sum \Delta h = -mgh. \tag{2.11}$$

Thus, the total work done by the gravitational force over any path from A to B is simply $W_{\text{AB}} = -mgh$. Similarly, over the return journey from B to A by *any* path, the work done by the gravitational force is $W_{\text{BA}} = +mgh$, so the round trip by any path is completed with zero net work done by gravity.

In general, conservative and non-conservative forces are defined as follows:

A force acting on a particle is said to be **conservative** if, when the particle moves around a closed path, the total work done by the force is zero, irrespective of the choice of closed path. Forces that do not satisfy this condition (i.e. those for which there is a closed path around which the work done is not zero) are said to be **non-conservative**.

An equivalent definition of a conservative force that may be shown to be a direct consequence of the first is the following:

A force acting on a particle is said to be conservative if the work that it does when the particle moves from point A to point B is independent of the path that the particle follows from A to B. Forces which do not satisfy this condition (i.e. those for which the work done between A and B is path-dependent) are said to be non-conservative.

From the argument we have given above, it is clear that the force of terrestrial gravitation satisfies these equivalent requirements and is therefore a conservative force. We have not proved it, but the same will also be true of the universal gravitational force we shall consider later, and of many other forces you will meet in physics. However, it's important not to make the mistake of thinking that every force is conservative. We have already noted that the muscular forces you would employ in moving a stone around a closed path will do work within your own body, and you certainly can't recover that. No amount of additional stone moving can replace the energy you expend. Similarly, while the stone is moving, its motion is always opposed by a drag force arising from air resistance. Precisely because it is opposed to the motion, this force will always do negative work, no matter which way the stone moves, so its contributions can never add up to zero, and, consequently it cannot be conservative.

Question 2.5   Explain why the frictional force on a sliding object cannot be conservative.   ■

3.2 Potential energy — energy due to configuration

We have just seen that lowering a body of mass m through a height h close to the Earth's surface means that the gravitational force on the body will *always* do a positive amount of work mgh, irrespective of the path followed. This work represents energy transferred to the body during its descent, by the gravitational force. This energy might show up as the increasing kinetic energy of a falling body, or as the energy dissipated in overcoming air resistance, or in many other ways, but whatever happens to it there can be no doubt that it is transferred to the descending body. (The fact that the gravitational force is conservative ensures that this will be so.) As a result, the body behaves as though the process of raising it to the height h had somehow enabled it to 'store' an amount of energy mgh that can be released by lowering the body again.

This stored energy, mgh, is referred to as the **gravitational potential energy** of the body, and may be represented by the symbol E_{grav}, so we may write:

$$E_{grav} = mgh. \tag{2.12}$$

The name 'gravitational potential energy' makes good sense. The energy is certainly associated with the conservative gravitational force, and it is potential energy in the sense that it represents the energy that is potentially available for release by returning the body to the ground.

Gravitational potential energy has many applications. The energy stored by water behind a dam is a case in point. The controlled release of that energy, by allowing the water to descend through turbines, produces hydroelectricity that is important in many regions of the world. The operation of a piledriver provides another example

Figure 2.13 Some examples of systems that use gravitational potential energy.

of gravitational potential energy in action, as does a long-case clock that relies on the gradual descent of a weight to keep the pendulum moving (Figure 2.13).

Equation 2.12 implies that $E_{\text{grav}} = 0$ when $h = 0$. This seems natural; it makes sense to say that a body sitting on the ground has no gravitational potential energy. However, it is important to realize that this is nothing more than an arbitrary choice on our part. Our earlier arguments certainly showed that the work done by gravity when a body is lowered is mgh, and this certainly represents a *change* in gravitational potential energy, but that doesn't compel us to say that E_{grav} will be zero at ground level. We implicitly chose to adopt that convention when we wrote down Equation 2.12, but we could have made a different choice. If, for instance, we had chosen to say that a body sitting on the ground has gravitational potential energy E_0, then a body of mass m at height h would have $E_{\text{grav}} = mgh + E_0$. Sometimes it is useful to adopt this alternative convention and to adjust Equation 2.12 accordingly. The point to remember is this:

> Only *changes* in gravitational potential energy are physically significant. Any formula that assigns particular values to the gravitational potential energy (such as $E_{\text{grav}} = mgh$) always involves making an arbitrary choice concerning the condition of zero gravitational potential energy.

In the case of Equation 2.12, working out the change in gravitational potential energy associated with a change in height is easy. Suppose that a body of mass m moves from some initial position in which it is at height h above the ground to some final position which is at height $h + \Delta h$, where Δh represents the change in height and may be positive or negative (depending on whether the body is raised or lowered). The change in gravitational potential energy as a result of the change in position can be found by subtracting the initial value of the gravitational potential energy from the final value. Thus,

$$\Delta E_{\text{grav}} = mg(h + \Delta h) - mgh \qquad (2.13)$$

i.e. $\qquad \Delta E_{\text{grav}} = mg\Delta h.$ $\qquad\qquad\qquad (2.14)$

So, under the conditions of terrestrial gravitation, the *change* in a body's gravitational potential energy is simply proportional to the *change* in its height.

Question 2.6   A cannon-ball of mass 10 kg is fired vertically upwards with an initial speed of 200 m s^{-1}.

(a) Use the equation of uniformly accelerated motion to calculate how high it goes before it stops. (Ignore air resistance and use $g = 9.8$ m s^{-2}.)

(b) Use $W = \boldsymbol{F} \cdot \boldsymbol{s}$ to evaluate the work done by gravity from the time the cannon-ball leaves the cannon until it reaches its highest point.

(c) Use Equation 2.14 to determine the change in the gravitational potential energy of the cannon-ball as it subsequently returns halfway to the ground from the highest point.

(d) What is the work done on the cannon-ball during this part of the descent?   ■

Now, all that we have said so far has applied specifically to a system consisting of the Earth and a body of mass m sufficiently close to its surface for us to treat the gravitational force on the body as a constant, $m\boldsymbol{g}$. However, similar ideas apply to any system composed of parts that interact via conservative forces, whatever those forces may be. It will be true, for example, that each configuration of a system of charged particles that interact by means of electrical forces can be associated with a definite amount of electrical potential energy, and it is also possible to associate a definite amount of gravitational potential energy with each configuration of a system whose parts interact via the force of universal gravitation. Both the electrical force between charged particles and the universal gravitational force are conservative forces, so potential energy may be associated with either system. In fact, we can make the following general statement:

> The *potential energy* of a body is the energy associated with the position of that body relative to other bodies with which it interacts via conservative forces.

Notice that although we speak of the potential energy 'of a body', and describe it informally as energy 'by virtue of position', it is really the energy of a *system* and it really depends on the *configuration* of the system. The gravitational potential energy of a stone increases when we move the stone away from the Earth, but it would increase just as much if we moved the Earth away from the stone. It is the *separation* of the Earth *and* the stone that is significant, not just the position of the stone.

No matter which conservative forces cause us to introduce potential energy, we can always say that *changes* in potential energy are physically significant, and that the change in potential energy in going from some initial configuration to some final configuration is equal to the work done *by those conservative forces* in returning the system from its final configuration to its initial one. We can represent this in terms of symbols by writing

$$\Delta E_{\text{pot}} = E_{\text{pot}}(\text{final}) - E_{\text{pot}}(\text{initial}) = W_{\text{cons}}(\text{final} \rightarrow \text{initial}). \qquad (2.15)$$

The subscript 'cons' is to remind us that it is only the work done by the conservative forces (such as gravitational or electrical forces) that is of any relevance. Any work you may do in changing the configuration, or any work done by dissipative forces such as those due to friction or air resistance, is irrelevant.

Question 2.7 In your own words, briefly explain why it is possible to associate potential energy with a conservative force such as terrestrial gravitation, but not with a dissipative force such as that due to friction. (*Hint*: Start from the statement that potential energy is energy by virtue of position.) ■

If we actually want to associate particular values of potential energy with particular configurations of the system, then we must first arbitrarily choose a configuration of zero potential energy. Having chosen this zero energy configuration we may then say that

> The potential energy of any configuration of a system is the work that would be done by the relevant conservative forces in going from that configuration to the configuration of zero potential energy.

This definition once again emphasizes the interplay of energy and work. The only potential energy formula we have been able to write down so far is $E_{grav} = mgh$, because it applies to cases where the gravitational force is constant, and we were therefore able to use the results of Section 2 to evaluate the work done. In order to discuss more general situations, including the potential energy of a spacecraft escaping from the Earth, we must first learn how to evaluate the work done by a changing force. An important example of this, is considered in the next subsection.

3.3 Strain potential energy

Raising a body above the ground is certainly not the only way of storing energy. If you have ever snapped an elastic band against your fingers or fired a catapult you will be aware that these simple devices also store energy. The same is true of a spring that has been stretched or compressed. In all these cases the energy is stored in a form referred to as **strain potential energy**. Fundamentally, this is actually a form of electrical potential energy, since it arises from the electrical attraction between atoms in the stretched material. However, you don't need to know anything about those electrical forces in order to analyse strain potential energy, as you are about to see.

When a spring is stretched or compressed, it is apparent that the force required varies with the amount of stretching or compressing. This is turned to good account in simple weighing devices, where the compression or extension of a spring is used as a measure of the applied force.

In this subsection we shall evaluate the change in strain potential energy when an ideal spring is stretched or compressed. This will involve learning how to calculate the work done by a varying force, a skill that will be of use later. The particular spring we are going to discuss is shown in Figure 2.14. It lies along the x-axis, has one end fixed, and, when unstretched, has its free end is at $x = 0$. Being ideal, the spring obeys Hooke's law

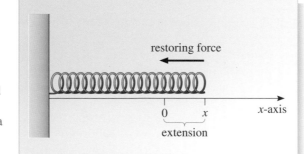

Figure 2.14 The spring lies along the x-axis. When unstretched its free end is at $x = 0$. When stretched (or compressed) there is a restoring force towards the origin described by $F_x = -k_s x$.

(see Chapter 1), so when the free end is moved to any position x, there is a linear restoring force tending to return it to its unstretched position:

$$F_x = -k_s x \qquad \text{(Eqn 1.26)}$$

where k_s is called the spring constant, and is characteristic of the material of the spring. This restoring force is a conservative force. The work that it does when the spring is stretched or compressed can be fully recovered when the spring is released.

If in this case we let the strain potential energy be zero when the spring is unstretched, then it follows from what was said in the last subsection that the strain potential energy when the spring is extended by some particular amount $x = X$, will be equal to the work done by the restoring force as the extension is reduced from $x = X$ to $x = 0$. Evaluating this amount of work presents us with a problem; as the extension is reduced the magnitude of the force diminishes, and we have not yet encountered any way of evaluating the work done by a varying force, even when it is aligned with the displacement.

The strategy adopted for calculating the work done is, in outline, to divide the extension into intervals small enough that the restoring force can be considered constant over each interval. The total work done by the restoring force during the return to equilibrium is then the sum of the work done during each small part of the return. (Since the restoring force is conservative, this will also be the negative of the work done by the restoring force during the process of extending the free end from $x = 0$ to $x = X$.)

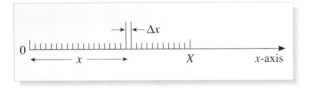

Figure 2.15 The extension x, between zero and X (the maximum extension for which we are calculating the energy), can be divided into many small intervals each of length Δx. There is no limit to how small Δx can be.

Figure 2.15 shows one of these small steps along the x-axis, at some general position x. The step is of length Δx, so the displacement of the free end as it takes this small step *towards* the equilibrium position is $s_x = -\Delta x$. Since the restoring force at this general position is $F_x = -k_s x$, it follows that the work done over this small step will be

$$\Delta W = F_x s_x = (-k_s x) \times (-\Delta x) = k_s x \Delta x.$$

Now for the clever bit! Note that this amount of work is equal to the area of a rectangle of height $k_s x$ and width Δx; just such a rectangle is shown in Figure 2.16. Its height is actually equal to the magnitude of the restoring force at the relevant value of x, so we have labelled the vertical axis to show that. The rectangle is supposed to be narrow enough that the force is essentially constant over the interval Δx, so it does not matter whether x refers to the low-x or high-x side of the rectangle.

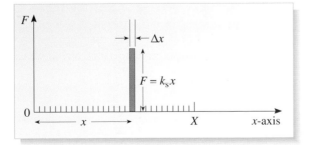

Figure 2.16 If Δx is small enough for F_x to be effectively constant over this interval, then the work done by the restoring force as the free end crosses this interval, heading towards $x = 0$, is represented by the area of the small coloured rectangle.

The total work done by the restoring force throughout the return from $x = X$ to $x = 0$ can be represented graphically by drawing similar little rectangles for each step and adding their respective areas together. This process is indicated in Figure 2.17. Note that the rectangles become shorter as we approach $x = 0$, because the magnitude of the restoring force decreases as we approach the origin. Since we have treated the force as constant across each small step this area is actually only an approximation to the total work done, but if the steps are small enough it should be a good one. We can indicate this approximation mathematically by using our summation sign again.

$$W \approx \sum k_s x \, \Delta x$$

where the symbol $\approx$ indicates an *approximate* equality, and the summation symbol tells us to add together the contributions $k_s x \Delta x$ from each of the small steps in the interval from $x = 0$ to $x = X$.

Figure 2.17 As the intervals Δx become smaller, it becomes evident that the sum of the areas of the rectangles (and hence the total work done) is the area of the coloured triangle, i.e. $\frac{1}{2} k_s X \times X = \frac{1}{2} k_s X^2$.

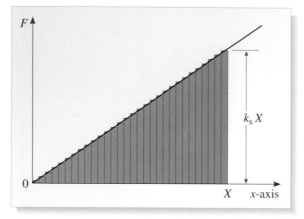

We can make the approximation more accurate by letting the steps get smaller and smaller. As you can see from Figure 2.17, if we do this, then the shaded area will get closer and closer to that of a triangle of base length X and height $k_s X$. The area of such a triangle is equal to $\frac{1}{2} k_s X^2$ (it's half the area of a rectangle of base X and height $k_s X$). So we can say that the accurate value of the total work done is

$$W = \frac{1}{2} k_s X^2.$$

Now, there was nothing special about the value X that we chose in arriving at this result. It could have been any value of x, so we can safely say that the total work done by the restoring force in returning an ideal spring from *any* extension x to its unstretched length is $\frac{1}{2} k_s x^2$. From which it follows that

> The strain potential energy of an ideal spring that has been extended by an amount x (positive or negative) is
> $$E_{str} = \frac{1}{2} k_s x^2. \tag{2.16}$$

This is the expression we were seeking. It allows us to work out the energy stored (as strain potential energy) in a stretched or compressed spring. Admittedly, the formula only applies to ideal springs, but it will also work for many real springs, or for stretched wires, etc. as long as they obey Hooke's law, and are not stretched too far.

Question 2.8 A simple catapult (Figure 2.18), is made from a spring that obeys Hooke's law and for which $k_s = 200 \, \text{N m}^{-1}$. The spring is stretched by 20 cm and then released. If all the stored energy is transferred to a projectile of mass 0.005 kg, at what speed will it travel as it leaves the catapult?

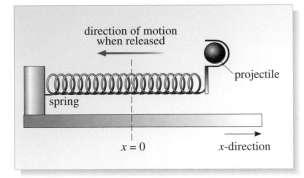

Figure 2.18 A simple catapult made from a spring that can be extended and released. When unextended, the free end of the catapult is at the origin, $x = 0$.

Question 2.9 What is the increase in the strain potential energy of a spring, with $k_s = 200 \, \text{N m}^{-1}$, when its extension is increased from 15 cm to 20 cm? Compare this with the increase in energy when the extension is increased from 20 cm to 25 cm. ■

3.4 The work done by a varying force

The graphical procedure that we have just used to calculate the work done by a linear restoring force can be applied to other forces, even when the magnitude of the force is not linearly related to the displacement. It should be noted that the forces we shall be discussing in this subsection *do not have to be conservative*. Any force may do work, though if the force is non-conservative then the work that it does cannot be immediately related to a change in potential energy.

Suppose a force F_x acts along the x-axis, and that its point of application moves from some point $x = A$ to some other point $x = B$. (It might, for example, be a contact force pushing a chest of drawers, or a frictional force opposing the motion of such a chest.) Also suppose that during this movement the force varies in the way indicated by the smooth curve in Figure 2.19. Then, by adding together a series of narrow rectangular areas, each of which represents the work done by a constant force over a small displacement of length Δx, and by considering what happens to that sum as the number of strips increases while the width of each is reduced, we can say that:

> The work done by the force F_x over the displacement from $x = A$ to $x = B$ is represented by the area under the graph of F_x against x, between A and B.

This is just a generalization of the result we obtained in the last subsection.

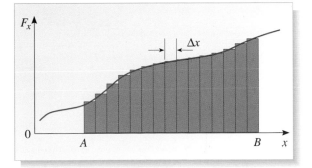

Figure 2.19 No matter how F_x varies with x, the work that it does over a displacement from $x = A$ to $x = B$, is still represented by the area under the force–displacement graph. As Δx becomes smaller and smaller, the total area of the rectangles approximates more and more closely to the area under the curve.

It is rather reassuring to know that it is possible to evaluate the work done by a varying aligned force by determining the area under an appropriate force–displacement curve. But it's also slightly unnerving since you might be asked to carry out such an evaluation. Apart from a few special cases, measuring the area under a curve is a very time-consuming and usually pretty inaccurate process. Nobody would do it if they could avoid it.

Fortunately, if we know exactly how F_x depends on x, i.e. if we know how to express F_x as a *function* of x, then there is a mathematical technique that will often allow us to evaluate the relevant area (and hence determine the work done) without needing to draw any graphs at all. This technique is called **integration** — a very appropriate name, since it suggests adding things together, which is precisely what we do when we take the sum of many small rectangular areas. Integration forms an important part of the subject known as **integral calculus**. As such it is a counterpart to the technique of *differentiation* (which forms part of *differential calculus*) that was reviewed in Chapter 1, and which is used to avoid having to evaluate gradients graphically.

In the notation of integral calculus, the work W done by a force F_x over a displacement from $x = A$ to $x = B$ is indicated by writing

$$W = \int_A^B F_x \, dx. \tag{2.17}$$

The right-hand side of this equation is said to be 'the **definite integral** of F_x, with respect to x, from A to B'. It can be thought of as a shorthand for 'the area under the F_x against x curve, between A and B' (Figure 2.20), and it consists of several distinct elements, each with its own name (see Figure 2.21). There is a large distorted S (called an *integral sign*) which indicates that we are dealing with integration and hence, in some sense, with a sum of vanishingly small terms. The integral sign is topped and tailed by the final and initial values of the displacement, B and A, (respectively called the *upper* and *lower limits of integration*) with the starting value at the bottom. The variable x (called the *integration variable* in this context) that changes continuously across the displacement is indicated by writing an *integration element* dx on the right of the integral, and the quantity that depends on the integration variable (the *integrand*) is written in the centre; in our case it is F_x. There's no need to remember all these terms, but the thing you should note is that taken together they provide all the mathematical information that is required to identify precisely the relevant graph (F_x against x) and the relevant part of the area under it (from $x = A$ to $x = B$) in order that the work done, W, can be evaluated.

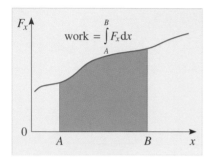

Figure 2.20 The work done by a force F_x over the displacement from $x = A$ to $x = B$, is equal to the definite integral of F_x from A to B.

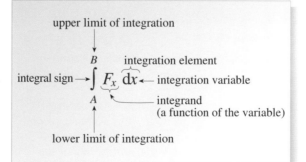

Figure 2.21 The elements of a definite integral. You do not need to remember these terms.

Of course, the important point about the definite integral is that it may be evaluated using algebra alone provided F_x is known as a function of x. You already know the result for one particular definite integral. From the graphical discussion in the last subsection you know that if $F_x = -k_s x$, and if $A = X$ and $B = 0$, then

$$\int_A^B F_x \, dx = \int_X^0 (-k_s x) \, dx = \tfrac{1}{2} k_s X^2. \tag{2.18}$$

We were able to arrive at the answer on the right by working out the triangular area under a sloping line, but to someone who was trained in the evaluation of definite integrals the result on the right would follow immediately (by algebra alone) from the expression in the middle.

If you have not previously met integration you are not expected to be able to follow these lines.

If you have already been taught how to integrate then you will recognize that if $F_x = kx^3$, for example, then

$$\int_A^B F_x \, dx = \int_A^B (kx^3) \, dx = \tfrac{1}{4}k(B^4 - A^4) \tag{2.19}$$

and, if $F_x = \dfrac{k}{x^2}$ then

$$\int_A^B F_x \, dx = \int_A^B \left(\frac{k}{x^2}\right) dx = k\left(\frac{1}{A} - \frac{1}{B}\right) \tag{2.20}$$

where k is a constant in both cases.

Although we shall not write it down, there is a precise general definition of the definite integral that allows mathematicians to evaluate definite integrals of a wide range of functions between given limits. However, this is rarely used by physicists, who generally know a few standard integrals and can work out many others by combining the ones they already know according to certain standard rules. (Increasingly though, physicists and engineers make use of computer packages that evaluate integrals for them.) This makes integration a valuable everyday tool, and many important laws and results are commonly expressed in terms of integrals.

Equation 2.17 is a good example of an important result that is most easily expressed in terms of an integral. Thus, rather than defining the work done by a varying aligned force as the area under a graph, it is actually more conventional to write the definition as follows:

> The work done by a variable force F_x (that is a function of x), over a displacement from $x = A$ to $x = B$ is
>
> $$W = \int_A^B F_x \, dx \tag{Eqn 2.17}$$

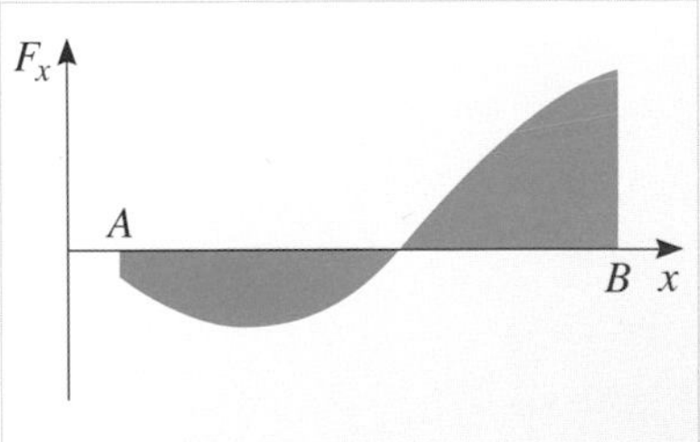

Figure 2.22 In this case, the work done by the force F_x over the displacement from A to B, is negative during the first part of the displacement and positive during the second part. If the positive and negative areas are equal in magnitude, then the total work done by the force will be zero.

Note that this result holds true even if F_x is negative over some part of the range of integration. In graphical terms (see Figure 2.22) any region where F_x is negative corresponds to an 'area under the curve' that is below the horizontal axis. Such an area is a signed quantity and is counted as negative since, if x increases during the displacement, a negative force will do a negative amount of work when its point of application moves through a positive displacement. (In the last subsection, when dealing with the relaxing spring, we were actually dealing with a negative force $(-k_s x)$ *and* a negative displacement (from X to 0), so the work done was positive, as we found.)

Finally, and mainly for the sake of completeness, we consider the work done by a varying force $\boldsymbol{F}$ as its point of application moves along some specified path in two or three dimensions. (As an example you might like to think of the work done on a roller-coaster by a resultant force arising from gravity, friction, air resistance, etc.) If

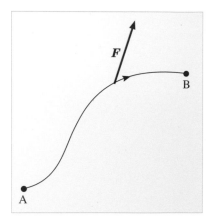

Figure 2.23 The work done by a force F as its point of application moves from A to B along a specified path in two or three dimensions.

the path starts at some initial point A and ends at some final point B, as indicated in Figure 2.23, then the work done by the force is

$$W = \int_A^B F \cdot ds \tag{2.21}$$

where the integration element ds indicates that the displacement is along the specified path, and the bold dot indicates that we are dealing with a scalar product of vectors.

3.5 Gravitational potential energy

In Section 3.2 we saw that the gravitational potential energy of a body of mass m at a height h above the ground can be written

$$E_{grav} = mgh \tag{Eqn 2.12}$$

provided that:

1 The body is sufficiently close to the surface that we can suppose the gravitational force acting on it to be the constant vector mg.

2 We adopt the convention that the gravitational potential energy of the body is zero when the body is on the ground (i.e. when $h = 0$).

Our main aim in this subsection is to find an expression for the gravitational potential energy of a body of mass m that is *not* necessarily close to the Earth. The gravitational force acting on such a body will be described by Newton's law of universal gravitation. If we use a polar coordinate system in which the radial coordinate r of the body's centre of mass is measured outward from the centre of the Earth (Figure 2.24), we can say that the gravitational force that the Earth exerts on the body is

$$F = F_r \hat{r} = \frac{-GmM_E}{r^2} \hat{r} \tag{2.22}$$

where F_r is the radial component of the gravitational force (the only component that is not zero in this case) and $\hat{r}$ is a radial unit vector pointing away from the centre of the Earth. Note that the combination of the outward unit vector $\hat{r}$ and a minus sign on the right-hand side of Equation 2.22, indicates that the gravitational force on the body is actually directed *towards* the Earth.

Figure 2.24 The gravitational force exerted by the Earth on a body of mass m. The only significant feature of the body's position is its distance from the centre of the Earth, which is determined by its radial position coordinate r.

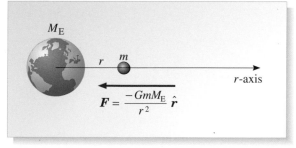

The universal gravitational force is a conservative force, so it may be associated with a gravitational potential energy. The general procedure for working out the potential energy of any given configuration of a system whose components (the Earth and the body in this case) interact via a conservative force was laid out in Section 3.2. It consists of two steps.

- First, choose a configuration of zero potential energy. (This is an arbitrary choice; it may be whatever we wish, since it is only *changes* in potential energy that are physically significant.)

- Second, calculate the potential energy of any given configuration of the system by determining the work done by the relevant conservative force (the gravitational force in this case) in going from the given configuration to the configuration of zero potential energy.

> When dealing with the gravitational potential energy arising from universal gravitation, it is conventional to choose the configuration of zero potential energy to be one in which the body is infinitely far away from the Earth.

This may seem a rather surprising choice at first; there's certainly no denying that it's different from the convention we adopted earlier when dealing with terrestrial gravitation. However, a body that is infinitely far away from the Earth is completely free from its gravitational pull, so it makes sense to say that such a body has no energy arising from its gravitational interaction with the Earth. In any event, that is the convention we shall adopt.

The gravitational potential energy of a body of mass m, at a distance r from the centre of the Earth, will be given by the work done by the gravitational force as the body is moved from its starting point to some other point infinitely far away. Since the gravitational force is conservative, the work done during this displacement will not depend on the path followed, only on the endpoints. Hence, we only need to consider the work done in moving along the simplest possible path; namely, one that leads radially outwards from the initial point to infinity. The precise location of the initial and final points is also irrelevant in this case, since it is only the body's initial and final *distance* from the centre of the Earth that will affect the amount of work done. Let's suppose that the body of mass m is initially located at a distance R from the centre of the Earth, so that initially $r = R$, and let us adopt the mathematical convention that infinity is represented by the symbol ∞, so that finally $r = \infty$. The work done by the universal gravitational force over the outward radial displacement from R to ∞ can then be written

$$W = \int_R^\infty F_r \, dr = \int_R^\infty \frac{-GmM_E}{r^2} \, dr. \tag{2.23}$$

The meaning of this definite integral is indicated graphically in Figure 2.25. It's just the area under the graph of F_r against r between $r = R$ and $r = \infty$.

Now, you may think that, because the value of F_r never quite reduces to zero, no matter how big r becomes, the shaded area 'under' the graph must be infinite. However, this is not so. The shaded area (even when r is extended out to infinity) turns out to be finite and may be shown to be the negative quantity

$$W = \int_R^\infty F_r \, dr = \frac{-GmM_E}{R}. \tag{2.24}$$

Remember, this is the work that would be done by the gravitational force, which points inwards, if the body was moved outwards (by whatever means) from R to ∞. So it represents the gravitational potential energy of a body of mass m at a distance R from the centre of the Earth. Using an argument similar to that employed when seeking an expression for strain potential energy, we can note that there is nothing special about the value R that we choose to characterize our starting point. We could

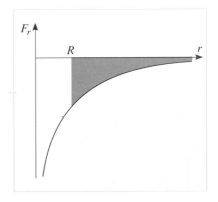

Figure 2.25 The curve shows the force F_r on a body of mass m at a distance r from the centre of the Earth. The fact that the force is negative indicates that it is directed towards the Earth. The work done by the gravitational force when the body is displaced from $r = R$ to $r = \infty$ is equal to the shaded area under the curve. Note that this area will be negative because it lies below the horizontal axis.

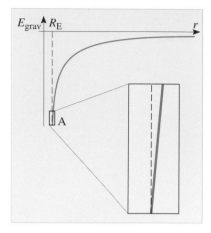

Figure 2.26 The (universal) gravitational potential energy E_{grav} $= -GmM_E/r$ plotted against r. Over a sufficiently small region, close to the Earth's surface, the curve can be approximated by a straight line (see boxed region).

just as easily have chosen any other value of r (provided it exceeded the radius of the Earth R_E). So, from Equation 2.24 we can say that:

> The gravitational potential energy of a body of mass m, the centre of mass of which is at a distance r from the centre of the Earth is
>
> $$E_{grav} = \frac{-GmM_E}{r} \quad \text{(provided } r > R_E\text{).}$$
> (2.25)

The graph of this gravitational potential energy function is shown in Figure 2.26. As you can see from the graph, the (universal) gravitational potential energy is *negative* at all points. This disagrees with our earlier result for the (terrestrial) gravitational potential energy of a body at height h above the Earth's surface, which was given by the *positive* quantity

$$E_{grav} = mgh,$$
(Eqn 2.12)

but such a disagreement about signs is to be expected because we were using a different convention about the zero of gravitational potential energy when we derived that earlier 'terrestrial' result. A more significant point concerns the consistency of the two results (Equations 2.12 and 2.25) in predicting *changes* in gravitational potential energy, since it is only changes in potential energy that are physically significant. If you look at the part of Figure 2.26 that corresponds to conditions close to the Earth's surface — the part enclosed in a box near point A — you should be able to see that the curve can be approximated by a straight line provided we don't examine it too closely, and provided we don't try to use the linear approximation over too great a range of values. If we identify a change in radial distance, r in Equation 2.25 with a change in height h in Equation 2.12, then the predictions of Equation 2.12 regarding *changes* in gravitational potential will approximately agree with the predictions of Equation 2.25 provided we restrict our attention to fairly modest values of Δh in the region close to the Earth's surface. The truth of this claim is left for you to explore in Question 2.10, but the overall conclusion is this:

> Within reasonable limits the gravitational potential energy predicted by terrestrial gravitation (Equation 2.12) closely agrees with that predicted by universal gravitation (Equation 2.25) apart from a fixed difference arising from the different configurations chosen to represent zero potential energy in the two cases.

Question 2.10 (a) Given that the radius of the Earth is $R_E = 6.378 \times 10^6$ m, and that the mass of the Earth is $M_E = 5.977 \times 10^{24}$ kg, estimate your own gravitational potential energy using both Equations 2.12 and 2.25. (b) Use both equations again to predict the amount by which your gravitational energy should increase if you board a plane and are flown to an altitude of 10 km. Work to four significant figures and comment on the consistency of your results. (c) How would your answers to part (b) be affected if you were to neglect air resistance? ■

One final point to end this particular discussion of gravitational potential energy. Equation 2.25 provides the expression we said we were seeking for the potential energy of a body due to its gravitational interaction with the Earth. However, since the expression was derived on the basis of Newton's law of universal gravitation it can be easily generalized. One such generalization is the following:

The gravitational potential energy of two particles of masses m_1 and m_2, separated by a distance r is

$$E_{\text{grav}} = \frac{-Gm_1m_2}{r}. \tag{2.26}$$

Other generalizations are possible. You may like to think of some for yourself.

3.6 Force as the negative gradient of potential energy

You have now seen several potential energy functions derived from the relevant conservative forces, and you have learned the general technique for deriving others. (Choose a configuration of zero potential energy and calculate the work done by the relevant conservative forces in going from any given configuration to the zero energy configuration.) You have also seen that when the configuration of a system changes, that change generally involves the storage of energy (positive or negative), as a result of which the potential energy changes by an amount

$$\Delta E_{\text{pot}} = W_{\text{cons}}(\text{final} \rightarrow \text{initial}) \tag{Eqn 2.15}$$

where $W_{\text{cons}}(\text{final} \rightarrow \text{initial})$ is the work that would be done by the conservative force in returning the system to its initial configuration.

Now, the total work done by a conservative force when its point of application moves around a closed path is zero, so

$$W_{\text{cons}}(\text{final} \rightarrow \text{initial}) + W_{\text{cons}}(\text{initial} \rightarrow \text{final}) = 0.$$

It follows that

$$W_{\text{cons}}(\text{final} \rightarrow \text{initial}) = -W_{\text{cons}}(\text{initial} \rightarrow \text{final}),$$

and that we may rewrite Equation 2.15 as

$$\Delta E_{\text{pot}} = -W_{\text{cons}}(\text{initial} \rightarrow \text{final}). \tag{2.27}$$

In other words, any change in potential energy that results from a change in configuration is *minus* the work done by the relevant conservative force.

Now suppose that this change in potential is the result of a body being displaced in one dimension by a small amount Δx while being subject to a conservative force F_x that acts in the x-direction. If Δx is sufficiently small we can treat F_x as approximately constant throughout the small displacement and thus arrive at the approximate equality

$$W_{\text{cons}}(\text{initial} \rightarrow \text{final}) \approx F_x \Delta x.$$

Substituting this into Equation 2.27 gives

$$\Delta E_{\text{pot}} \approx -F_x \Delta x, \tag{2.28}$$

from which it follows that

$$F_x \approx \frac{-\Delta E_{\text{pot}}}{\Delta x}. \tag{2.29}$$

This relationship will become more accurate as Δx becomes smaller, and it implies (in the limit as Δx becomes vanishingly small) that the value of F_x at any particular value of x will be given by *minus* the gradient of the E_{pot} against x graph at the relevant value of x. As you will recall from the discussion of derivatives in Chapter 1, this is most easily expressed in mathematical terms by writing

$$F_x = \frac{-\mathrm{d}E_{\text{pot}}}{\mathrm{d}x}. \tag{2.30}$$

This equation provides a very powerful link between a potential energy function and the conservative force responsible for it. It can be easily modified to deal with cases where the relevant displacement is in some direction other than the *x*-direction. It can also be extended to cover three-dimensional situations in which the potential energy is a function of *x*, *y* and *z*, and the corresponding conservative force may have three non-zero components. We shall not pursue the rather complicated mathematics of this three-dimensional generalization, but it is worth noting that the vector $\boldsymbol{F} = (F_x, F_y, F_z)$ representing the conservative force at any position will point in the direction in which the associated potential energy is decreasing most rapidly and will have a magnitude which is proportional to that rate of decrease. This justifies the general claim that, even in three dimensions:

> Given a conservative force that may act at any point (such as the gravitational force), its value at any particular point is *minus* the gradient of the related potential energy at that point.

Equation 2.30 is the mathematical expression of this statement in one dimension. You may check the correctness of Equation 2.30 by using it to analyse some familiar examples where you already know both the force and the potential energy.

Question 2.11   When an ideal spring of spring constant k_s is extended by an amount *x*, its strain potential energy is $E_{str} = \frac{1}{2} k_s x^2$. Use Equation 2.30 together with the guidance on evaluating derivatives given in Section 4.1 of Chapter 1, to show that the restoring force exerted by the spring is given by Hooke's law.

Question 2.12   When a body of mass *m* is located at a modest height *h* above the surface of the Earth, its gravitational potential energy is $E_{grav} = mgh$. Explain the relevance of Equation 2.30 to the analysis of this situation, and find the corresponding expression for the gravitational force on the object due to the Earth.

Question 2.13   When a particle of mass *m* is located at a distance *r* from the centre of the Earth (where *r* is greater than the radius of the Earth) its gravitational potential energy is $E_{grav} = -GmM_E/r$, where M_E is the mass of the Earth. Explain the relevance of Equation 2.30 to the analysis of this situation, and find the corresponding expression for the gravitational force on the object due to the Earth. You will find it useful to know that $\dfrac{d}{dr}\left(\dfrac{1}{r}\right) = -\left(\dfrac{1}{r}\right)^2$.

Question 2.14   What specific feature of Figure 2.26 can be determined from Equation 2.30 and the fact that the weight of a body of mass *m* close to the Earth's surface has magnitude *mg* and is directed downwards, towards the centre of the Earth? ■

We shall make much use of this important idea, that a conservative force can be expressed as *minus* the gradient of a related potential energy, later in the course when we investigate the physics of fields in the book *Static fields and potentials*.

3.7 The conservation of mechanical energy

Any force (conservative or not) that acts on a particle may do work and thereby transfer energy to or from that particle. However, according to the work–energy theorem of Section 2.2, the change in the translational kinetic energy of a particle will be equal to the work done by the resultant force that acts upon it, so we may write

$$\Delta E_{trans} = W_{res}. \tag{2.31}$$

We also know that the change in the potential energy of a particle is equal to *minus* the work done on that particle by the relevant conservative force that acts upon it during the change, so we may write

$$\Delta E_{\text{pot}} = -W_{\text{cons}}(\text{initial} \rightarrow \text{final}). \tag{2.32}$$

Equation 2.32 includes an explicit reference to the initial and final configurations because it concerns conservative forces and those configurations entirely determine the work done in that case. No such mention of initial and final configurations is made in Equation 2.31 because the resultant force may be non-conservative, in which case the work done will depend on the path followed and hence on all the configurations between the initial and final ones. However, if we deliberately restrict our attention to situations in which the only forces acting on a particle are conservative, then we can also say that $W_{\text{res}} = W_{\text{cons}}(\text{initial} \rightarrow \text{final})$, and hence, from Equations 2.31 and 2.32 that

$$\Delta E_{\text{pot}} = -\Delta E_{\text{trans}} \qquad \text{(conservative forces only)}$$

i.e. $\qquad \Delta E_{\text{pot}} + \Delta E_{\text{trans}} = 0. \qquad \text{(conservative forces only)} \tag{2.33}$

Thus, for a particle subject only to conservative forces, the sum of the changes in kinetic and potential energy will be zero.

A simple example of this is provided by a stone thrown vertically upwards from the ground and moving under the influence of terrestrial gravity. In the absence of air resistance, the stone will slow down and lose kinetic energy as it rises. At the same time, it will gain potential energy. Over any part of the ascent the (positive) change in gravitational potential energy when added to the (negative) change in kinetic energy will produce a total change of zero, because the gravitational force is conservative (Figure 2.27). Contrast this with what happens when a stone slides across a rough horizontal surface. The stone soon slows down and stops. In this case, the gravitational potential energy is unchanged but the kinetic energy has undergone a negative change, so the sum of the changes is not zero. This has occurred because the stone is subject to a non-conservative frictional force, so Equation 2.33 does not apply.

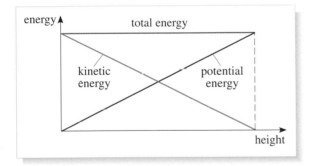

Figure 2.27   The kinetic energy and the gravitational potential energy of a stone thrown vertically upwards from the Earth. Note that the sum of the two energies is constant and equal to the kinetic energy at the time of launch.

According to Equation 2.33 a particle may alter its kinetic energy or its potential energy, or both, but, provided all the forces that act on the particle are conservative, the sum of the changes must be zero. As a consequence, the sum of the particle's kinetic energy and potential energy, $E_{\text{trans}} + E_{\text{pot}}$, must be constant. That sum, which is known as the particle's total **mechanical energy**, must have the same value at the beginning and end of any process — provided that only conservative forces act during the process. This important result is known as the law of **conservation of mechanical energy** and may be stated formally as follows:

Provided the only forces which act on a particle are conservative forces, then the mechanical energy of that particle will remain constant

$$E_{\text{mech}} = E_{\text{trans}} + E_{\text{pot}} = \text{constant}. \tag{2.34}$$

This principle often provides a means of relating the speed of a particle to its position, and plays an important part in the solution of many problems in mechanics. As you will see later, it can be modified to cover systems of particles and rigid bodies (that might be rotating), but provided we ignore rotation we can apply it to some extended bodies as it stands.

Question 2.15   A cannon-ball of mass 10 kg is fired vertically upwards with an initial speed of 200 m s^{-1}. (The same cannon-ball was discussed in Question 2.6.)

(a) What is its initial translational kinetic energy?

(b) Use the conservation of mechanical energy to calculate the maximum height that the cannon-ball will reach in the absence of air resistance.

(c) Ignoring air resistance, work out the speed of the cannon-ball when it has covered half the distance back to the ground.

(d) What will be the speed of the cannon-ball just before it hits the ground?

(e) When the cannon-ball hits the ground it is (almost) immediately brought to rest. What do you deduce about the nature of the forces that stop it?   ■

Although mechanical energy is tremendously useful in helping to simplify calculations involving force and motion, its true importance only began to be perceived in the mid-nineteenth century. Around 1850, a number of scientists realized that the concept of energy had applications well beyond the confines of mechanics. It was just as meaningful, for instance, to talk about electrical energy or thermal energy as it was to talk about mechanical energy. Fundamental to this realization was the recognition that energy in general is a *conserved quantity*, which is to say that provided we take account of all the energy contributions

> The total amount of energy in any isolated system is always constant.

This immensely powerful statement is known as the **law of conservation of energy**. Its discovery was of such fundamental importance in the development of physics that it led to a substantial reorganization of the whole subject. A discussion of its full significance belongs elsewhere (see the next book in the series, *Classical physics of matter*), but as far as force and motion are concerned it implies that when mechanical energy is not conserved in an isolated system (due to the action of a non-conservative force such as friction) the mechanical energy that is 'lost' always reappears in some other form (such as thermal energy), elsewhere in the system.

3.8   The role of energy in predicting motion

The last subsection included an example of the way in which energy can be used to predict the speed of a body. Similar calculations, based on the conservation of mechanical energy, can be carried out in many situations. It is always the case that the same information may be obtained by considering the relevant forces, formulating the appropriate equation of motion and solving it. However, the force-based approach can easily become very complicated, whereas the energy-based approach is generally much simpler if it can be applied. Energy is therefore of great practical value in the prediction of motion.

An example of this is the one that was mentioned at the start of this chapter, the calculation of the minimum speed with which a projectile must be launched if it is to escape completely from the Earth. Provided we ignore air resistance, the calculation of this **escape speed**, v_{escape}, is pretty straightforward since it may be based on the conservation of mechanical energy. Using Equation 2.25 to evaluate the projectile's potential energy at the launch site (where $r = R_E$), the initial mechanical energy will be:

$$E_{mech} = E_{trans} + E_{pot} = \frac{1}{2}mv_{escape}^2 - \frac{GmM_E}{R_E}.$$

Interpreting 'escape' to mean 'arriving at infinity (the point of zero potential energy) with no kinetic energy left', means that the final value of the projectile's mechanical energy must be

$$E_{mech} = E_{trans} + E_{pot} = 0.$$

But conservation of mechanical energy tells us that these two expressions for E_{mech} must be equal, so

$$\frac{1}{2}mv_{escape}^2 - \frac{GmM_E}{R_E} = 0$$

which implies that the escape speed is

$$v_{escape} = \sqrt{\frac{2GM_E}{R_E}}. \tag{2.35}$$

Question 2.16 Evaluate the escape speed from the Earth, given that Newton's gravitational constant $G = 6.67 \times 10^{-11} \, \text{N m}^2 \, \text{kg}^{-2}$, $M_E = 5.98 \times 10^{24} \, \text{kg}$ and $R_E = 6.38 \times 10^6 \, \text{m}$. ∎

The relative simplicity of this calculation underlines the power of energy-based methods, but fails to properly indicate their scope. In 1788, nearly thirty years before the term 'energy' was given the current meaning, the French scientist Joseph Louis Lagrange (Figure 2.28a) developed a general approach to the prediction of motion that made no direct use of forces at all, though it was mathematically equivalent to the Newtonian approach. Lagrange's method concentrated on a quantity later called the Lagrangian, which we now recognize as the difference between the kinetic and potential energies of the system under investigation. Applying Lagrange's method is not simple, but it was his proud boast that it took away the need to draw diagrams and effectively reduced mechanics to a branch of mathematics. Not all physicists would see this as an advantage, but most would admit the elegance and practical utility of Lagrange's so-called 'analytical mechanics'. Additional developments by Sir William Rowan Hamilton (Figure 2.28b) further emphasized the importance of energy and introduced another quantity, now called the Hamiltonian, that effectively represents the sum of the kinetic and potential energies. Almost a century later it was the concept of the Hamiltonian and the methods of Hamiltonian mechanics that were to provide the vital link between classical mechanics and quantum mechanics, an area of study where forces and Newton's laws are of very little use, but energy is crucial.

(a)

(b)

Figure 2.28 (a) Joseph Louis Lagrange (1736–1813) the founder of Lagrangian mechanics. Lagrange headed the commission that introduced the metric system in 1793. (b) Sir William Rowan Hamilton (1805–65) (on the left) the founder of Hamiltonian mechanics. Hamilton was appointed Astronomer Royal for Ireland in 1827, while still an undergraduate.

4 Power

4.1 Power — the rate of energy transfer

When comparing the capabilities of two machines we are often concerned with the *rate* at which work is done, that is, the amount of work done per second. This quantity is called the *power*. Since doing work transfers energy we may also say that:

Power is the rate at which work is done and energy is transferred.

Figure 2.29 James Watt (1736–1819) was a Scottish engineer who played a major role in the development of the steam engine by cooling the used steam in a condenser separate from the main cylinder. Steam engines and other devices of his invention were successfully built by him in partnership with Matthew Bolton, and were vital to the industrial revolution. The SI unit of power, the watt, is named after him.

The SI unit of energy is the joule (J), the corresponding unit of power is the joule per second (J s⁻¹). This unit is more usually called the **watt** (W), in honour of James Watt (Figure 2.29), thus:

$$1 \text{ watt} = 1 \text{ W} = 1 \text{ J s}^{-1}.$$

As you will almost certainly be aware, ordinary domestic light bulbs are sold according to their power consumption and are typically rated at 60 W, 100 W and 150 W. A typical electric kettle consumes 3 kW ($1 \text{ kW} = 10^3 \text{ W}$) (other examples are given in Figure 2.30). In all these cases it should be noted that the power 'consumption' is actually the rate of energy transfer. These devices 'convert' energy from one form to another. In a similar spirit, your own body consumes energy even when at rest; this is what constitutes your basal metabolic rate and is typically about 85 W in a young man, slightly less in a young woman.

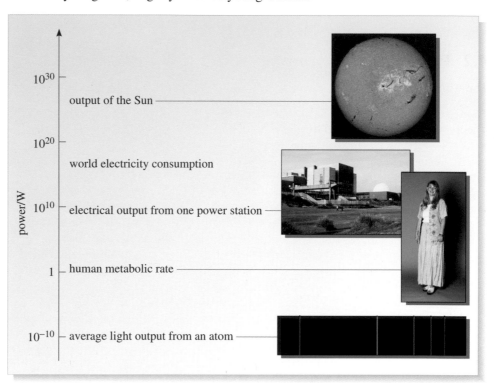

Figure 2.30 Some physically interesting powers.

Suppose you pull a heavy trolley through a certain displacement, and consequently do 400 J of work in 10 s. Your average rate of doing work during the displacement would be 400 J/10 s = 40 W. This is the *average power* you supply during the movement.

In general, the average power supplied when an amount of work ΔW is done in a time Δt is defined by:

$$\langle P \rangle = \frac{\Delta W}{\Delta t}, \tag{2.36}$$

where we have used the angle brackets $\langle \ \rangle$ to indicate the average of the quantity (P) enclosed. Sometimes, rather than the average power we may want to know the instantaneous rate of energy transfer at some particular time. This is called the *instantaneous power*, and is what is usually meant by the unqualified term 'power'. It may be thought of as the gradient of a graph of energy transfer against time and may be written in terms of a derivative as follows

$$P = \frac{dW}{dt}.$$
(2.37)

This is usually described as the limit of $\frac{\Delta W}{\Delta t}$ as Δt tends to zero.

From Equation 2.36 we see that:

$$\Delta W = \langle P \rangle \Delta t.$$

This means that we could use the watt second (W s) instead of the joule (J) as an SI unit of work or energy. In practice, the watt second unit is rarely used, but the **kilowatt hour** (kW h) is commonly used as the commercial unit for the measurement of electrical energy consumption. One kilowatt hour is the energy transferred or work done in one hour when the power is 1 kW, so:

$$1 \text{ kW h} = 10^3 \text{ J s}^{-1} \times 3600 \text{ s} = 3.600 \times 10^6 \text{ J} = 3.600 \text{ MJ}.$$

4.2 Power and vectors

The work done by a constant force $\boldsymbol{F}$ over a small displacement $\Delta\boldsymbol{s}$ is:

$$\Delta W = \boldsymbol{F} \cdot \Delta\boldsymbol{s}.$$
(2.38)

The average power delivered by that force over the time Δt is:

$$\langle P \rangle = \frac{\Delta W}{\Delta t} = \frac{(\boldsymbol{F} \cdot \Delta\boldsymbol{s})}{\Delta t} = \boldsymbol{F} \cdot \frac{\Delta\boldsymbol{s}}{\Delta t}.$$

The instantaneous power is given by the limit of this as Δt tends to zero, so

$$P = \frac{dW}{dt} = \boldsymbol{F} \cdot \left(\frac{d\boldsymbol{s}}{dt}\right) = \boldsymbol{F} \cdot \boldsymbol{v}.$$
(2.39)

Since $\boldsymbol{F} \cdot \boldsymbol{v} = Fv \cos\theta$, an immediate consequence of this is that the rate of energy transfer by a force $\boldsymbol{F}$ acting on a body which is moving in a direction perpendicular to $\boldsymbol{F}$ must be zero, since $\theta = 90°$ in that case and $\cos 90° = 0$. Some other consequences are explored in the following questions.

Question 2.17   The three engines of an airliner can develop a total take-off thrust (force) of magnitude 6.72×10^6 N. If the speed at take-off is 90 m s^{-1}, calculate the power developed at take-off.

Question 2.18   A horse pulls a barge along a canal at a constant speed of 2.0 m s^{-1}. The force due to the tension in the tow rope is 300 N and the rope makes an angle of 30° to the direction in which the barge travels. What power does the horse supply? ■

5   Energy in oscillating systems

Oscillating systems are of great interest and importance, whether they are weights on springs, atoms in solids, chimneys that sway in the wind, or bridges that shake themselves to destruction. In this section we examine several such systems from the point of view of energy and power, starting with the simplest harmonic oscillators and progressing to more complicated but more realistic systems in which both conservative and non-conservative forces are at work. Along the way you will be introduced to the important mathematical concept of the *exponential function* ($\exp(x) = e^x$) that appears in various guises throughout *The Physical World*.

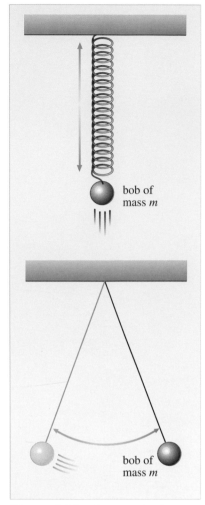

Figure 2.31 Some typical examples of simple harmonic oscillators; a bob on a spring, and the bob of a pendulum (for small oscillations).

5.1 The energy of a simple harmonic oscillator

You will recall from Section 4.2 of Chapter 1 that simple harmonic motion (s.h.m.) arises when a particle's acceleration is directed towards a fixed point (called the *equilibrium position*) and is proportional to its displacement from that point (see Figure 2.31). If the simple harmonic motion is along the *x*-axis, and if the equilibrium position is at $x = 0$, then the displacement, velocity and acceleration of the oscillator at any time t are given by

$$x(t) = A\sin(\omega t + \phi) \tag{Eqn 1.50}$$

$$v_x(t) = \frac{dx}{dt} = A\omega\cos(\omega t + \phi) \tag{Eqn 1.51}$$

$$a_x(t) = \frac{d^2x}{dt^2} = -A\omega^2\sin(\omega t + \phi) = -\omega^2 x \tag{Eqn 1.52}$$

where the constants A and ϕ represent the *amplitude* (i.e. maximum displacement) and the *initial phase* of the motion. The constant ω is called the *angular frequency* of the oscillator and is related to the time period T required for one complete *cycle* of oscillation by $T = 2\pi/\omega$.

Such motion, in which $a_x = -\omega^2 x$, is always the result of a *linear restoring force* of the form

$$F_x = -kx,$$

where the constant k is called the *force constant*, and is related to the angular frequency by $\omega = \sqrt{k/m}$. This general linear restoring force is conservative and is just a simple generalization of the restoring force we encountered when deriving an expression for strain potential energy in Section 3.3. Consequently, whatever the cause of the restoring force for any particular simple harmonic oscillator, we can always say that the potential energy associated with a displacement x will be

$$E_{pot} = \tfrac{1}{2}kx^2. \tag{2.40}$$

This potential energy is plotted against displacement in Figure 2.32.

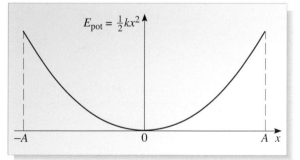

Figure 2.32 The potential energy of a simple harmonic oscillator plotted against displacement from equilibrium. The maximum value of the displacement is called the amplitude and is denoted by A.

When the restoring force F_x is provided by a spring of spring constant k_s, the potential energy will be strain potential energy and the force constant will be equal to the spring constant, but there are other possibilities. For example, if the oscillator is the bob of a simple pendulum, the potential energy will be the gravitational potential energy and the force constant will be $k = mg/l$, where m is the mass of the bob, g is the magnitude of the acceleration due to gravity, and l is the length of the pendulum.

In any case, whatever the precise nature of the oscillator may be, we can say that its total mechanical energy at any time is

$$E_{mech} = E_{trans} + E_{pot} = \tfrac{1}{2}mv^2 + \tfrac{1}{2}kx^2 \tag{2.41}$$

where v and x represent the instantaneous values of speed and displacement at the relevant time. Noting that v^2 will always have the same value as v_x^2, we can use Equations 1.50 and 1.51 to rewrite this expression for the mechanical energy at time t as

$$E_{mech} = E_{trans} + E_{pot}$$
$$= \tfrac{1}{2}mA^2\,\omega^2\cos^2(\omega t + \phi) + \tfrac{1}{2}kA^2\sin^2(\omega t + \phi). \tag{2.42}$$

Recalling that $\omega = \sqrt{k/m}$, we see that Equation 2.42 implies

$$E_{mech} = \tfrac{1}{2}kA^2\cos^2(\omega t + \phi) + \tfrac{1}{2}kA^2\sin^2(\omega t + \phi)$$
$$= \tfrac{1}{2}kA^2[\cos^2(\omega t + \phi) + \sin^2(\omega t + \phi)]. \tag{2.43}$$

But, for any value of θ, $\sin^2\theta + \cos^2\theta = 1$, so Equation 2.43 implies that

$$E_{mech} = \tfrac{1}{2}kA^2. \tag{2.44}$$

Hence the total mechanical energy of the oscillator is constant (as we should expect of a system that only involves a conservative restoring force) and its value is determined by the force constant k and the square of the amplitude, A^2. For any given oscillator, k will be a constant, so we may say that:

> For a given oscillator, the total mechanical energy in any particular case of simple harmonic motion will be proportional to the square of the amplitude of that motion.

So, for a given oscillator, doubling the amplitude of the motion requires that its total mechanical energy should be quadrupled. Tripling the amplitude implies a nine-fold increase in energy.

Although the total mechanical energy of any particular example of s.h.m. is constant, the individual contributions from the kinetic and potential energy are changing all the time. This is most easily seen by drawing graphs of those contributions. Choosing to set $\phi = 0$ for simplicity, which we are free to do, we see from Equations 2.42 and 2.43 that

$$E_{trans} = \tfrac{1}{2}kA^2\cos^2(\omega t) \tag{2.45}$$

and $\qquad E_{pot} = \tfrac{1}{2}kA^2\sin^2(\omega t). \tag{2.46}$

The graphs of these functions, together with the total mechanical energy, $E_{mech} = kA^2/2$, are plotted against time in Figure 2.33. As the oscillation progresses the constant total energy is divided between potential energy and kinetic energy. There is a cyclical transformation of potential energy to kinetic energy, and back again, as the mass oscillates.

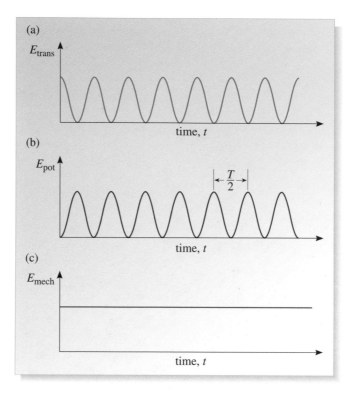

Figure 2.33 The variation with time of (a) the translational kinetic energy, (b) the potential energy, and (c) the total mechanical energy of a simple harmonic oscillator. (T is the period of the oscillation.)

The same situation, for the particular case of a particle on a spring, is illustrated somewhat more physically in Figure 2.34. At the maximum displacement, when the particle is momentarily at rest (snapshot 1, or 5, in Figure 2.34), the strain potential energy has its maximum value, and the kinetic energy is zero. In snapshot 2, as the particle moves towards its unextended position, strain energy is transformed into kinetic energy as the speed increases. In snapshot 3, the spring is unextended, so its strain potential energy is zero. However, the speed of the particle is highest at this point, so the kinetic energy here is a maximum, and is equal to the total mechanical energy. As the particle continues to oscillate (snapshots 4 and 5), it slows down as kinetic energy is transformed back into strain potential energy. These energy transformations are reversed as the oscillation swings back to the configuration shown in snapshot 1, and they are then repeated in successive cycles.

An interesting point to note about Figures 2.33 and 2.34 is that both kinetic energy and potential energy vary with period $T/2$, i.e. just half the period of the oscillation itself. The physical reason for this is clear; the potential energy will attain its maximum value twice per cycle, once when the displacement is $x = A$ and then again when the displacement is $x = -A$. The mathematical reason is also fairly easy to

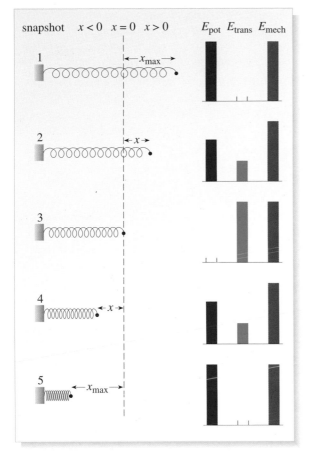

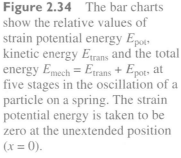

Figure 2.34 The bar charts show the relative values of strain potential energy E_{pot}, kinetic energy E_{trans} and the total energy $E_{mech} = E_{trans} + E_{pot}$, at five stages in the oscillation of a particle on a spring. The strain potential energy is taken to be zero at the unextended position ($x = 0$).

understand; whereas the displacement is described by a sine function with period T, which is negative for a half of each cycle, the potential energy is described by a $\sin^2$ function that is always positive and has period $T/2$ (see Figure 2.35).

Question 2.19 Sketch an energy–displacement graph similar to Figure 2.32, but including curves that correspond to the total mechanical energy and the kinetic energy as well as the potential energy. State the convention you have adopted regarding the zero value of the potential energy. ∎

5.2 The energy of a damped harmonic oscillator

From your everyday experience, you cannot fail to have noticed that the amplitude of most oscillators does not remain constant with time. If you construct a pendulum from a bob and a piece of string and then set it swinging, the amplitude will gradually become smaller and smaller, and the bob will stop swinging altogether if left for long enough. This gradual reduction in amplitude shows that the oscillator is losing mechanical energy. Any oscillating system that loses mechanical energy in this way is said to be **damped**.

● What might have happened to the mechanical energy that the oscillator lost due to damping?

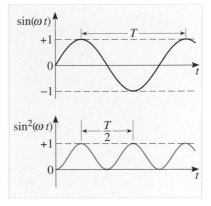

Figure 2.35 Graphs of $\sin(\omega t)$ and $\sin^2(\omega t)$ plotted against t. The period of the former is T, while that of the latter is $T/2$, where $T = 2\pi/\omega$.

○ Typically, air resistance, viscosity or friction dissipate the mechanical energy of an oscillating (or vibrating) system, usually resulting in an increase in thermal energy elsewhere. For instance, as the amplitude of a swinging pendulum decreases, the surrounding air might become very slightly warmer. ∎

Nearly all oscillating systems are subject to some damping, even though it may be very small. Sometimes damping is a nuisance, but it can also be desirable. The damping of vibrations is of great technological importance. Many engineers spend their time designing systems to damp vibrations as efficiently as possible.

Motor vehicle suspension systems (see Figure 2.36) use damped oscillations to smooth out the effect of bumps in a road. Riding in a car without a suspension system is very uncomfortable, but riding in one that uses undamped oscillators, where every bump sets the car and its passengers into continuous vertical oscillations, would be even more unpleasant and dangerous. Chemical balances and other weighing devices also provide examples of systems where the correct degree of damping is important — too much damping and the balance will take a long time to settle at a reading, too little damping and the balance will oscillate about the reading. Sound too is an oscillatory phenomenon; music and speech are transmitted through air in the form of oscillations in the air pressure. The energy contained in these oscillations can be absorbed by walls and other materials, which can thereby deaden sound and prevent unwanted noise. Choosing the best way to damp sound oscillations is important when sound-proofing rooms or preventing echoes in concert halls. Damping also plays an important role in electrical circuits, as you will learn later.

As all of these examples show, understanding the nature of damped oscillations is an important matter with many practical applications. However, before we can give a more quantitative description of how damping affects oscillatory motion we need to introduce an important mathematical concept — *exponential decay*.

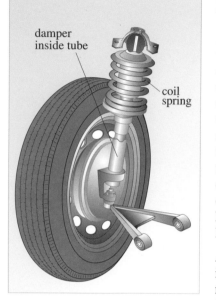

Figure 2.36 A typical part of a car suspension system — a MacPherson strut.

Study comment *If you have encountered exponential processes before, most of the material covered in the next few pages will be familiar and you should proceed quickly to Question 2.20. If you have not met exponential decay before, work through the material leading to Question 2.20 carefully — exponential decay is a very important concept.*

Exponential decay

Earlier, in Chapter 1, we noted that air resistance to a moving body can give rise to a resistive force, called aerodynamic drag, with a magnitude proportional to the speed of the body. If we consider such a body moving in the x-direction with velocity v_x, we can describe this force by the equation

$$F_x^{\text{air}} = -bv_x \tag{2.47}$$

where b is a positive constant that depends on the size and shape of the body, and the minus sign indicates that the aerodynamic drag points in the opposite direction to the body's instantaneous velocity.

If F_x^{air} is the only force acting on the body, Newton's second law tells us that

$$F_x^{\text{air}} = ma_x$$

where m is the mass of the body and a_x is its acceleration. Using Equation 2.47 to eliminate F_x^{air} from this equation, we have

$$a_x = -\left(\frac{b}{m}\right)v_x. \tag{2.48}$$

Given that m and b are positive constants, the minus sign in this equation indicates that the acceleration acts in the opposite direction to the velocity and hence tends to slow the body, just as you would expect.

Figure 2.37 shows the effect that this acceleration would have on a body initially travelling with velocity $v_x = 0.1 \text{ m s}^{-1}$, for a particular value of b/m. Similar, though not identical curves, would have resulted whatever value of b/m had been used. In all such cases the velocity reduces (or decays) towards zero in a highly characteristic way.

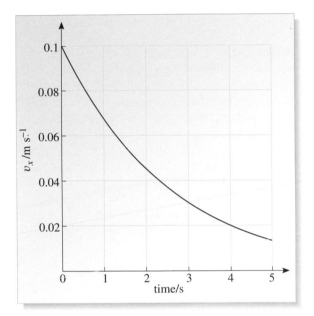

Figure 2.37 The velocity of a small body moving in the x-direction subject only to a force arising from air resistance, $F_x^{\text{air}} = -bv_x$. The initial speed of the body was 0.1 m s^{-1}.

We can gain more insight into this special kind of decay by using the fact that acceleration is the rate of change of velocity. This allows us to rewrite Equation 2.48 as a *differential equation*:

$$\frac{dv_x}{dt} = -\left(\frac{b}{m}\right) v_x. \tag{2.49}$$

As usual, let's try to 'read' this equation and see what it is telling us. The first thing it tells us is that if v_x is positive then the rate of change of v_x must be negative. The effect of this is clear in Figure 2.37; at every value of t the velocity v_x is positive and at each of those values of t the gradient of the curve (the graphical equivalent of the rate of change) is indeed negative — it's downhill all the way as t increases. The second thing Equation 2.49 tells us is that the value of the gradient at any time t will be proportional to the value of v_x at that time. This too influences Figure 2.37. As t increases and v_x decreases, the gradient of the curve also decreases, so the curve gets flatter and flatter as a result. Of course, detailed measurements would be required to confirm that the plotted curve really does have a gradient that is accurately proportional to the value of v_x at every point, but, faced with the alternative of actually carrying out some of those measurements, I hope you will accept my assurance that it does.

The kind of decay process shown in Figure 2.37 is just one example of a whole class of processes called *exponential decay processes*. Such processes are not restricted to moving bodies; other examples of exponential decay include the decline in the number of atoms of a given type in a sample of radioactive material, and the decrease in air pressure with height during a steady ascent through the Earth's atmosphere. The feature that distinguishes an exponential decay from any other kind of decay process is this:

Exponential decay is a process in which the instantaneous rate of change of the decaying quantity is at all times proportional to the instantaneous value of that decaying quantity.

This statement implies that if we are considering the behaviour of some general quantity v that depends on t, then, in order for it to decay exponentially, it must be the case that at every instant

$$\frac{dv}{dt} \propto v. \tag{2.50}$$

This proportional relationship can be turned into an equation by introducing a constant of proportionality on the right-hand side. But remember that we are describing a decay process, so we want v to approach zero as t increases. This means that the constant of proportionality must be negative. (A positive proportionality constant would lead to exponential growth which we shall discuss later.) Hence, we see that if v decays exponentially then

$$\frac{dv}{dt} = (\text{a negative constant}) \times v.$$

For reasons that will soon become clear, this negative constant is usually written in the form $-1/\tau$, where τ (the Greek letter tau) is a positive constant called the *time constant* of the decay. Adopting this convention, the differential equation describing the exponential decay of the general quantity v is

$$\frac{dv}{dt} = -\left(\frac{1}{\tau}\right)v. \tag{2.51}$$

Equation 2.49 was a particular example of this general relationship, in which v was represented by v_x and τ was represented by m/b.

From a mathematical point of view, Equation 2.51 is said to be a *first-order* differential equation since the only derivative that it contains, dv/dt, is a first derivative. The general solution to Equation 2.51 will be an expression for v as a function of t. It will involve the given constant τ, but it will also introduce an arbitrary constant corresponding to whatever initial value we choose for v, just as Figure 2.37 involved the initial velocity $0.1\,\mathrm{m\,s^{-1}}$.

There are many equivalent ways of writing the general solution to Equation 2.51. Each would constitute a mathematical representation of the general exponential decay process. The one that we shall use is the simplest and most conventional. It is

$$v(t) = v_0 e^{-t/\tau} \tag{2.52}$$

where the parenthetical t on the left-hand side is simply there to remind us that v is a function of (i.e. depends on) the variable t. The form of that function is spelt out on the right-hand side of Equation 2.52. As expected, it contains the given constant τ that appeared in Equation 2.51, and the arbitrary constant v_0 that represents the value of v when $t = 0$, but it also contains another constant, represented by e, which stands for a particular number. The constant e is a mathematical constant, rather like π. For most purposes its value may be taken to be 2.718, but it is actually a non-recurring decimal whose value is known to extraordinarily high precision. Your calculator will certainly have a built-in function that will enable you to evaluate the result of raising e to some power. This may appear as an 'e^x' button, or you may have to access it by some combination of keys such as 'INV' followed by 'ln', or '2ndF' followed by '$\log_e$'. In any case, make sure you know how to access this function on your calculator, and, when you do, use it to evaluate e^1. If you have pressed the right keys your calculator will display something like

$$e = 2.718\,281\,828.$$

The most striking feature of Equation 2.52 is that, the only place where the variable t appears on the right-hand side of the equation is in the power or *exponent* of e. For this reason we say that $v(t)$, as defined by Equation 2.52, is an **exponential function** of t. Exponential functions of this sort are very common throughout mathematics, physics and engineering. As a class of functions they stand alongside the trigonometric functions (e.g. $\sin(\omega t)$ and $\cos(\omega t)$) and the polynomial functions (e.g. $At + Bt^2$ and $At + Bt^2 + Ct^3$) that you have already met. Like those other functions, exponential functions have a number of characteristic properties that make them particularly valuable when it comes to modelling the physical world.

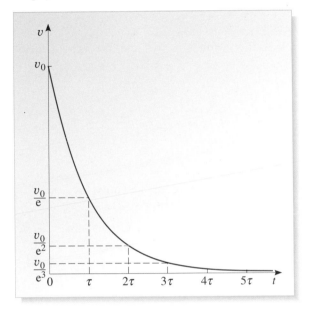

Figure 2.38 The graph of the function $v(t) = v_0 e^{-t/\tau}$.

Figure 2.38 shows the graph of the exponential function in Equation 2.52. It is really just a generalization of the particular example of exponential decay we plotted in Figure 2.37. We shall now examine some of the characteristic features of exponential decay by comparing the function given in Equation 2.52 with the corresponding graph in Figure 2.38.

The first thing to notice is that if we set $t = 0$ in Equation 2.52, then we find

$$v(0) = v_0 e^{-0} = v_0$$

as expected. (You shouldn't be surprised that $e^0 = 1$, any number raised to the power 0 is equal to 1.) In Figure 2.38, v_0 is duly shown as the value of v at $t = 0$.

The next thing to learn is the significance of the time constant τ. You may be able to discover this for yourself by answering the following question.

● Use Equation 2.52 to write down expressions for the value of v at time $t = \tau$, $t = 2\tau$, $t = 3\tau$ and $t = 4\tau$. What will be the value of v at $t = 100\tau$?

○ The required values are $v(\tau) = v_0/e$, $v(2\tau) = v_0/e^2$, $v(3\tau) = v_0/e^3$ and $v(4\tau) = v_0/e^4$. At $t = 100\tau$, the value of v will be $v(100\tau) = v_0/e^{100}$. ∎

Can you see the significance of τ? Figure 2.38 shows some of the values that you were asked to write down. The point to note is that each additional increment in time by an amount τ corresponds to a further decrease in v by a *factor* of 1/e. (The value of 1/e is 0.368 to three decimal places.) Thus we can say that:

The **time constant** τ of an exponential decay is the time required for the decaying quantity to reduce its value by a factor of 1/e.

Note that this statement is true no matter when we start timing. Any increase in time by an amount τ corresponds to a 1/e decrease in v because

$$v(t + \tau) = v_0 e^{-(t + \tau)/\tau} = v_0 e^{-(t/\tau)-1}.$$

We can then use the relation $e^{a + b} = e^a \times e^b$, which is always true, to rewrite this as

$$v(t + \tau) = v_0 e^{-t/\tau} e^{-1} = v(t) e^{-1} = \frac{v(t)}{e}.$$

The time constant τ is important because it tells us about the rate of the decay; for a given v_0 a large time constant indicates a slow decay. Figure 2.39 shows a number of exponential decays starting from the same value $v = v_0$, but corresponding to different values of τ. Increasing τ results in a slower decay.

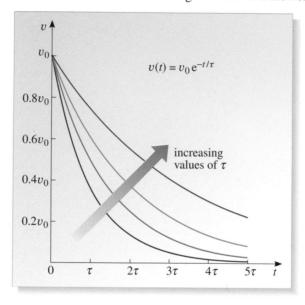

Figure 2.39 Examples of exponential decay for various values of τ and a fixed (though arbitrary) initial value v_0 of some general quantity v. Note that larger values of τ correspond to slower decays.

Question 2.20 Three exponential decays are described by: $y_1 = 10e^{-12t}$, $y_2 = 20e^{-6t}$, $y_3 = 5e^{-t/6}$. (In this particular question, y_1, y_2, y_3 and t are all dimensionless variables without units.) (a) Which has the largest value at $t = 0$? (b) Which decays most rapidly as t increases?

Question 2.21 Given a graph, such as Figure 2.38, that shows the exponential decay of a quantity, devise a step-by-step procedure that will enable you to determine the relevant time constant from the graph. Write down your procedure and apply it to Figure 2.37 to determine the time constant ($\tau = m/b$ in this case) that was used to plot the graph. ■

Damped harmonic motion

Armed with an understanding of exponential decay processes, we can now return to our discussion of damped harmonic motion and its energy. For the speeds encountered in many simple oscillators, the damping forces due to air resistance may be considered as being proportional to the speed and they always act in the direction that opposes the motion. This implies that in one dimension, the total force acting on the oscillator at any time is given by

$$F_x = -kx - bv_x \tag{2.53}$$

where v_x is the velocity, x is the displacement, k is the force constant and b is another constant related to the strength of the damping. Since the restoring force is no longer proportional to the displacement as it was in the undamped case, the motion will no longer be simple harmonic motion, and it will not obey the s.h.m. equation (Equation 1.70). Instead, the oscillator's equation of motion will be

$$\frac{d^2 x}{dt^2} = \frac{-k}{m} x - \frac{b}{m} \frac{dx}{dt}. \tag{2.54}$$

(A similar equation was briefly discussed towards the end of Chapter 1.) You are not expected to solve this equation, but it is worth noting that in the case of **light damping**, that is when b/m is small, its solution describes the oscillator's displacement as a function of time as follows

$$x(t) = (A_0\, e^{-t/\tau}) \sin(\omega t + \phi) \tag{2.55}$$

where A_0 and ϕ are arbitrary constants determined by the initial conditions we choose to impose. The constant τ is equal to $2m/b$ and the angular frequency ω may be taken to have the same value as in the undamped case, i.e. $\omega = \sqrt{k/m}$. (Equation 2.55 becomes increasingly inaccurate as the damping increases, so it should only be regarded as providing an approximate description of the motion.)

We can interpret Equation 2.55 as describing a damped oscillatory motion of the form

$$x(t) = A(t) \times \sin(\omega t + \phi) \tag{2.56}$$

where the amplitude $A(t)$ is not constant, but decays exponentially with time according to

$$A(t) = A_0 e^{-t/\tau} \tag{2.57}$$

with a time constant, $\tau = 2m/b$, that is inversely proportional to the **damping constant**, b. This is reasonable — a more heavily damped system (larger b) will take a short time to decay (implying a small value of τ) whereas a very lightly damped system (small b) will go on oscillating much longer (implying a larger value of τ).

The kind of behaviour described by Equations 2.56 and 2.57 is shown graphically in Figure 2.40. The exponentially decaying amplitude, $A(t)$, forms an envelope that restricts the harmonic (i.e. sinusoidal) oscillations either side of the equilibrium position. As $A(t)$ decreases the whole oscillation gradually dies away. When notes are played on a piano, the strings exhibit damped harmonic motion of just this kind. The result is that we hear the notes getting softer and softer over a period of time as their mechanical energy is dissipated and the vibration of the strings gradually decays away.

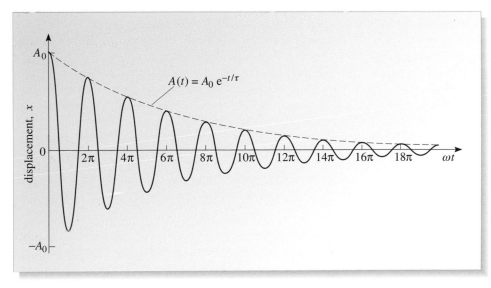

Figure 2.40   A lightly damped oscillator, released from rest at a positive displacement A_0 from its equilibrium position $x = 0$, at time $t = 0$, will exhibit the damped harmonic motion shown in the graph and described by Equation 2.55. Note that the displacement x has been plotted against ωt, for convenience, and that the specified initial conditions imply that the initial phase of the motion is $\phi = \pi/2$ in this case.

The mechanical energy of a damped oscillator decreases with time, so its calculation is inevitably more complicated than that of the constant energy of an undamped oscillator. However, if the damping is so light that there is little change in the amplitude from one oscillation to the next, then we can continue to use the result we obtained for a simple harmonic oscillator ($E_{mech} = \frac{1}{2}kA^2$), provided we recognize that E_{mech} is now a function of time and that the constant A is to be replaced by the time-dependent amplitude, $A(t)$ of Equation 2.57. Hence, for a sufficiently lightly damped harmonic oscillator, we can say that

$$E_{mech} = \tfrac{1}{2}kA^2 = \tfrac{1}{2}kA_0^2\,\mathrm{e}^{-2t/\tau} \; . \tag{2.58}$$

This implies that the mechanical energy decays exponentially, with a time constant $\tau/2 = m/b$ which is just half of that for the amplitude. (Note that the exponent in Equation 2.58 is $-2t/\tau = -t/(\tau/2)$; that's why the decay constant is $\tau/2$ in this case, rather than τ.)

In situations where the damping is somewhat greater, though still light enough for many oscillations to occur, a useful way of describing the decay of an oscillator's energy is in terms of its *quality*, or **Q-factor**. This is calculated by dividing the total energy stored in the oscillator at any time by the energy lost per oscillation and multiplying the result by 2π, so

$$Q = \frac{2\pi \times \text{total stored energy}}{\text{average energy loss per oscillation}}. \tag{2.59}$$

A damped oscillator with a high Q-factor will complete many oscillations before most of its energy is dissipated. A low Q oscillator will complete very few. Table 2.1 lists the approximate Q-factors of some physically interesting damped harmonic oscillators.

Table 2.1 Approximate Q-factors of some physically interesting oscillators.

Oscillator	Q-factor
vibrating atom	10^7
tuning fork	10^4
piano string	10^3
Earth (due to earthquakes)	200 to 1500

In describing the lightly damped harmonic oscillator we have assumed that the damping has no effect on the oscillator's angular frequency ω. However, as pointed out earlier, this is not strictly correct. The presence of damping, even light damping, actually increases the period slightly and thus decreases the angular frequency (since, by definition, $T = 2\pi/\omega$). This increase is usually so small that we can indeed neglect it except when the damping is quite large. Nonetheless, in recognition of the fact that the angular frequency of the damped oscillator really is somewhat different from that of the undamped (simple harmonic) oscillator, it is customary to refer to the quantity $\sqrt{k/m}$ as the **natural frequency** of the oscillator and to give it the special symbol ω_0.

When a system is subject to sufficiently **heavy damping**, oscillations do not occur at all. In this case, the displacement changes exponentially almost from the outset, as shown in Figure 2.41.

Heavy damping is put to practical use in designing systems for noise isolation, in car speedometers and in the closing mechanisms on some swing doors. You may be able to think of other examples.

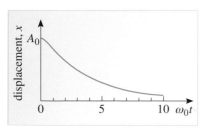

Figure 2.41 A heavily damped oscillator, released at a displacement A_0 from its equilibrium position $x = 0$, returns to equilibrium without any oscillation at all. ω_0 is the natural frequency of the system.

At intermediate levels of damping there is an important special case known as **critical damping**. When this occurs the system approaches equilibrium especially rapidly, as indicated in Figure 2.42. Critical damping is particularly useful in the design of balances and other measuring devices where you want the reading to be reached quickly but you don't want the meter to oscillate about the reading. It is also used in some car suspension systems.

5.3 The energy of a driven damped harmonic oscillator

If the oscillations of a damped oscillator are to be maintained at a given amplitude, then energy must be transferred to the oscillator to compensate for the energy lost due to damping. The simplest way to supply this energy is to apply an additional force to the oscillator that does work on the oscillator at the required rate. Such a force is called a *driving force* and an oscillator that is subject to both a driving force and a damping force is said to be a **driven damped harmonic oscillator**.

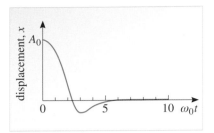

Figure 2.42 A critically damped oscillator approaches equilibrium very quickly with no more than one overshoot. ω_0 is the natural frequency of the system.

The classic example of a driven damped oscillator is a child on a swing being pushed by another child. As every child learns, the effectiveness of a given push depends crucially on when it is applied. Pushing at the swing's own natural frequency is the most efficient way of transferring energy to the swing. Knowing this can help us to devise efficient driving systems for oscillators that we wish to maintain such as those in clocks and watches; it can also help us to avoid unwanted oscillations such as those that caused the Tacoma Narrows suspension bridge to shake itself to pieces on a windy day in November 1940 (Figure 2.43).

Figure 2.43 An oscillator driven to destruction. The Tacoma Narrows Bridge, Washington, USA, collapsing in dramatic fashion due to the effects of wind-induced oscillations.

Driving forces can be of many kinds, but one of the simplest yet most informative to consider is one that is itself periodic and which varies sinusoidally with time. In one dimension, one such force can be described by an expression of the form

$$F_x^{\text{drive}} = F_0 \cos(\Omega t) \tag{2.60}$$

where F_0 represents the maximum magnitude of the force, and Ω is its angular frequency (usually called the **driving frequency**). Notice that this force spends half its time acting in the direction of increasing x and the other half of its time acting in exactly the opposite direction. (You might think that this means it can have no overall effect, but you will soon learn otherwise.) Also note that the driving frequency Ω does not necessarily have anything to do with the properties of the system that is being driven. We are free to choose the driving frequency to have whatever value we want, just as a child is, in principle, free to push a swing as frequently as he or she wishes.

Now, with the addition of the driving force in Equation 2.60, the total force on the oscillator becomes

$$F_x = -kx - bv_x + F_0 \cos(\Omega t)$$

and it follows that the equation of motion of the driven damped oscillator we discussed in the last section will become

$$\frac{d^2x}{dt^2} = \frac{-k}{m}x - \frac{b}{m}\frac{dx}{dt} + \frac{F_0}{m}\cos(\Omega t). \tag{2.61}$$

As usual, you are not expected to solve this equation, only to note some of the features of its solution. Perhaps the first thing to note is that on this occasion the solution, $x(t)$, can be written as the sum of two terms

$$x(t) = x_{dec} + x_{dri} \tag{2.62}$$

where x_{dec} represents the kind of decaying damped motion that was discussed in the last subsection, while x_{dri} represents a new kind of motion, due to the driving force, that will persist after the decaying motion has died away. (x_{dec} is often described as the *transient* motion, while x_{dri} is said to be the *steady state* motion.) If we wait long enough for the decaying motion to die away, so that only the steady state motion remains, then we will find that the motion is described by

$$x(t) = A\sin(\Omega t + \phi). \tag{2.63}$$

This looks like simple harmonic motion with amplitude A, angular frequency Ω and initial phase ϕ, but there are some important differences from the 'natural' s.h.m. that we studied in Section 5.1. First, the angular frequency Ω is that of the driving force, not the angular frequency of the (undriven) damped oscillator ω, nor the natural frequency $\omega_0 = \sqrt{k/m}$ that the oscillator would have in the absence of driving and damping. Second, the constants A and ϕ that appear in Equation 2.63 are *not* arbitrary constants determined by the initial conditions of the motion. (The arbitrary constants of the driven damped harmonic motion would have helped determine the transient motion that has decayed away.) Instead, the A and ϕ in Equation 2.63 are completely determined by the constants that appear in the equation of motion, Equation 2.61.

● List and briefly describe the constants in the equation of motion of the driven damped harmonic oscillator.

○ m, the mass of the oscillator

k, the force constant of the restoring force

b, the damping constant

F_0, the maximum magnitude of the periodic driving force

Ω, the angular frequency of the driving force (usually called the driving frequency, for convenience). ■

The expression that relates the steady state amplitude A to these constants is complicated, and you certainly shouldn't try to remember it, but you can learn a lot from it, especially when it is written in the following way

$$A = \frac{F_0/m}{\sqrt{(\omega_0^2 - \Omega^2)^2 + (\Omega b/m)^2}} \tag{2.64}$$

where $\omega_0 = \sqrt{k/m}$ is the natural frequency of the oscillator. The significance of Equation 2.64 is best understood in graphical terms. It tells us that, for a given oscillator, with a given damping constant b, and a given strength of driving force F_0, the steady state amplitude A will depend on the driving frequency Ω, as indicated in Figure 2.44.

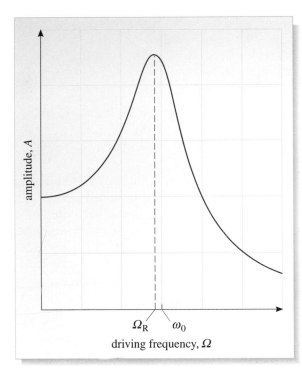

Figure 2.44 The amplitude of the steady state oscillation for a driven damped harmonic oscillator as a function of the driving frequency. The particular value of Ω at which the amplitude attains its maximum is called the resonant frequency and is denoted Ω_R. Its value is close to (but slightly less than) the value of the oscillator's natural frequency ω_0.

As you can see, the amplitude has a maximum value that occurs when the driving frequency is quite close to the oscillator's natural frequency ω_0. The precise value of Ω at which this maximum occurs is denoted Ω_R and is called the **resonant frequency** of the oscillator. A damped oscillator that is driven at this particular angular frequency is said to be in a state of **resonance**, and will have the maximum attainable amplitude.

Both the resonant frequency of an oscillator, and the maximum amplitude to which it corresponds depend on the degree of damping. As Figure 2.45 indicates, reducing the damping constant of a given driven damped oscillator will allow the maximum

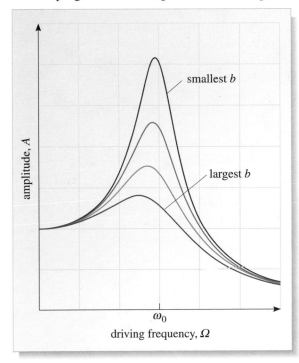

Figure 2.45 Reducing the damping constant b of a given oscillator changes the relationship between the amplitude and the driving frequency, causing the resonant frequency to approach the natural frequency and increasing the maximum amplitude.

steady state amplitude to increase and cause the resonant frequency to approach the natural frequency even more closely. So, for a lightly damped oscillator, resonance will occur when the driving frequency is almost identical to the natural frequency $\omega_0 = \sqrt{k/m}$, and may result in a very substantial amplitude.

When a damped oscillator is driven at its natural frequency, it can be shown that the phase constant ϕ in Equation 2.63 is zero, so the oscillator's steady state velocity will then be

$$v_x = \frac{\mathrm{d}x}{\mathrm{d}t} = A\Omega\cos(\Omega t). \tag{2.65}$$

This means that the velocity and the driving force (Equation 2.60) will be in phase, both attaining their greatest magnitude at the same time (as the oscillator moves through its equilibrium position) then, later, both reversing direction at the same time. It is under these conditions that a driving force of given maximum magnitude F_0 will provide the greatest power $P = F_x v_x$, and transfer the maximum amount of energy to the oscillator during each cycle of its oscillation. Of course, we are talking about an oscillator that is in its steady state, so all of this transferred energy must be dissipated again by the damping force during each cycle, but the conclusion is clear:

> The energy that sustains a driven damped harmonic oscillator in a steady state is supplied entirely by the driving force. The power transferred by that force will be a maximum when the driving frequency is equal to the natural frequency of the oscillator, and, for a lightly damped oscillator, this will also ensure that the oscillator is in (or very near) a state of resonance in which its amplitude is a maximum.

Although we have been discussing resonance in the context of a particular system with a highly specific driving force, it is actually a very common phenomenon that has many important applications in both the natural and the technological worlds. Some of these applications are discussed in Box 2.1.

Box 2.1   Resonance in action

Nature makes good use of resonance; there are many examples of it in the biophysics of living organisms and in animal behaviour. For instance, during *photosynthesis* (the process whereby plants use the energy of sunlight to power chemical processes that enable them to store energy), certain frequencies of light cause atoms within plant cells to resonate. In this way, the energy transferred from the light to the plant is maximized, to the benefit of the plant. Some possibly better known, though hardly more important examples of resonance, can be found in the production of sound by various animals. The air inside the horn-shaped burrow constructed by mole crickets resonates at the frequency of the vibration produced by the cricket. This results in the sound produced by the crickets being amplified by a factor of twenty or so. Similarly, howler monkeys and some birds, frogs and toads have large vocal sacs (Figure 2.46). When the vocal chords

vibrate, the air trapped inside the vocal sac resonates and the sound made naturally by the animal is amplified.

Figure 2.46   The resonant cavity of a frog.

Similar principles are sometimes employed in musical instruments to amplify the naturally weak sounds produced by strings and reeds. Resonance is also exploited in the electrical circuits used to tune-in to radio signals of a given frequency. You will learn about this in greater detail later in the course.

Resonance is not always benign and beneficial. The destruction of the Tacoma Narrows suspension bridge (Figure 2.43) has already been mentioned. In that case the wind, which reached no more than 68 km h⁻¹, forced the bridge to vibrate in much the same way that the curved metal slats of a venetian blind flutter when air rushes over them. The exact mechanism is controversial, but evidence suggests that a complex series of wind vortices provided the driving force. They transferred energy to the swaying bridge at its natural frequency and the amplitudes of the vertical vibrations and then of the twisting vibrations

got so large that the roadway ultimately ripped from its hangers and plunged into the water below. You may have seen the classic film footage of this remarkable event.

A more recent and more tragic example of resonance-induced destruction was the collapse of a section of the upper deck of the Nimitz freeway onto the lower deck following an earthquake whose epicentre was about 100 km away (Figure 2.47). The collapse caused extensive damage with the loss of 67 lives.

Figure 2.47 The collapse of the Nimitz freeway, San Francisco Bay, California, USA in October 1989.

Question 2.22 Figure 2.48 shows the damped vibrations of the end of a ruler that is projecting over the edge of a table.

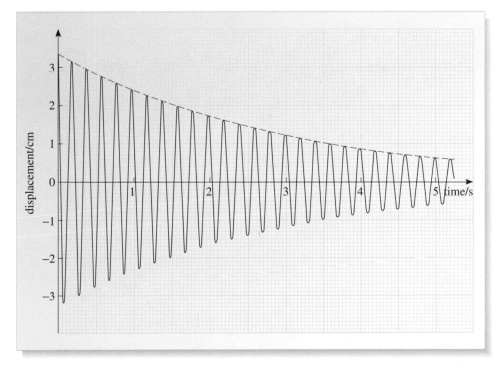

Figure 2.48 Damped vibrations of a ruler, for use with Question 2.22.

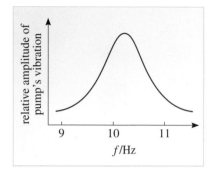

Figure 2.49 The frequency response of a pump mounted on springy feet, for use with Question 2.23.

(a) Does the amplitude of the vibrations decay exponentially, and if so, what is the time constant τ of the decay?

(b) Suppose that the ruler and table were placed in a large chamber, and the air was then pumped out. How would the displacement–time graph differ from that shown in Figure 2.48?

(c) What is the natural (angular) frequency of the ruler, ω_0, when vibrating in the way shown in Figure 2.48?

(d) Describe the relationship between the resonant frequency of the ruler, its natural frequency, and the driving (angular) frequency of an applied periodic force that will produce oscillations of maximum amplitude.

Question 2.23   A pump is mounted on springy rubber feet to reduce the transmission of vibrations to the surroundings. The response curve for this system is shown in Figure 2.49. Unfortunately, the pump rotates at 10 Hz, which is close to the resonant frequency of the system, and so large amplitude vibrations are set up. Suggest three modifications that could be made to reduce the amplitude of the vibrations. ∎

5.4 A note on exponential functions

We end this section with a mathematical note on exponential functions. This summarizes and extends the discussion of exponential decay contained in Section 5.2 and is mainly intended to provide handy reference material.

An **exponential process** is one in which the rate at which some quantity is changing at any time is proportional to the value of the quantity at that time.

If we denote the changing quantity by v, then the definition of an exponential process implies that

$$\frac{dv}{dt} = \alpha v \qquad (2.66)$$

where α is a constant.

Equation 2.66 is a first-order differential equation and its general solution is given by the following *exponential function* of t:

$$v(t) = v_0 e^{\alpha t} \qquad (2.67)$$

where v_0 is an arbitrary constant representing the initial value of v, i.e. the value at $t = 0$, and e is a mathematical constant with the value e = 2.718 281 828 to nine decimal places.

If α is negative, Equation 2.67 describes *exponential decay* and it is customary to write $\alpha = -1/\tau$. If α is positive then Equation 2.67 describes *exponential growth* and it is usual to write $\alpha = 1/\tau$. In either case the positive constant τ is called the *time constant* of the exponential process and represents the time required for the varying quantity to decrease or increase by a factor of 1/e or e, respectively.

Some examples of exponential processes are shown in Figure 2.50.

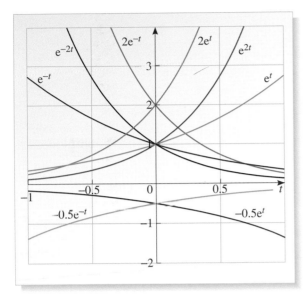

Figure 2.50   Some examples of exponential growth and exponential decay.

The following relationships hold true for any values of x and y:

$$e^x e^y = e^{x+y}, \quad (e^x)^y = e^{xy}, \quad e^{-x} = 1/e^x$$

$$e^0 = 1, \quad e^1 = e, \quad e^{-1} = 1/e.$$

Any function of the form $y = a^x$, where a is a positive constant, may be written as $y = e^{kx}$ since it is always possible to find a constant k such that $e^k = a$, and we can then write $a^x = (e^k)^x = e^{kx}$.

Given a positive quantity x, the **logarithm to the base e** of x is the power to which e must be raised to obtain x. So, if $x = e^z$, then $z = \log_e x$. The **logarithmic function** defined by $y(x) = \log_e x$ is said to be the *inverse* of the exponential function since it 'undoes' the effect of the exponential function in the sense that $\log_e e^x = x$. This is why some calculators provide access to the built-in exponential function via the key combination 'INV' followed by '$\log_e$' (some calculators use 'ln' in place of '$\log_e$': 'ln' stands for **natural logarithm**, a synonym for 'log to the base e').

A quantity that decays (or grows) exponentially decreases (or increases) by equal factors in equal intervals of time, irrespective of when those intervals begin. In the case of exponential decay we have already noted that the time constant τ is the time required for a quantity v to decrease by a factor of $1/e$, but it is generally true that in *any* fixed interval of time, Δt, the quantity v will *always* decrease by the factor $1/e^{\Delta t/\tau}$. An important implication of this is that the time required for any exponentially decaying quantity to reduce its value by a factor of $\frac{1}{2}$ is $\tau \log_e 2 = 0.693\tau$ (to three decimal places). This quantity is known as the *half-life* of an exponential decay process. Similarly, any exponentially growing quantity will double its value in any time interval of length $\tau \log_e 2$, which is accordingly called the *doubling time* of an exponential growth process.

The simplest way to define the number e is in terms of differential calculus; e is the unique number with the property that

$$\frac{de^x}{dx} = e^x \quad \text{or, more generally, if } a \text{ is a constant,} \quad \frac{de^{ax}}{dx} = ae^{ax}.$$

These definitions are consistent with Equations 2.66 and 2.67.

6 Closing items

6.1 Chapter summary

1   The energy of a system is a measure of its capacity for doing work. The SI unit of work and of energy is the joule (J), where $1\,\text{J} = 1\,\text{kg}\,\text{m}^2\,\text{s}^{-2} = 1\,\text{N}\,\text{m}$.

2   The translational kinetic energy of a body of mass m and speed v is $E_{\text{trans}} = \frac{1}{2}mv^2$.

3   The work done on any body by a force is the energy transferred to or from that body by the force. When a non-zero resultant force acts on a particle, the work done by that force is related to the change in the particle's translational kinetic energy by the work–energy theorem

$$W = \Delta E_{\text{trans}} = \tfrac{1}{2}mv^2 - \tfrac{1}{2}mu^2.$$

4   The work done by a constant force $\boldsymbol{F}$ on a body that undergoes a displacement $\boldsymbol{s}$ is defined by the scalar product, $W = \boldsymbol{F} \cdot \boldsymbol{s} = Fs\cos\theta$. If the force is not constant, then this product is replaced by the limit of an appropriate sum, which may be expressed as a definite integral. In the case of a force that varies in strength but always acts along the x-axis this integral takes the form

$$W = \int_A^B F_x\,\mathrm{d}x$$

which may be interpreted as the area under the graph of F_x against x between $x = A$ and $x = B$. In three dimensions the work done generally depends on the particular path that the body moves along and can be represented by the integral

$$W = \int_A^B \boldsymbol{F} \cdot \mathrm{d}\boldsymbol{s}.$$

5   A conservative force is one where the total work done by the force is zero for any round trip, or equivalently, where the work done by the force is independent of the path connecting the start and end points. Other forces are non-conservative forces.

6   A potential energy may be associated with each conservative force that acts on a body or between a system of bodies. The potential energy, E_{pot}, associated with any particular configuration is the work that would be done by the relevant conservative force in going from that configuration to an agreed reference configuration that has been arbitrarily assigned zero potential energy. Because of the arbitrary nature of this reference configuration, only changes in potential energy are physically significant.

7   The change in potential energy when a system goes from some initial configuration to some final configuration is *minus* the work done by the relevant conservative force during that change. (Note that this is *not* generally equal to the work done by any external forces that bring about the change, since those forces may be non-conservative.)

$$E_{\text{pot}}(\text{final}) - E_{\text{pot}}(\text{initial}) = \Delta E_{\text{pot}} = -W_{\text{cons}}(\text{initial} \to \text{final}).$$

8   The gravitational potential energy of a body of mass m at a small height h above the Earth's surface is given by

$$E_{\text{grav}} = mgh \quad (E_{\text{grav}} = 0 \text{ at the Earth's surface}).$$

The gravitational potential energy of a body of mass m at a distance r from the Earth's centre is given by

$$E_{\text{grav}} = \frac{-GmM_{\text{E}}}{r} \quad (E_{\text{grav}} = 0 \text{ at } r = \infty).$$

The strain potential energy of an ideal spring extended by an amount x from its unextended state, is given by

$$E_{str} = \tfrac{1}{2} k_s x^2 \quad (E_{str} = 0 \text{ at } x = 0).$$

9   If the potential energy associated with a particular conservative force is a function of the single variable x, then the only non-zero component of the force will be F_x, and its value at any point will be given by *minus* the gradient of the E_{pot} against x graph at that point

$$F_x = \frac{-dE_{pot}}{dx}.$$

10 For systems in which only conservative forces act, the total mechanical energy, E_{mech}, is conserved

$$E_{mech} = E_{trans} + E_{pot} = \text{constant}$$

i.e. $\Delta E_{pot} + \Delta E_{trans} = 0.$

This equation provides the means of linking the speed of a body to its position in such a system.

11 The law of conservation of mechanical energy can be extended to cover all forms of energy, leading to the law of conservation of energy.

12 Power is defined as the rate at which work is done and energy is transferred. The SI unit of power is the watt (W), where $1\,W = 1\,J\,s^{-1}$.

The instantaneous power delivered by a force $\boldsymbol{F}$ acting on a body moving with velocity $\boldsymbol{v}$ is given by

$$P = \frac{dW}{dt} = \boldsymbol{F} \cdot \boldsymbol{v}.$$

13 A simple harmonic oscillator, subject to a force $F_x = -kx$, has potential energy $E_{pot} = \tfrac{1}{2} kx^2$, where $x\,(= A\sin(\omega t + \phi))$ is the oscillator's instantaneous displacement from its equilibrium position , and k is the force constant $(\omega = \sqrt{k/m})$. The total mechanical energy of such an oscillator is constant, and is given by $E_{mech} = \tfrac{1}{2} kA^2$. In such a system, potential energy is converted into kinetic energy and back again with a period equal to half that of the motion.

14 A lightly damped harmonic oscillator, subject to a force $F_x = -kx - bv_x$, may be regarded as having an amplitude that decays exponentially with time, as described by the equation $A(t) = A_0 e^{-t/\tau}$, where the time constant of the decay is $\tau = 2m/b$. If the damping is sufficiently light, the energy of such an oscillator will also decay exponentially with time constant $\tau/2 = m/b$. (In exponential decay, the decaying quantity will decrease by equal factors in equal intervals of time.)

15 A driven damped harmonic oscillator, subject to a force $F_x = -kx - bv_x + F_0\cos(\Omega t)$, will eventually enter into a steady state in which it exhibits simple harmonic motion at the driving (angular) frequency Ω. The amplitude of this motion depends on the value of Ω and the damping constant b, but for a fixed value of b the amplitude will be a maximum when the driving frequency is equal to the resonant (angular) frequency Ω_R, which is generally quite close to the oscillator's natural (angular) frequency $\omega_0 = \sqrt{k/m}$. This condition of maximum amplitude response to a periodic driving force is referred to as resonance and has many important applications.

16 The energy that sustains a driven damped harmonic oscillator in a steady state is supplied entirely by the driving force. The power transferred by that force will be a maximum when the driving frequency is equal to the natural frequency of the oscillator, and, for a lightly damped oscillator, this will also ensure that the oscillator is in (or very near) a state of resonance.

6.2 Achievements

Having completed this chapter, you should be able to:

A1 Understand the meaning of all the newly defined (emboldened) terms introduced in this chapter.

A2 Write down expressions for the work done by a force on a body when the body undergoes a specified displacement or moves along a specified path.

A3 Write down and justify an expression for the work done by a force in stretching a spring that obeys Hooke's law.

A4 State the work–energy theorem, derive an equation which represents that theorem in sufficiently simple cases, and use that equation to solve problems involving work done and kinetic energy changes.

A5 Distinguish between conservative and non-conservative forces and explain why the concept of potential energy is meaningful only in relation to conservative forces.

A6 Derive expressions for gravitational potential energy near the Earth's surface and the potential energy of a stretched or compressed spring in sufficiently simple cases. (This might involve writing down integrals, but you are not expected to evaluate those integrals unless you can do so graphically.)

A7 Derive an expression which represents the law of conservation of mechanical energy and solve problems using this conservation principle.

A8 Recall the relation between a conservative force and the negative gradient of the associated potential energy function.

A9 Explain the meaning of the term power and describe the use of this quantity in comparing the performances of machines.

A10 Recall and derive an expression relating power, force and velocity.

A11 Write down and use a general expression for the gravitational potential energy of two point-like masses separated by a given distance. Use this expression to justify a formula for the escape speed from the Earth.

A12 Obtain an expression for the total mechanical energy of a simple harmonic oscillator. Describe the periodic transformation of potential energy into kinetic energy for such an oscillator and solve simple problems involving the energy of vibrating objects.

A13 Recognize the mathematical description of exponential decay, $v = v_0 e^{-t/\tau}$ and understand the meaning of the symbols. Recognize cases of exponential decay and use their mathematical description to solve a variety of problems relating to those decays.

A14 Describe the behaviour of a damped harmonic oscillator for various levels of damping, and, in the case of sufficiently light damping, relate the exponential decay of the mechanical energy to that of the amplitude.

A15 Describe the steady state behaviour of a driven damped harmonic oscillator (with a sinusoidal driving force) and explain the significance of resonance to such an oscillator. Describe the role of the driving force in maintaining the oscillations of such a system, especially when close to resonance.

A16 Describe and discuss the nature and importance of the condition of resonance in nature and technology, including the relevance of damping.

6.3 End-of-chapter questions

Question 2.24   Explain the meaning of the term 'work done' and determine the work done by a force of magnitude 5 N acting on a particle which undergoes a displacement of 3 m, if the angle between the force and the displacement is: (a) 0°, (b) 53.1°, (c) 90°.

Question 2.25   A force does the same amount of work in stretching springs A and B by extensions $2x$ and x, respectively. If both springs obey Hooke's law, what is the ratio of the spring constants of springs A and B?

Question 2.26   Show that the work done by a constant resultant force on a particle is equal to the change in the translational kinetic energy of the particle.

Question 2.27 A single force acts on a particle initially at rest, causing it to accelerate and supplying power at the constant rate of 20 W. What is the magnitude of the force when the speed of the particle is 8 m s^{-1}?

Question 2.28   Explain the difference between a conservative and a non-conservative force. Explain the meaning of the term potential energy. How are these concepts related?

Question 2.29   The potential energy function associated with a central force (i.e. one directed towards a fixed point O) is given by $E_{pot} = kr$, where r is the distance from O, and k is a constant. Find an expression for the force in terms of r and k.

Question 2.30   A block of mass m projected with speed u across a rough horizontal surface comes to rest after travelling a distance d. Explain what happens to the kinetic energy lost by the block and derive an expression for the magnitude of the (constant) frictional force acting on the block.

Question 2.31   The Moon has a mass of 7.35×10^{22} kg and radius 1.74×10^{6} m. Calculate the escape speed from the lunar surface.

Question 2.32   Plot, on the same graph, the amplitude of a damped harmonic oscillator against time for the three cases: (a) heavily damped, (b) lightly damped and (c) critically damped. Suggest an application for each of the cases. ■

Chapter 3   Linear momentum and collisions

Figure 3.1   Ernest Rutherford (1871–1937), on the right, the discoverer of the atomic nucleus, together with Hans Geiger.

Alpha-particles were eventually identified as the nuclei of helium atoms.

1   Discovering the atomic nucleus — an example

Early in the twentieth century, J. J. Thomson (1856–1940), already famous as the discoverer of the electron, proposed that the positive charge associated with most of the mass of an atom was smeared throughout the volume of the atom with electrons 'dotted about' within it. This became known as Thomson's 'plum pudding' model of the atom.

In 1909, at the University of Manchester, England, Hans Geiger (1882–1945) and Ernest Marsden (1889–1970), two young assistants of the Nobel laureate Ernest Rutherford (Figure 3.1), conducted an experiment that caused Rutherford to reject Thomson's model of the atom and to replace it with his own 'nuclear' model, a refinement of which is still in use today. The experiment that so profoundly altered Rutherford's view of the atom was essentially a subatomic *collision* experiment, and the analysis of it, that Rutherford performed, made essential use of the concept of *momentum*.

In their experiment, Geiger and Marsden arranged for positively charged **alpha-particles** — subatomic particles with about four times the mass of a hydrogen atom — coming from a radioactive source, to strike an extremely thin gold foil. The subsequent behaviour of the alpha-particles was determined by looking for the tiny flashes of light that were produced when they struck a zinc sulfide screen — a primitive sort of particle detector. What Geiger and Marsden found, as a result of hours spent sitting in the dark recording flashes, was totally counter to their expectations. Some of the alpha-particles suffered substantial deflections as they passed through the thin foil and a few of them, about one in every 8000, actually bounced back in the direction from which they came.

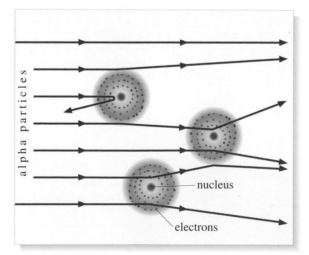

Figure 3.2   Alpha-particles 'colliding' with gold atoms in Rutherford's interpretation of the results of Geiger and Marsden. Most of the alpha-particles are deflected only slightly, if at all, but a few bounce back as they collide head-on with dense atomic nuclei.

When these results were reported to Rutherford, he was astonished. He later said that it was as if 'you had fired a 15-inch naval shell at a piece of tissue paper and the shell came right back at you'. If Thomson's plum pudding model had been correct, all the relatively heavy alpha-particles would have passed through the foil, suffering deflections of no more than a few degrees. The fact that this did not happen caused Rutherford to propose that all of the atom's positive charge, and most of its mass, is concentrated in a very tiny central nucleus. The negatively charged electrons, according to Rutherford, moved around the nucleus, held in place by their electrical attraction towards it. In order to fit the experimental data, the nuclear radius had to be less than one ten-thousandth of the atomic radius, yet the nucleus also had to account for 99.9% of the atom's mass (see Figure 3.2). Despite these exotic sounding properties, atomic nuclei do exist (every atom in your body contains one!) and nuclear physics is now a major research field.

The rebounding alpha-particles that Geiger and Marsden observed were the result of essentially head-on collisions between alpha-particles and individual gold nuclei. The forces between these objects cause the observed change in the motion, but the analysis of the collision is most easily performed by combining the law of conservation of energy with a related law of *conservation of momentum*. By using these principles Rutherford realized that the only sensible interpretation of the observations was that the alpha-particles were colliding with some more massive body than themselves, hence the need for a nucleus within the atom. A collision between an alpha-particle and a 'plum pudding' cloud of positive charge, or a low mass electron, could not have caused the observed deflections and rebounds. In this chapter you will learn about the concept of momentum and its role in analysing collisions. You will also learn about the importance of momentum in everyday life and the varied role that collisions play in the world around us.

2 Linear momentum

2.1 The linear momentum of a body

Newton's laws are useful for predicting motion in situations where known forces act for known periods of time. However, in physics, as in life, we are often faced with situations in which predicting motion is vital, even though the relevant forces are not known in sufficient detail for Newton's laws to be applied. Games such as snooker and pool provide an excellent example of this. When two snooker balls collide, they are in contact for a very short time during which the forces between them are rapidly varying. Predicting the outcome of such a collision using Newton's laws would be very difficult if not impossible, yet snooker players intuitively predict the outcomes of such collisions all the time, often with astonishing accuracy. The physical quantity that makes such predictions possible is known as *momentum*.

Loosely speaking, the momentum of a body is a measure of the difficulty of stopping that body. It is apparent that a locomotive travelling at $60\,\mathrm{m\,s^{-1}}$ 'takes more stopping' than a similar locomotive travelling at $30\,\mathrm{m\,s^{-1}}$. Likewise, a heavy locomotive travelling at $60\,\mathrm{m\,s^{-1}}$ 'takes more stopping' than a light locomotive travelling at that speed. The importance of the physical quantity that combines the mass and velocity of a body was perceived in the fifteenth and sixteenth centuries, and the concept was developed mathematically by Galileo and others in the seventeenth century. Newton called it the 'quantity of motion' of the body, but is now called linear momentum and is defined as follows:

> The **linear momentum** of a body of mass m whose centre of mass is moving with velocity $\boldsymbol{v}$ is
>
> $$\boldsymbol{p} = m\boldsymbol{v}. \tag{3.1}$$

For the rest of this chapter, the word 'linear' will be dropped and we will simply refer to the '**momentum**' of a body. In Chapter 4, however, you will see that bodies can also possess *angular momentum*, and you must not confuse the two.

Now mass is a positive scalar quantity, and velocity is a vector quantity that can be specified in three dimensions in terms of its components $\boldsymbol{v} = (v_x, v_y, v_z)$. It therefore follows from Equation 3.1 that the momentum of a body is also a vector quantity, and that it points in the same *direction* as the body's velocity. So,

$$\boldsymbol{p} = m\boldsymbol{v} = (mv_x, mv_y, mv_z)$$

and we can write the (scalar) magnitude of the momentum in any of the following ways

$$p = |\boldsymbol{p}| = |m\boldsymbol{v}| = m|\boldsymbol{v}| = mv$$

where v is the speed of the body, given by $v = \sqrt{v_x^2 + v_y^2 + v_z^2}$.

A vector equation is always equivalent to three scalar equations involving the components of the vectors. In the case of $\boldsymbol{p} = m\boldsymbol{v}$, if we let $\boldsymbol{p} = (p_x, p_y, p_z)$, the three equations are

$$p_x = mv_x, \quad p_y = mv_y, \quad p_z = mv_z.$$

This is illustrated, for two dimensions, in Figure 3.3.

It follows from all these expressions for momentum, its magnitude, and its components, that the SI unit of momentum is the $\mathrm{kg\,m\,s^{-1}}$. Rather unusually for such an important unit, this has no special name.

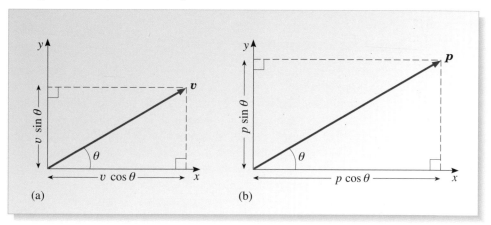

Figure 3.3 (a) The velocity $\boldsymbol{v}$ of an object moving in the xy-plane has magnitude v and is directed at an angle θ to the x-axis. The x-component is $v_x = v \cos \theta$, and the y-component is $v_y = v \sin \theta$. (b) The momentum $\boldsymbol{p}$ of the object is in the same direction as $\boldsymbol{v}$ and has components $p_x = p \cos \theta = mv_x = mv \cos \theta$, and $p_y = p \sin \theta = mv_y = mv \sin \theta$.

Since momentum is a vector, it is well suited to the analysis of motion that is not restricted to a single straight line. This is important, because when bodies moving in different directions collide, they can, and often do, change their direction of motion. Momentum vectors often play a vital role in determining these directions. The following questions involve momentum as a vector and will help you to familiarize yourself with the concept. Do not hesitate to draw a diagram as part of your answer if you think it will help.

Question 3.1 A body of mass 5.0 kg is moving north-east at $20\,\mathrm{m\,s^{-1}}$. (a) What is its momentum? (b) What are the components of momentum in the northerly direction and in the easterly direction? (c) What would be the corresponding components if the body moved with the same speed in the *north-west* direction?

Question 3.2 A ball of mass 0.5 kg moving horizontally in the positive x-direction with a speed of $2.0\,\mathrm{m\,s^{-1}}$ rebounds from a vertical wall. After the collision, the ball is found to be moving in the negative x-direction at $2.0\,\mathrm{m\,s^{-1}}$. What is the *change* in the ball's momentum?

Question 3.3 In a game of volleyball, player 1 near the back of the court passes the ball *horizontally* to her team-mate player 2, who is standing near the net. As it approaches player 2, the ball is moving uniformly at $3.0\,\mathrm{m\,s^{-1}}$. Player 2 taps the ball with a *vertical* force so that it achieves a *component* of momentum of $0.8\,\mathrm{kg\,m\,s^{-1}}$ in the vertical direction immediately afterwards. If the mass of the ball is 400 g, determine the magnitude and direction of its resulting momentum? (*Hint*: Draw a vector diagram!) ■

Question 3.2 illustrates an especially important point. Due to its vector nature, the momentum of an object can change simply due to a change in the *direction* of a body's motion, even though the *magnitude* of the velocity, and hence of the momentum, remains unchanged.

2.2 Momentum and Newton's second law

Newton's second law, as formulated in Chapter 1, implies that for a body of fixed mass m moving in one dimension (the x-direction), the product of mass and acceleration is equal to the resultant force in that direction, i.e.

$$F_x = ma_x.$$

The instantaneous acceleration a_x is the rate of change of the corresponding velocity (the gradient of the v_x against x graph) so it may be expressed as a derivative, and we may write

$$F_x = m\frac{\mathrm{d}v_x}{\mathrm{d}t}.$$

According to the rules of differentiation (as reviewed in Chapter 1), the derivative of a constant times a function is equal to the constant times the derivative of the function. So, given that v_x is a function of t, and that m is a constant, we can write

$$m\frac{\mathrm{d}v_x}{\mathrm{d}t} = \frac{\mathrm{d}}{\mathrm{d}t}(mv_x)$$

consequently $F_x = \dfrac{\mathrm{d}}{\mathrm{d}t}(mv_x) = \dfrac{\mathrm{d}}{\mathrm{d}t}(p_x).$

Expressed in words, this equation says that the x-component of force is equal to the rate of change of the x-component of momentum. Similar equations can be found in the same way for the y- and z-components of force, giving

$$F_y = \frac{\mathrm{d}p_y}{\mathrm{d}t} \quad \text{and} \quad F_z = \frac{\mathrm{d}p_z}{\mathrm{d}t},$$

and we can combine these three equations for the force components F_x, F_y and F_z into the single vector equation

$$(F_x, F_y, F_z) = \left(\frac{\mathrm{d}p_x}{\mathrm{d}t}, \frac{\mathrm{d}p_y}{\mathrm{d}t}, \frac{\mathrm{d}p_z}{\mathrm{d}t}\right).$$

This is an important general result that can be summarized as follows:

> The total force acting on a body is equal to the rate of change of momentum of that body
>
> $$F = \frac{dp}{dt}.$$ (3.2)

Although we have arrived at this equation by using $F = ma$ and by assuming that m is constant, it can be shown that Equation 3.2 is, in fact, a *more general formulation of Newton's second law* of motion than $F = ma$. It can be applied to systems of many particles, and, in some cases, can be used when the mass of the system varies, as in the case of a rocket that accelerates by expelling the fuel that it carries and thus continuously changes its mass. Because of its greater generality, many physicists prefer to refer to Equation 3.2 as 'Newton's second law' rather than $F = ma$.

In cases where the mass of a body does change with time, a direct mathematical consequence of Equation 3.2 is the following relationship between force and acceleration

$$F = ma + v\frac{dm}{dt}.$$ (3.3)

As you can see, the right-hand side of the familiar $F = ma$ is supplemented by an additional term that points in the direction of the velocity and is proportional to dm/dt, the rate of change of the mass. In cases where the mass of a body remains constant this extra term would be zero, which is why we have not considered it before. However, if m is *not* constant, Equation 3.3 shows that the relationship between force and acceleration is not as simple as we might have thought it to be. The inclusion of the $v(dm/dt)$ term in Equation 3.3 has some strange consequences. For instance, if the mass of a body is changing, then a resultant force acting on that body does not necessarily cause it to accelerate; if the mass is increasing at just the right rate, the effect of the force may be to maintain a constant velocity rather than to cause a change of velocity. Similarly, even in the absence of any resultant force, the velocity of a body may change with time if its mass is changing.

Question 3.4   What force is required to push a truck across a frictionless surface at a constant velocity of $5\,\text{m s}^{-1}$ in a given direction if sand is being poured into the truck at a constant rate of $6\,\text{kg s}^{-1}$?   ■

2.3 Impulse and impulsive forces

When snooker balls collide, or when a ball is struck by a bat, all we are usually interested in is the extent to which one object will be deflected as a result of its interaction with the other. We generally want to know where a ball will go, not the forces that sent it there. From a scientific point of view, what we are really interested in when we hit a ball with a bat is the change in the ball's momentum as a result of its brief interaction with the bat.

The quantity that represents the change of momentum of a body as a result of such a brief interaction is called the **impulse**. In many situations, the impulse delivered by a force is all we really want to know about the force. Forces that can be treated in this way are known as **impulsive forces**; they are usually short and abrupt.

If a constant force $\boldsymbol{F}$ acts on a body for a short time Δt, Equation 3.2 tells us that the body's momentum will change by an amount $\Delta\boldsymbol{p}$, where

$$\boldsymbol{F} = \frac{\Delta\boldsymbol{p}}{\Delta t}.$$

It follows that the momentum of the body will change by an amount

$$\Delta\boldsymbol{p} = \boldsymbol{F}\,\Delta t \qquad\qquad (3.4)$$

so, $\boldsymbol{F}\,\Delta t$ will be the impulse delivered by the constant force $\boldsymbol{F}$ over the short time Δt.

The same result applies when $\boldsymbol{F}$ represents the average force that acts over a time Δt, as shown in Example 3.1.

Example 3.1

While waiting at traffic lights, a car and its 80 kg driver are suddenly accelerated to a velocity of $10\,\text{m s}^{-1}$ as a result of a rear-end collision. Find (a) the change in momentum of the driver, and (b) the average force exerted on the driver's back by the seat of the car, assuming that the time during which the force acted was 0.2 s.

Solution

Let the direction in which the car moves be the x-direction.

(a) The driver's initial momentum *before* the collision is given by

$$p_x(\text{initial}) = mv_x(\text{initial}) = 80\,\text{kg} \times 0\,\text{m s}^{-1} = 0\,\text{kg m s}^{-1}.$$

The driver's final momentum *after* the collision is given by

$$p_x(\text{final}) = mv_x(\text{final}) = 80\,\text{kg} \times 10\,\text{m s}^{-1} = 800\,\text{kg m s}^{-1}.$$

Therefore,

$$\begin{array}{ccc} \text{change in} \\ \text{momentum} \end{array} = \begin{array}{c} \text{momentum} \\ \text{after collision} \end{array} - \begin{array}{c} \text{momentum} \\ \text{before collision} \end{array}$$

$$= 800\,\text{kg m s}^{-1} \quad - \quad 0\,\text{kg m s}^{-1}$$

$$= 800\,\text{kg m s}^{-1}.$$

(b) Since, at any time, the instantaneous force on the driver is equal to the rate of change of the driver's momentum, it must be the case that

$$\begin{array}{c} \text{average force} \\ \text{on the driver} \end{array} = \begin{array}{c} \text{average rate of change} \\ \text{of the driver's momentum.} \end{array}$$

It follows that if the driver's momentum changes by an amount Δp_x over a time Δt then

$$\begin{array}{c} \text{average force} \\ \text{in the }x\text{-direction} \end{array} = \frac{\Delta p_x}{\Delta t} = \frac{800\,\text{kg m s}^{-1}}{0.2\,\text{s}} = 4000\,\text{N}.$$

In reality the force that produces a given impulse often varies in quite a complicated way. Figure 3.4 shows the variation of the upward component of force, F_x, on a rubber ball during a typical bounce. Arguments similar to those used to evaluate the work done by a varying force in the last chapter, show that in this case the impulse is directed upwards, and may be represented by the area under the curve shown in Figure 3.4. It follows that we can write the impulse delivered by the varying force as a definite integral

$$\Delta \boldsymbol{p} = \int_{T_1}^{T_2} \boldsymbol{F} \, \mathrm{d}t$$

where $t = T_1$ and $t = T_2$ represent the times at which contact with the ground begins and ends.

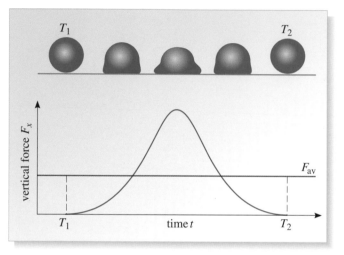

Figure 3.4 The vertical force on a rubber ball as it hits the floor is shown as a function of time. The equivalent average force is also shown. The upward impulse is represented by the area under the curve, between T_1 and T_2, in each case. The value of the F_{av} ensures these two areas are equal. (The extent of the deformation has been exaggerated.)

In practice, it is often much easier to measure a momentum change and hence deduce the corresponding impulse than it is to calculate the impulse from the force. However, having determined the impulse delivered by a force, it is always possible to determine the constant force $\boldsymbol{F}_{av}$ that would be required to deliver the same impulse over a given time Δt, simply by requiring

$$\boldsymbol{F}_{av} \, \Delta t = \int_{T_1}^{T_2} \boldsymbol{F} \, \mathrm{d}t. \tag{3.5}$$

This procedure can give surprising insights into the rapidly varying forces that arise in various impulsive interactions. For example, in baseball the bat and ball are typically in contact for one- or two-thousandths of a second, but the impulse delivered to a baseball might well have a magnitude of 6 to 12 kg m s^{-1}. The average force required to deliver the same impulse over the same time would have a magnitude of about 6000 N. Such large forces are not unusual.

2.4 The momentum of a system of bodies

A major reason for the importance of momentum is that it is *conserved* in a wide range of systems. We shall consider momentum conservation in the next subsection, but first we will consider what it means to speak of the 'momentum of a system'.

A system is simply that portion of the Universe we choose to study in any particular investigation. The momentum of a system is the vector sum of the momenta of its parts. (Momenta is the plural of momentum.)

In the case of a system consisting of two bodies, with masses m_1 and m_2, moving with respective velocities $\boldsymbol{v}_1$ and $\boldsymbol{v}_2$, the total momentum of the system is

$$\boldsymbol{P} = \boldsymbol{p}_1 + \boldsymbol{p}_2 = m_1\boldsymbol{v}_1 + m_2\boldsymbol{v}_2.$$

Notice we use a capital $\boldsymbol{P}$ to denote the sum of the momenta. Since each individual momentum is a vector, the addition must be carried out vectorially (using the triangle rule, for example, or in terms of components), so the total momentum $\boldsymbol{P}$ will also be a vector with magnitude and direction.

More generally, for a system containing a number of bodies, the total momentum of the whole system at any instant is just the vector sum of all the individual momenta at that instant. Thus:

$$\boldsymbol{P} = \boldsymbol{p}_1 + \boldsymbol{p}_2 + \boldsymbol{p}_3 + \ldots = m_1\boldsymbol{v}_1 + m_2\boldsymbol{v}_2 + m_3\boldsymbol{v}_3 + \ldots.$$

We can conveniently write this in terms of the summation symbol we used in Chapters 1 and 2. Thus

$$\boldsymbol{P} = m_1\boldsymbol{v}_1 + m_2\boldsymbol{v}_2 + m_3\boldsymbol{v}_3 + \ldots = \sum_i (m\boldsymbol{v})_i \tag{3.6}$$

where $(m\boldsymbol{v})_i = m_i\boldsymbol{v}_i$ and the symbol $\sum_i$ (read as 'the sum over i') tells us to add up all terms of the form immediately following the symbol for every allowed value of i, (in this case that means for $i = 1, 2, 3$ and so on). If the quantities being added together are vectors, then the method of addition must take that into account, so that their sum is also a vector.

In most cases, the easiest way of carrying out the vector sum is in terms of components. If we know the components of each of the individual velocities then we can express the x-, y- and z-components of $\boldsymbol{P}$ as follows

$$P_x = \sum_i (mv_x)_i \qquad P_y = \sum_i (mv_y)_i \qquad P_z = \sum_i (mv_z)_i. \tag{3.7}$$

Question 3.5   Two trains, each of mass 5.0×10^4 kg are on the same straight track. What is the total momentum of the system (a) if the trains are both travelling in the same direction at 90 km h^{-1}, and (b) if the trains are both travelling at 90 km h^{-1} but are approaching each other head on?   ■

2.5 Conservation of momentum

The identification of 'force' with 'rate of change of momentum' as expressed in Equation 3.2 implies that if $\boldsymbol{F} = \boldsymbol{0}$ then $\mathrm{d}\boldsymbol{p}/\mathrm{d}t = \boldsymbol{0}$. Thus:

If no resultant force acts on a body then its momentum does not change.

Now this statement is little more than an expression of Newton's first law of motion in terms of momentum. However, by extending it to systems of bodies and incorporating the consequences of Newton's third law, we can generalize it to produce a law of conservation of momentum that can be just as useful in predicting motion as the conservation of mechanical energy we discussed in the last chapter. This subsection concerns the justification and formulation of the law of *conservation of momentum*. Its applications will occupy much of the rest of this chapter.

Momentum conservation applies to systems consisting of two or more bodies that interact with each other but *not* with their surroundings. The forces that act on such bodies are said to be **internal forces** because the reaction forces with which they must be paired according to Newton's third law, also act *within* the system.

- State Newton's third law of motion.

○ If body A exerts a force on body B, then body B exerts a force on body A. These two forces are equal in magnitude but opposite in direction. (Less formally; 'to every action on a body, there is an equal but opposite reaction on some other body'.) ■

The internal forces in any system may be contrasted with the **external forces** that may act on that system. A force acting on a system is said to be *external* if the reaction to it acts outside the system. So, for example, in a system consisting of just two particles (see Figure 3.5), those particles may exert gravitational forces on each other which are then internal forces of that system. However, if both particles are also acted upon by the force due to the Earth's gravitational attraction, that would be an external force since the reaction to it is the pull that the particles exert on the Earth, which is certainly outside of the two-particle system.

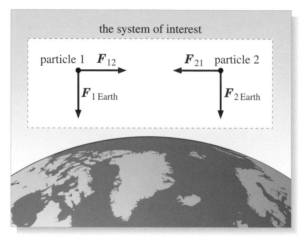

Figure 3.5 A system consisting of two particles. The system includes internal forces arising from the mutual gravitational interaction of the particles, and external forces due to their gravitational interaction with the Earth.

In most systems, the various parts of the system are subject to a mix of internal and external forces. Any system that is completely free of external forces is said to be an **isolated system**.

Now, within an isolated system, each of the bodies making up the system may have momentum, and the momentum of any of those bodies may change with time due to the action of internal forces. However, no matter how the momentum of any part of the system changes due to an internal force, a compensating change must occur elsewhere in the system due to the reaction to that force. (Remember, we are dealing with an *isolated* system, so the reaction to any force must also act *within* the system.) As a result, no matter how the momentum of any part of the system changes, there can be no change to the total momentum of all the system's parts. This, in essence, is the proof of the following important principle:

> The law of **conservation of linear momentum**:
>
> The total linear momentum of any *isolated* system is constant.

Remember that the 'total linear momentum' referred to here is the *vector sum* of the momenta of the system's parts. So, if P is the total momentum of an *isolated* system, the principle of momentum conservation can be expressed algebraically by saying

$$P = \text{constant vector} \quad \text{or,} \quad P(\text{initial}) = P(\text{final}). \tag{3.8}$$

The fact that this is a vector relationship means we can also express it in terms of individual momentum components, thus, for example

$$\sum_i \left(mv_x(\text{initial})\right)_i = \sum_i \left(mv_x(\text{final})\right)_i \tag{3.9a}$$

$$\sum_i \left(mv_y(\text{initial})\right)_i = \sum_i \left(mv_y(\text{final})\right)_i \tag{3.9b}$$

$$\sum_i \left(mv_z(\text{initial})\right)_i = \sum_i \left(mv_z(\text{final})\right)_i \tag{3.9c}$$

The law of conservation of momentum was suggested by the English cleric and mathematician John Wallis (1616–1703), though it was significantly clarified and extended by the Dutch scientist Christiaan Huygens (1629–1695) who conducted experiments with colliding balls.

● When a ball bounces from a wall, its momentum changes, and is therefore not conserved. Does this constitute a violation of the law of conservation of momentum? Explain your answer.

○ No, the ball is not an isolated system when it strikes the wall, so there is no reason why its momentum should be conserved. However, the wall is fixed to the Earth, and a system consisting of the ball plus the wall plus the Earth would be an isolated system as far as the bouncing ball is concerned. It is the momentum of this system that is conserved. The change in momentum of the ball is equal in magnitude and opposite in direction to the change in momentum of the wall plus the Earth. However, since the Earth is so massive, you will not see any recoil of the wall when the ball strikes it; the required momentum change of the Earth corresponds to only an imperceptible change in its velocity. ■

Question 3.6   When you walk along the road with constant velocity you have constant momentum. When you stop walking you have zero momentum. Where has your momentum gone?   ■

2.6 The role of momentum in predicting motion

It is important to recognize that the law of conservation of momentum holds even when we do not fully understand the details of the forces under consideration. This fact makes it a powerful tool in solving problems and predicting motion. Example 3.2 illustrates its value in analysing the radioactive decay of a nucleus.

Example 3.2

A radioactive nucleus at rest decays by forming a 'daughter nucleus' and an alpha-particle (an alpha-particle is actually a helium nucleus; it consists of two protons and two neutrons bound together). Suppose the mass of the daughter nucleus is 54 times that of the alpha-particle. What is the velocity of the daughter nucleus if the velocity of the alpha-particle is $v_x(\text{alpha}) = 1.0 \times 10^7 \, \text{m s}^{-1}$ along the positive x-axis?

Solution

The nucleus and its decay products constitute an isolated system, so momentum must be conserved in the decay. The total momentum before the decay is zero, so the total momentum afterwards must also be zero. This means that the vector sum of the momenta of the two particles resulting from the decay must be zero. To satisfy this condition the two momenta must have equal magnitudes but their directions must be opposite to each other. Since the alpha-particle travels along the positive x-axis, the daughter nucleus must also travel along the x-axis, but in the opposite direction. Despite knowing this we shall simply call its velocity v_x(daughter), and let the mathematics take care of the sign (which we expect to be negative). Letting m be the mass of the alpha-particle and conserving momentum along the x-axis, Equation 3.9a gives:

$$mv_x(\text{alpha}) + 54mv_x(\text{daughter}) = 0.$$

So, $v_x(\text{daughter}) = -v_x(\text{alpha})/54 = -1.0 \times 10^7 \,\text{m s}^{-1}/54 = -1.9 \times 10^5 \,\text{m s}^{-1}.$

The negative sign confirms that the daughter nucleus recoils in the opposite direction to the alpha-particle, as expected.

Here is another one-dimensional example (Example 3.3) illustrating the power of momentum conservation. This time it concerns two bodies that collide and stick together. It's almost the reverse of a decay problem. (Later, we will extend our discussion to two- and three-dimensional motion and to collisions where the bodies do not stick together.)

Example 3.3

A body of mass 2 kg moving with a speed of $6 \,\text{m s}^{-1}$ along the x-axis collides with a stationary body of mass 4 kg. The bodies stick together on impact and move off as one combined body along the x-axis with velocity v_x. Find the value of v_x, assuming this is an isolated system.

Solution

In applying the law of conservation of momentum to the isolated system, we equate the total momentum before the collision to the total momentum after the collision. Noting that the mass of the final composite body is $m_1 + m_2$, and using Equation 3.9a,

$$m_1 u_{1x} + m_2 u_{2x} = (m_1 + m_2)v_x$$

where u_{1x} and u_{2x} are the initial velocities of the bodies.

It follows that

$$v_x = \frac{(m_1 u_{1x} + m_2 u_{2x})}{(m_1 + m_2)}.$$

So, $v_x = \dfrac{(2\,\text{kg} \times 6\,\text{m s}^{-1} + 4\,\text{kg} \times 0\,\text{m s}^{-1})}{(2\,\text{kg} + 4\,\text{kg})},$

i.e. $v_x = \dfrac{12\,\text{kg m s}^{-1}}{6\,\text{kg}} = 2\,\text{m s}^{-1}.$

In problems that involve more than one dimension particular care must be taken over the directions of motion of the bodies involved. This means that in applying conservation of momentum it is important to remember the vector nature of momentum. In our discussions so far we have prepared the ground for this by stressing the vector nature of the problem and by insisting on the use of component notation, even in one-dimensional problems.

Problems that require you to evaluate a velocity or a momentum cannot be fully answered unless you state the directions as well as the magnitudes of the required vectors. In order to do so you may well have to start your solution by introducing some appropriate convention regarding the labelling of directions. It may also be necessary to draw a clearly labelled diagram to show what is happening. Do not hesitate to take either or both of these steps if you need to do so. (See Example 3.4.)

Example 3.4

A 2000 kg van, travelling due east at $20 \, \text{m s}^{-1}$ collides, on an icy patch of road, with a 1000 kg car travelling due north at $30 \, \text{m s}^{-1}$. The vehicles lock together on impact and then move off as one body over the frictionless surface. Find the common velocity of the vehicles after the collision.

Solution

In this case, let east be the x-direction and north be the y-direction. We can then draw a diagram (Figure 3.6) showing the initial individual momenta of the van, $\mathbf{p}_1 = m_1\mathbf{u}_1$, and the car, $\mathbf{p}_2 = m_2\mathbf{u}_2$, and their vector sum $\mathbf{P} = \mathbf{p}_1 + \mathbf{p}_2$. Since there are no external forces in the xy-plane that operate on the system, the total momentum $\mathbf{P}$ and its components $P_x = p_1$ and $P_y = p_2$ will both be conserved.

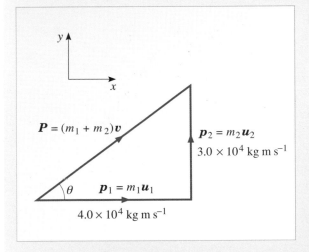

Figure 3.6 The resultant momentum of two vehicles moving east and north is constructed from the vector sum of the momenta of the vehicles.

It follows that the momentum components of the composite body that results from the collision will be

$$P_x = P \cos \theta = p_1 = m_1 u_1 = 4.0 \times 10^4 \, \text{kg m s}^{-1}$$

and $$P_y = P \sin \theta = p_2 = m_2 u_2 = 3.0 \times 10^4 \, \text{kg m s}^{-1}$$

where $$P = \sqrt{P_x^2 + P_y^2} = 5.0 \times 10^4 \, \text{kg m s}^{-1}.$$

and $$\theta = \arctan(P_y/P_x) = \arctan(3/4) = 36.9°.$$

Thus, since the composite body has mass 3000 kg, its velocity will have magnitude

$$v = P/(m_1 + m_2) = 5.0 \times 10^4 \, \text{kg m s}^{-1}/3000 \, \text{kg} = 16.7 \, \text{m s}^{-1}$$

and it will be directed 36.9° north of east.

The arctan function may appear on your calculator as $\tan^{-1}$, or you may have to key 'INV' followed by 'tan'.

If you prefer, you could express the final velocity in terms of its components, thus avoiding evaluating θ:

$$\mathbf{v} = (v_x, v_y) = (4.0 \times 10^4, 3.0 \times 10^4) \, \text{kg m s}^{-1}/3000 \, \text{kg} = (13.3, 10.0) \, \text{m s}^{-1}.$$

Question 3.7   Two blocks of masses 0.3 kg and 0.2 kg are moving directly towards one another along a frictionless, horizontal surface with speeds 1.0 m s^{-1} and 2.0 m s^{-1}, respectively. If the blocks stick together on impact, find their final velocity.

Question 3.8   Consider the following scenario. The population of the Earth (5.6×10^9 people in September 1994) all simultaneously begin walking at a speed of 1.0 m s^{-1} in the direction of the Earth's instantaneous motion through space. *Estimate* the effect on the Earth. (Take the mass of the Earth as 6.0×10^{24} kg.) What would happen when everybody stopped walking?   ■

3   Collisions and conservation laws

In a scientific context, the term **collision** means a brief but powerful interaction between two particles or bodies in close proximity. The analysis of collisions is of fundamental importance in physics, particularly in nuclear and particle physics, and the techniques used to analyse collisions are well established and widely used. They are also very firmly rooted in the basic conservation principles (or 'conservation laws' as they are sometimes known), particularly those of momentum and energy.

In this section we consider various aspects of collisions and conservation laws, including some of the modifications that must be made to Newtonian principles when we seek to apply them to subatomic particles that typically collide at 'relativistic' speeds, i.e. at speeds close to the speed of light.

3.1   Elastic and inelastic collisions

In the collision problems tackled so far we have always either started with a single body (Example 3.2), or finished up with a single body (Examples 3.3 and 3.4). The reason for this self-imposed limitation is that such problems can be solved by applying momentum conservation alone. The analysis of more general collisions, however, requires the use of other principles in addition to momentum conservation. To illustrate this, we now consider a one-dimensional problem in which two colliding bodies with known masses m_1 and m_2, and with known initial velocities u_{1x} and u_{2x} collide and then separate with final velocities v_{1x} and v_{2x}. The problem is that of finding the two unknowns v_{1x} and v_{2x}. Conservation of momentum in the x-direction provides only one equation linking these two unknowns:

$$m_1 u_{1x} + m_2 u_{2x} = m_1 v_{1x} + m_2 v_{2x} \tag{3.10}$$

which is insufficient to determine both unknowns.

In the absence of any detailed knowledge about the forces involved in the collision, the usual source of an additional relationship between v_{1x} and v_{2x} comes from some consideration of the translational kinetic energy involved in the collision. The precise form of this additional relationship depends on the nature of the collision.

Collisions may be classified by comparing the total (translational) kinetic energy of the colliding bodies before and after the collision. If there is no change in the total kinetic energy, then the collision is an **elastic collision**. If the kinetic energy after the collision is less than that before the collision then the collision is an **inelastic collision**. In some situations (e.g. where internal potential energy is released) the total kinetic energy may even increase in the collision; in which case the collision is said to be a **superelastic collision**.

In the simplest case, when the collision is elastic, the consequent conservation of kinetic energy means that

$$\tfrac{1}{2}m_1 u_{1x}^2 + \tfrac{1}{2}m_2 u_{2x}^2 = \tfrac{1}{2}m_1 v_{1x}^2 + \tfrac{1}{2}m_2 v_{2x}^2. \tag{3.11}$$

This equation, together with Equation 3.10 will allow v_{1x} and v_{2x} to be determined provided the masses and initial velocities have been specified. We consider this situation in more detail in the next subsection.

Real collisions between macroscopic objects are usually inelastic but some collisions, such as those between steel ball bearings or between billiard balls, are very nearly elastic. Collisions between subatomic particles, such as electrons and/or protons, commonly are elastic. The kinetic energy which is lost in an inelastic collision appears as energy of a different form, (e.g. thermal energy, sound energy, light energy, etc.), so that the total energy is conserved. The collisions which we have dealt with so far, in which the bodies stick together on collision and move off together afterwards, are examples of **completely inelastic collisions**. In these cases the maximum amount of kinetic energy, consistent with momentum conservation, is lost. (Momentum conservation usually implies that the final body or bodies must be moving and this inevitably implies that there must be some final kinetic energy; it is the remainder of the initial kinetic energy, after this final kinetic energy has been subtracted, that is lost in a completely inelastic collision.)

3.2 Elastic collisions in one dimension

In this subsection we examine the outcomes of various elastic collisions in one dimension. These are essentially particular cases of the general elastic collision described by Equations 3.10 and 3.11 in the last subsection. Cataloguing one-dimensional elastic collisions may sound like a rather esoteric pastime, but as you will see, you have probably witnessed many collisions of this type, and may even have paid handsomely for the privilege.

We begin with an example of a one-dimensional elastic collision between two particles of identical mass (Example 3.5). An incoming projectile strikes an initially stationary target with the result that both particles move off along the same line. Our aim is to determine the final velocity of each particle after the collision.

Example 3.5

A particle of mass m moves along the x-axis with velocity u_{1x} and collides *elastically* with an identical particle at rest. What are the velocities of the two particles after the collision?

Solution

Let the final velocities be v_{1x} and v_{2x}. Conservation of momentum along the x-axis gives

$$mu_{1x} = mv_{1x} + mv_{2x} \tag{3.12}$$

and conservation of kinetic energy for this elastic collision gives

$$\tfrac{1}{2}mu_{1x}^2 = \tfrac{1}{2}mv_{1x}^2 + \tfrac{1}{2}mv_{2x}^2. \tag{3.13}$$

By eliminating common factors, Equation 3.12 can be simplified to give

$$u_{1x} = v_{1x} + v_{2x}$$

i.e. $\quad v_{2x} = u_{1x} - v_{1x} \tag{3.14}$

and Equation 3.13 can be treated similarly to give

$$u_{1x}^2 = v_{1x}^2 + v_{2x}^2. \qquad (3.15)$$

Rearranging Equation 3.15 gives

$$v_{2x}^2 = u_{1x}^2 - v_{1x}^2$$

the right-hand side of which may be rewritten using the general identity $a^2 - b^2 = (a - b)(a + b)$, thus

$$v_{2x}^2 = (u_{1x} - v_{1x})(u_{1x} + v_{1x}).$$

Dividing both sides of this last equation by v_{2x}, and using Equation 3.14 to simplify the resulting right-hand side gives

$$v_{2x} = u_{1x} + v_{1x}.$$

Comparing this expression for v_{2x} with that in Equation 3.14 shows that

$$v_{1x} = 0 \quad \text{and, thus} \quad v_{2x} = u_{1x}.$$

The result of Example 3.5 will be familiar to anyone who has seen the head-on collision of two bowls on a bowling green. The moving one stops, and the one that was initially stationary moves off with the original velocity of the first. In effect, the bowls exchange velocities.

● In Example 3.5, the two masses were equal. Predict qualitatively (i.e. without calculation) what would happen with the target mass m_2 at rest if: (a) $m_1 \gg m_2$, (b) $m_2 \gg m_1$? (The symbol $\gg$ should be read as 'is very much greater than'.)

○ Experience should tell you (a) that a high-mass projectile fired at a low-mass target would be essentially unaffected by the collision, whereas (b) a low-mass projectile fired at a massive target would bounce back with unchanged speed. ■

We now consider the general one-dimensional elastic collision between particles of mass m_1 and m_2 which move with initial velocities u_{1x} and u_{2x} before the collision and final velocities v_{1x} and v_{2x} after the collision. As stated earlier, the outcome of collisions of this kind is determined by Equations 3.10 and 3.11. We shall not write down the details (though you might like to work them out for yourself) but by arguments similar to those used in Example 3.5, the following result can be obtained.

If the initial velocity of particle 1 *relative* to particle 2 is taken to be

$$u_{12x} = u_{1x} - u_{2x}$$

and if the final velocity of particle 1 *relative* to particle 2 is taken to be

$$v_{12x} = v_{1x} - v_{2x},$$

then, as a result of an elastic collision

$$v_{12x} = -u_{12x}.$$

In other words:

In a one-dimensional *elastic* collision between two particles the relative velocity of approach is the negative of the relative velocity of separation

$$v_{1x} - v_{2x} = -(u_{1x} - u_{2x}).$$

Combining this result (which incorporates the conservation of energy) with Equation 3.10 (which expresses conservation of momentum), leads to the following expressions for the final velocities:

$$v_{1x} = \frac{u_{1x}(m_1 - m_2) + 2m_2 u_{2x}}{m_1 + m_2} \quad \text{(elastic)} \tag{3.16}$$

$$v_{2x} = \frac{u_{2x}(m_2 - m_1) + 2m_1 u_{1x}}{m_1 + m_2} \quad \text{(elastic)} \tag{3.17}$$

Note: you are not expected to memorize Equations 3.16 and 3.17!

It is interesting to examine these results for v_{1x} and v_{2x} in a few special cases, including some that have been mentioned earlier. The cases are illustrated in Figure 3.7, and have many familiar sporting applications.

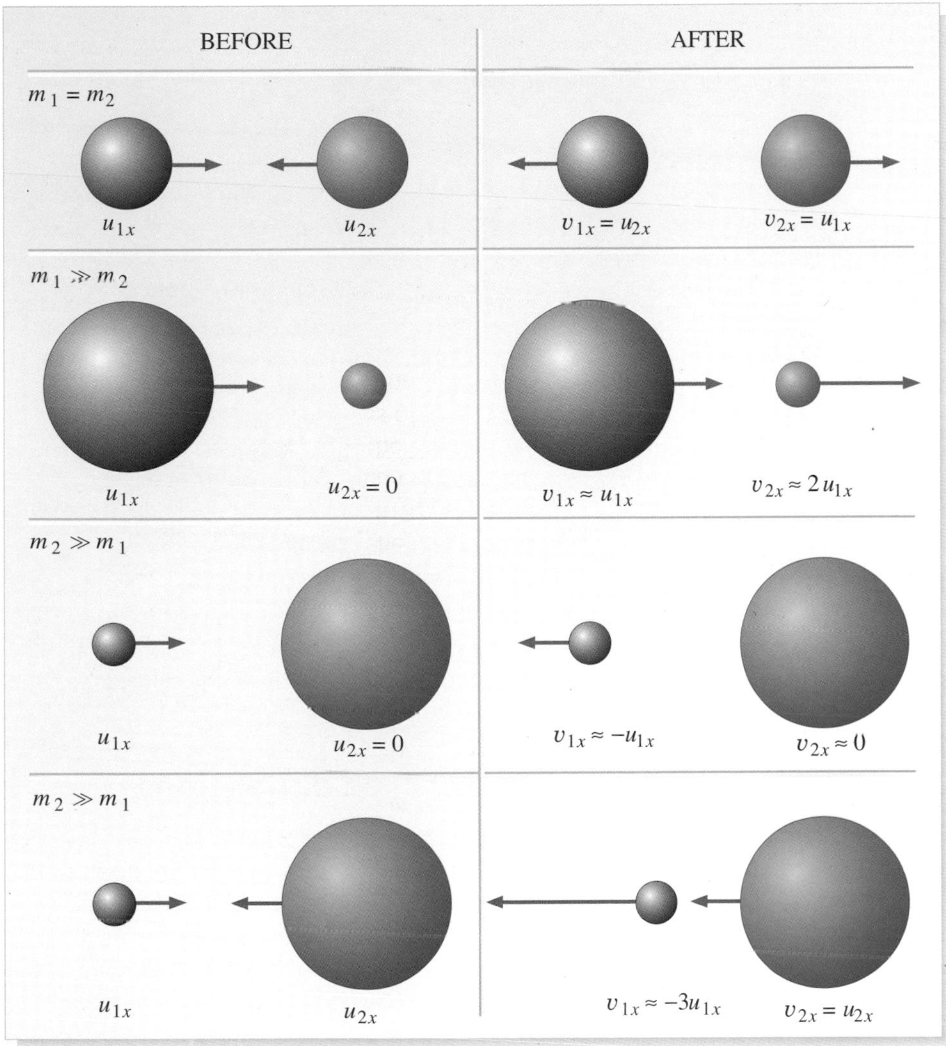

BEFORE	AFTER
$m_1 = m_2$	

Figure 3.7 Four special cases seen in collision problems. See text for complete descriptions.

1   $m_1 = m_2$

If two objects of equal mass collide, Equations 3.16 and 3.17 give $v_{1x} = u_{2x}$ and $v_{2x} = u_{1x}$. This means that *the particles simply exchange velocities* on collision. For example, if particle 2 is at rest initially ($u_{2x} = 0$), then finally particle 1 is at rest ($v_{1x} = 0$) and particle 2 moves with the initial velocity of particle 1 ($v_{2x} = u_{1x}$). We saw this result earlier in this subsection; it is a familiar tactic in bowls and snooker.

2    $m_1 \gg m_2$; $u_{2x} = 0$

If m_2 is very small compared with m_1, then $m_1 + m_2 \approx m_1$ and $m_1 - m_2 \approx m_1$. With $u_{2x} = 0$, Equations 3.16 and 3.17 then give $v_{1x} \approx u_{1x}$ and $v_{2x} \approx 2u_{1x}$. (The symbol $\approx$ is read as 'is approximately equal to'.)

Thus the motion of the high-mass particle is virtually unchanged by the collision but the low-mass particle moves off with a velocity of *twice* that of the high-mass particle. Tennis players serving will be familiar with this case.

3    $m_2 \gg m_1$; $u_{2x} = 0$

If m_1 is very small compared with m_2 (which is stationary), then Equations 3.16 and 3.17 lead to $v_{1x} \approx -u_{1x}$ and $v_{2x} \approx 0$. Therefore, the low-mass particle rebounds with almost unchanged speed while the high-mass particle remains essentially at rest. Golfers whose ball hits a tree will recognize this situation.

4    $m_2 \gg m_1$; $u_{2x} \approx -u_{1x}$

If m_1 is negligible compared with m_2, and the two bodies approach head-on with equal speeds then Equations 3.16 and 3.17 lead to $v_{1x} \approx -3u_{1x}$ and $v_{2x} \approx u_{2x}$. This shows that the low-mass particle bounces back with *three* times its initial speed, while the high-mass particle continues essentially unaffected by the collision. This case will be recognized by a batsman playing cricket or by a tennis player returning a serve.

The results for these four special cases accord with common experience. The results quoted above under points 2, 3 and 4 give an upper limit to the speed that can be imparted to a ball hit by a club, bat or racket.

Question 3.9   A neutron of mass m rebounds elastically in a head-on collision with a gold nucleus of mass $197m$ that is initially at rest. What fraction of the neutron's initial kinetic energy is transferred to the recoiling gold nucleus? Repeat this calculation when the target is a carbon nucleus at rest and of mass $12m$.

Question 3.10   A tennis player returns a service in the direction of the server. The ball of mass 50 g arrives at the racket of mass 350 g with a speed of 45 m s^{-1} and the racket is travelling at 10 m s^{-1} at impact. Calculate the velocity of the returning ball, assuming elastic conditions. ■

3.3   Elastic collisions in two or three dimensions

The laws of conservation of momentum and energy that we used to analyse elastic collisions in one dimension are also used to analyse elastic collisions in two or three dimensions. We simply treat the motions in each dimension as independent, and apply conservation of momentum separately along each Cartesian coordinate axis. Kinetic energy conservation continues to provide one additional equation relating the squares of the particle speeds. Since we have been careful to use vector notation throughout, this extension to two or three dimensions is easily made.

Elastic collisions between two particles of equal mass, with one particle at rest

Consider the elastic collision of two identical bodies of mass m, one at rest and the other approaching with velocity $\boldsymbol{u}_1$. The particles are no longer confined to move in one dimension, so our x-component equation (Equation 3.10), embodying conservation of momentum, becomes a full vector equation:

$$m\boldsymbol{u}_1 = m\boldsymbol{v}_1 + m\boldsymbol{v}_2.$$

The law of conservation of energy (Equation 3.11) stays in scalar form

$$\tfrac{1}{2} m u_1^2 = \tfrac{1}{2} m v_1^2 + \tfrac{1}{2} m v_2^2.$$

These can be simplified to:

$$\boldsymbol{u}_1 = \boldsymbol{v}_1 + \boldsymbol{v}_2 \qquad (3.18)$$

and $\qquad u_1^2 = v_1^2 + v_2^2. \qquad (3.19)$

These equations are most easily interpreted by a diagram. Figure 3.8 shows how the three vectors $\boldsymbol{u}_1$, $\boldsymbol{v}_1$ and $\boldsymbol{v}_2$ are related to one another. Equation 3.18 tells us that *all three velocity vectors must lie in a single plane*, and that they must form a closed triangle. Equation 3.19 tells us that the triangle must be a *right-angled triangle*, since its sides obey Pythagoras' theorem. The implication of this is striking, it means that the angle between $\boldsymbol{v}_1$ and $\boldsymbol{v}_2$ must be 90°.

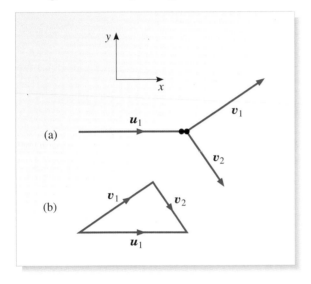

Figure 3.8 (a) Elastic collision between particles of equal mass, with one at rest; (b) the corresponding vector triangle.

> Following the elastic collision of two identical particles, one of which is initially at rest, the final velocities of the two particles will be at right angles.

This is a simplifying feature of equal-mass collisions in two or three dimensions, analogous to the simple result of the exchange of velocities, which we found in one dimension.

You may have noticed that this result docs not tell us exactly where the bodies go after the collision. *Any* pair of final velocities which can be represented by Figure 3.8 will be equally satisfactory, and there are an infinite number of these. The reason for this is that we have said nothing about the shape or size of the bodies, or just how they collide. We usually need to have additional information of this kind if we are to determine unique final velocities in such cases. (Figure 3.9 shows the outcome of a particular collision in which spherical bodies make contact at a specific point. The location of this point is the sort of additional information required to determine unique values for v_1 and v_2.)

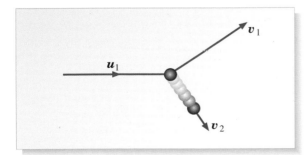

Figure 3.9 When ball 1 strikes ball 2, the reaction forces at the contact point ensure that ball 2 is propelled away along the line of centres, as in snooker.

Question 3.11 For the case illustrated in Figure 3.8 (two bodies of equal mass, one of which is initially at rest), if the moving body has an initial speed of $10\,\text{m s}^{-1}$, and is deflected through $20°$ in the collision, find the magnitudes and directions of the velocities $\boldsymbol{v}_1$ and $\boldsymbol{v}_2$.

Question 3.12 In the same situation (Figure 3.8), if, instead of the conditions in Question 3.11, the speed of the moving body is reduced from $10\,\text{m s}^{-1}$ to $6\,\text{m s}^{-1}$ by the collision, find the final velocities. ■

When the masses of the two colliding particles are unequal the algebraic manipulations required to solve elastic collision problems become rather complicated, but no new physics is involved in the solution so we will not pursue such problems here.

3.4 Inelastic collisions

We now extend our discussion to include inelastic cases, where the total kinetic energy changes during the one-dimensional collision.

Completely inelastic collisions

First we return to our original case, where the two particles stick together on impact; this is an example of a completely inelastic collision, which occurs with the maximum loss of kinetic energy consistent with conservation of momentum. As a simple example, suppose we have two bodies of equal mass, with one initially at rest. If the initial velocity of the other is u_x and the initial momentum is mu_x, the final momentum must be the same so, since the mass has been doubled, the final velocity is $u_x/2$ and the final kinetic energy is therefore

$$\frac{1}{2} \times 2m \left(\frac{u_x}{2} \right)^2 = \frac{mu_x^2}{4}.$$

Since the initial kinetic energy was twice as large as this, it follows that half the original kinetic energy has been lost (mainly as thermal energy), during the collision.

For the more general case where the colliding masses are unequal, but they stick together at collision, we still have $v_{1x} = v_{2x} = v_x$ and so momentum conservation implies that

$$m_1 u_{1x} + m_2 u_{2x} = (m_1 + m_2)v_x$$

and provides a full solution of the problem (a value for v_x), without recourse to energy.

The general case of an inelastic collision in one dimension

To complete the picture, let us mention the general case where two particles collide but where the transfer of kinetic energy into other forms is less than that for the completely inelastic case. This problem has no general solution without more information, such as the fraction of kinetic energy converted. Such problems have solutions which lie between those for the two extremes of elastic and completely inelastic collisions but they must be tackled on an individual basis, using the general principles of conservation of momentum and energy.

You will see that in all these calculations we have not needed to invoke the rather complicated forces involved in the interaction of the two particles, but rather have been able to solve the problems using only the principles of conservation of momentum and energy. This is a great simplification and illustrates the power of using conservation principles whenever possible.

3.5 Collisions all around us

Collisions are occurring around us all the time and on all size scales, from the very largest to the very smallest.

On an astronomical scale, entire galaxies can collide and merge. Figure 3.10 shows a galaxy that is thought to have resulted from such a process of collision and merger. Our own galaxy, the Milky Way, in which the Sun is just one of about 10^{11} stars, is thought to have absorbed a number of small companion galaxies in a similar way. One such 'victim', a partly disrupted dwarf galaxy in Sagittarius, was only discovered in the late 1990s, as this book was being prepared.

A large body of scientific evidence now exists that supports the idea that a major asteroid or comet impact occurred in the Caribbean region at the boundary of the Cretaceous and Tertiary periods in Earth's geological history (about 65 million years ago). Such an impact, depicted in Figure 3.11, is suspected of being responsible for the mass extinction of many species of plants and animals, including the large dinosaurs. The correctness of this hypothesis is still not certain, but the possibility of such a collision is very real; it is becoming increasingly clear that the Earth orbits the Sun in a sort of cosmic shooting gallery. The collision of the fragments of comet Shoemaker–Levy 9 with the planet Jupiter in 1994 (Figure 3.12) was just one of the many side-shows in this gallery. The most dangerous asteroids and comets, those capable of causing major regional or global disasters, are extremely rare, impacting on the Earth perhaps once every 250 000 years or so. Nevertheless, a great deal of media and scientific attention has been focused on strategies of defence. These include the use of nuclear devices and high-speed collisions to deflect or fragment such an object heading for Earth. Calculations using the principles we have discussed in this chapter have shown that high-speed *inelastic* collisions, at $20\,000\,\mathrm{km\,h^{-1}}$ or so, of 15 m wide projectiles with a kilometre-wide asteroid would only change the speed of the main body of the asteroid by about $1\,\mathrm{km\,h^{-1}}$. Such collisions would have to take place years before impact in order for enough deflection to take place to avert catastrophe.

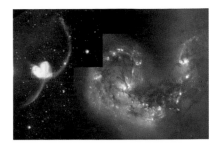

Figure 3.10 A pair of colliding galaxies in the constellation of Corvus. A Hubble Space Telescope image.

Figure 3.11 An artist's impression of a large asteroid impacting on the Earth causing global extinction of a number of species of plants and animals.

Figure 3.12 An artist's depiction of a fragment of the Shoemaker–Levy 9 comet heading towards its collision with Jupiter in July, 1994.

The growing number of cars on today's roads makes it increasingly likely that each driver will be involved in at least one collision during their lifetime. It is very important then that the cars on the road are constructed to a high safety standard. The mythical car that will guard its occupants from all injury in every single type of crash does not exist. But crashworthiness tests on vehicles can reveal how well a vehicle protects people in particular kinds of inelastic collision. Such tests show how well the body of the car can dissipate the energy around the occupants' compartment. Among the most important of the tests is the *frontal offset crash test* at 40 mph. In this test, shown in Figure 3.13, 40% of the total width of each vehicle strikes a barrier on the driver's side. The barrier's deformable face is made of aluminium honeycomb which makes the forces of the impact similar to those in real offset crashes between two identical vehicles, each going just slower than 40 mph. The offset impacts complement full-width frontal tests at 35 mph, shown in Figure 3.14. The full-width test is especially demanding of restraint systems while frontal offsets are more demanding of vehicles' front-end structure, or crumple zone.

Figure 3.13 A frontal offset crash test which probes the integrity of a vehicle's front-end structure.

Figure 3.14 A full-width frontal crash test which is especially demanding of a vehicle's restraint system.

● People who conduct crash tests on vehicles are sure to point out that crash test ratings are only meaningful when comparing vehicles in the same weight class. Why?

○ Since the crash tests are performed at a fixed speed for all vehicles, different masses mean that different momenta are used. It would be unfair to compare the collision of a large truck with that of a small car. ■

In case you think it's a long time since you personally were involved in a collision, you should be aware that even the air that you breathe has its properties regulated by the innumerable collisions that occur every second between the molecules in the air. The air pressure that helps to keep your lungs inflated and enables you to breathe is a result of the rate at which momentum is transferred between the molecules in the air and lung tissue.

On the even smaller subatomic scale, we have already mentioned the role that collisions and momentum analysis played in the discovery of the nucleus. Collisions continue to be of importance in nuclear physics, but they are even more significant in subnuclear physics. Collision experiments, usually at very high energies, are almost synonymous with the experimental investigation of elementary particles such as protons, and their supposedly fundamental constituents, the quarks and gluons. These investigations are carried out with the aid of purpose-built **particle accelerators**, such as the ones at The European Centre for Particle Physics (CERN) or Fermilab in the USA (Figure 3.15). Sophisticated detectors allow the energies and momenta of the emerging particles to be measured, aiding the identification of the particles and the analysis of their behaviour. The results give an indication of the underlying structure of the colliding particles, and have revealed the existence of forms of matter that would still be unknown and possibly even unsuspected were it not for collision experiments.

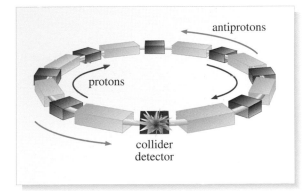

Figure 3.15 A highly schematic diagram of a large particle accelerator.

3.6 Relativistic collisions

The high-energy collision experiments carried out at CERN, Fermilab and other such facilities, involve particles that travel at speeds close to that of light. Under such circumstances the definitions of momentum and translational kinetic energy, that play such an important role in Newtonian mechanics, reveal certain shortcomings. It is still the case that translational kinetic energy is conserved in an elastic collision and that the momentum of an isolated system is always conserved, but the Newtonian expressions

$$\boldsymbol{p} = m\boldsymbol{v} \quad \text{and} \quad E_{\text{trans}} = \tfrac{1}{2}mv^2$$

are now recognized as approximations, valid only at low speeds (i.e. at speeds much less than the speed of light), to more complicated expressions that work at any speed, up to the speed of light. The breakthrough that led to this realization was the development of Einstein's special theory of relativity in 1905. Special relativity will be dealt with in detail in *Dynamic fields and waves*; here we shall just quote a few of its well-established results concerning momentum and energy. You are not expected to memorize the formulae that follow, but you should be able to recognize them when you see them again, and you should certainly aim to understand their significance.

According to Einstein's theory the momentum of a particle with mass m and velocity $\boldsymbol{v}$ is given by

$$\boldsymbol{p} = \frac{m\boldsymbol{v}}{\sqrt{1 - \dfrac{v^2}{c^2}}} \tag{3.20}$$

where v is the speed of the particle and c is the speed of light in a vacuum. The speed of light in a vacuum, $c = 3.0 \times 10^8 \,\mathrm{m\,s^{-1}}$, plays an important role throughout special relativity. Among other things it represents an upper limit to the speed of any particle.

Equation 3.20 implies that the momentum of a particle increases more rapidly with increasing speed than the Newtonian relation ($\boldsymbol{p} = m\boldsymbol{v}$) predicts. This is shown in Figure 3.16, where the behaviour of the Newtonian and relativistic definitions of momentum magnitude are compared. You can see the good agreement at low speed, but you can also see the increasing discrepancy as the speed increases. Note that the relativistic definition does not extend beyond $v = c$. This reflects the fact that in special relativity it is impossible to accelerate a particle with mass to the speed of light, as you will soon see.

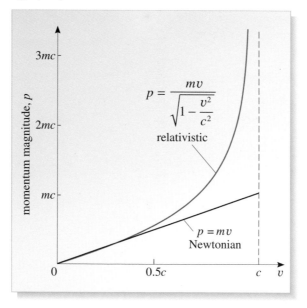

Figure 3.16 The magnitude of the momentum of a particle of mass m plotted against the particle's speed v according to Newtonian mechanics and special relativity. The Newtonian relation closely approximates that of relativity for values of v that are small compared with the speed of light, c.

One of the most celebrated aspects of special relativity is Einstein's discovery of **mass energy**, the energy that a particle has by virtue of its mass. The mass energy of a particle of mass m (sometimes called the *rest mass* in this context) is given by

$$E_{\mathrm{mass}} = mc^2. \tag{3.21}$$

The reason for mentioning this relation here is that it plays a part in determining the kinetic energy of a particle. How is this? Well, according to special relativity the total energy (including the mass energy) of a particle of mass m travelling with speed v is

$$E_{\text{tot}} = \frac{mc^2}{\sqrt{1 - \dfrac{v^2}{c^2}}}. \tag{3.22}$$

Since this quantity is the sum of the translational kinetic energy and the mass energy of the particle it follows that, according to the theory of relativity, the translational kinetic energy of a particle of mass m and speed v is

$$E_{\text{trans}} = \frac{mc^2}{\sqrt{1 - \dfrac{v^2}{c^2}}} - mc^2. \tag{3.23}$$

Unlikely as it may seem, this expression actually agrees very closely with the Newtonian expression for translational kinetic energy ($\frac{1}{2}mv^2$) when v is small compared with c. The relativistic and Newtonian definitions of translational kinetic energy are compared in Figure 3.17. The figure also indicates one reason why it is impossible to accelerate a particle to the speed of light; doing so would require the transfer of an unlimited amount of energy to the particle.

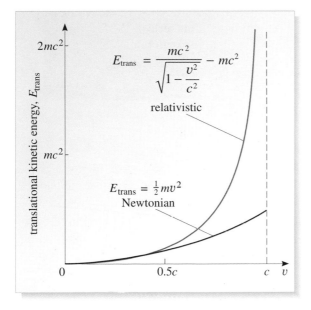

Figure 3.17 The translational kinetic energy of a particle of mass m plotted against the particle's speed v according to Newtonian physics and special relativity. The Newtonian relation closely approximates that of relativity for values of v that are small compared with the speed of light, c.

In analysing high-speed relativistic collisions it is the relativistic expressions for momentum and energy that must be used rather than their Newtonian counterparts. In an elastic collision, all of the quantities we have just defined will be conserved; momentum, rest energy, kinetic energy and total energy. However, many high-energy collisions are actually inelastic, and in a high-energy inelastic collision the only quantities that are certain to be conserved are the momentum and total energy. In a general inelastic collision neither kinetic energy, nor rest energy are necessarily conserved. This means that in an inelastic collision it is quite possible for particles to be created or destroyed, thereby increasing or decreasing the mass energy. However, the conservation of total energy means that any change in mass energy must be accompanied by a compensating change in the kinetic energy. Thus particles may be created, but only at the expense of kinetic energy.

The need for kinetic energy in order to create particles explains why advances in particle physics often require the construction of powerful new particle accelerators. Increasing the kinetic energy of the colliding particles increases the mass of the particles that may be created in the collision and thus opens up the possibility of creating previously undiscovered forms of matter. Figure 3.18 shows the tracks of particles created in one such 'ultra-relativistic' collision.

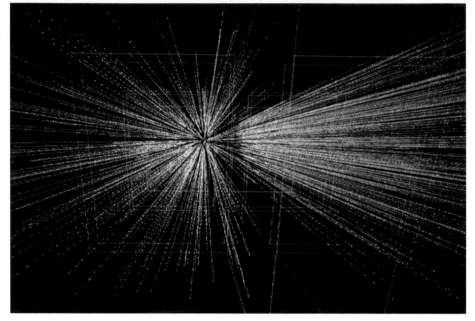

Figure 3.18 Tracks of particles coming from an 'ultra-relativistic' inelastic collision at CERN.

Question 3.13 This subsection has made much use of the phrase 'high-speed collision'. Which of the following collision speeds would indicate a high collision speed in the sense that we have been using the term here?

$$10^2\,\text{m s}^{-1},\ 10^4\,\text{m s}^{-1},\ 10^6\,\text{m s}^{-1},\ 10^8\,\text{m s}^{-1}.\ \blacksquare$$

4   Closing items

4.1   Chapter summary

1   In classical Newtonian mechanics the momentum of a body of mass m and velocity $\boldsymbol{v}$ is given by $\boldsymbol{p} = m\boldsymbol{v}$.

2   The total force acting on a body is equal to the rate of change of momentum of the body

$$\boldsymbol{F} = \frac{\mathrm{d}\boldsymbol{p}}{\mathrm{d}t}.$$

This is a more general statement of Newton's second law of motion than is $\boldsymbol{F} = m\boldsymbol{a}$, since the latter only applies to situations in which the mass is constant.

3   The impulse delivered by a constant force $\boldsymbol{F}$ over a time Δt is $\boldsymbol{F}\,\Delta t$, and this is equal to the change in momentum of the body to which the impulse is applied.

4   The momentum of a system of bodies is the (vector) sum of the momenta of the bodies that comprise the system.

5   According to the law of conservation of momentum, the total momentum of any isolated system is constant.

6   The law of conservation of momentum may be used in the solution of a variety of problems. It is commonly used in the analysis of collision problems, often in association with some aspect of energy conservation.

7   Collisions in which kinetic energy is conserved are said to be elastic. Collisions in which kinetic energy is not conserved are said to be inelastic.

8   In a one-dimensional elastic collision of two particles the relative velocity of approach is the negative of the relative velocity of separation. In such a collision, whatever the masses and initial velocities of the particles, their final velocities may always be determined on the basis of conservation of (kinetic) energy and conservation of momentum.

9   Problems involving motion in more than one dimension can only be solved fully if some details of the collision are known. An example is the elastic collision of two identical particles, one of which is initially at rest. Here, the final velocities are always at right angles.

10 Collisions play a vital role in the physical world, and in various fields of scientific research.

11 In high-energy collisions, where collision speeds approach that of light in a vacuum ($c = 3.0 \times 10^8 \, \text{m s}^{-1}$), it is generally necessary to use the relativistic definitions of momentum and translational kinetic energy

$$\boldsymbol{p} = \frac{m\boldsymbol{v}}{\sqrt{1 - \dfrac{v^2}{c^2}}} \quad \text{and} \quad E_{\text{trans}} = \frac{mc^2}{\sqrt{1 - \dfrac{v^2}{c^2}}} - mc^2 \, .$$

(for $v \ll c$ these approximate the expressions $\boldsymbol{p} = m\boldsymbol{v}$ and $E_{\text{trans}} = \frac{1}{2}mv^2$.) In such collisions, particles may be created (implying that mass energy increases) but total energy must be conserved, so there must be a corresponding decrease in kinetic energy.

4.2 Achievements

Having completed this chapter, you should be able to:

A1 Understand the meaning of all the newly defined (emboldened) terms introduced in the chapter.

A2 Determine the momentum of a body or system of bodies, given the mass and velocity of each body belonging to the system.

A3 State Newton's second law of motion in terms of momentum, and apply it in a variety of situations, including those that involve impulsive forces.

A4 State the law of conservation of momentum and justify it using Newton's second and third laws of motion.

A5 Describe the essential features of elastic and inelastic collisions, and give examples of each.

A6 Use the law of conservation of momentum, and (when appropriate) the law of conservation of kinetic energy, to solve a variety of simple collision problems.

A7 Recognize the expressions for momentum and energy that arise in special relativity, recall their relationship to the corresponding Newtonian expressions, and explain their implications for the creation of new particles in high-speed inelastic collisions.

4.3 End-of-chapter questions

Question 3.14   Find the ratio of the magnitudes of the momenta associated with a rifle bullet of mass 30×10^{-3} kg moving at $400\,\mathrm{m\,s^{-1}}$ and a brick of mass 1 kg moving at $10\,\mathrm{m\,s^{-1}}$.

Question 3.15   Explain qualitatively how it is possible for a body on which a resultant force acts to move with constant velocity.

Question 3.16   A particle of mass m moves along the x-axis with velocity u_x and collides head-on with a stationary particle of mass $2m$. If in the collision the first particle is brought to rest, find the velocity of the second particle. This second particle then collides head-on with a stationary third particle of mass $3m$. If the second and third particles stick together on impact, find their subsequent velocity after the collision.

Question 3.17   Two balls of the same mass travelling at the same velocity strike a window perpendicularly. The first bounces back but the second breaks the window. Explain how this could be so.

Question 3.18   A car travelling at $15\,\mathrm{m\,s^{-1}}$ goes out of control and hits a tree, with the car and driver coming to rest in 0.6 s. What is the initial momentum of the driver, whose mass is 70 kg? What is the magnitude of the average force exerted on the driver during the crash, and how does this force compare with the magnitude of the driver's weight?

Question 3.19   A spacecraft approaches the planet Jupiter, passes round behind it, and departs in precisely the opposite direction to that from which it approached. Initially, when the spacecraft is at a large distance from the planet, the spacecraft is moving in the negative x-direction at a speed of $10\,\mathrm{km\,s^{-1}}$, and Jupiter is moving in the positive x-direction at $13\,\mathrm{km\,s^{-1}}$. Assuming elastic conditions, find the final speed of the spacecraft when it is once again at a large distance from the planet. Use the approximation that the mass of the spacecraft is very small compared with the mass of Jupiter. (Note, answering this question involves quite a lot of algebra; consult the answer if you get stuck.)

Question 3.20   Two steel balls travelling at the same speed collide elastically head-on. One of them is stationary after the collision. If its mass is 900 g, what is the mass of the other?

Question 3.21   A block of stone at a quarry is blown up into three separate pieces of masses 10 000 kg, 8000 kg and 6000 kg. The 10 000 kg piece moves off in the positive x-direction at a speed of $8.00\,\mathrm{m\,s^{-1}}$, and the 8000 kg piece moves in the negative y-direction at $5.00\,\mathrm{m\,s^{-1}}$. Find (a) the momentum of the third piece, and (b) the energy of the explosion, which may be assumed to equal the total kinetic energy after the explosion. ■

Chapter 4 Torque and angular momentum

1 Precession — an example of rotational dynamics

As the Earth traces out its year-long orbit around the Sun, it also spins on its axis, creating the 24-hour cycle of day and night (Figure 4.1). This is well known, but there are other aspects of the Earth's rotation that are not so familiar. Its *precession* is one of them.

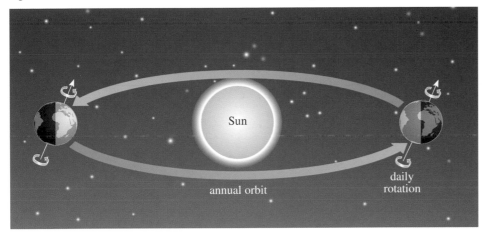

Figure 4.1 The Earth's orbital and rotational motion, the cause of years and days, respectively. The orientation of the rotation axis is essentially unaffected by the Earth's orbital motion. (Not to scale.)

One result of the Earth's daily rotation is the apparent nightly movement of the stars. Any observer in the Earth's Northern Hemisphere will find that the night sky appears to rotate, with a 24-hour period, around a fixed point. This rotational motion of the stars is easily recorded by pointing a camera towards the night sky and leaving its shutter open for several hours. The movement of the stars causes them to leave trails that have the form of circular arcs, as shown in Figure 4.2.

Figure 4.2 A long exposure photograph of the northern night sky, revealing the motion of the stars around the North Celestial Pole.

Examining a sequence of such photographs, taken with different exposure times, reveals that all the stars appear to move with a constant angular speed of 15° per hour in the anticlockwise sense. This observed motion has long been recognized as a reflection of the Earth's own rotation. The 15° per hour angular speed implies a 24-hour period for a complete 360° rotation.

The centre point of the rotation in Figure 4.2 is known to astronomers as the North Celestial Pole, and is the point on the northern sky towards which the Earth's axis of rotation points. (A corresponding South Celestial Pole can be observed from the Earth's Southern Hemisphere.) There is a moderately bright star close to the North Celestial Pole that appears as a bright spot at the centre of Figure 4.2. This is Polaris, the North Pole star (it's actually about 0.75° from the North Celestial Pole) in the otherwise undistinguished constellation of Ursa Minor.

The fact that the North Pole star is a good indicator of the direction of true (geographic) north shows that the Earth's axis of rotation maintains an essentially fixed orientation in space, despite the Earth's orbital motion about the Sun. The reason for this constancy of orientation will be explained later, but an equally interesting discovery is that the orientation is actually not quite constant. Detailed observations show that the Earth's rotation axis is very gradually sweeping out a cone, causing the North Celestial Pole to move in a circle on the sky, with an angular diameter of 47° and a period of 25 800 years. Thus, if you were to observe the night sky about 13 000 years from now, you would find that the nearest bright star to the North Celestial Pole would be Vega, in the constellation of Lyra, a star that is currently about 50° from Polaris (Figure 4.3).

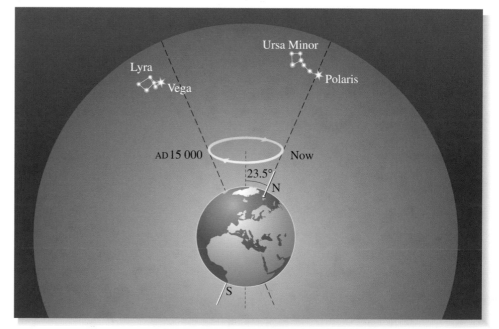

Figure 4.3 Due to the precession of the Earth's rotation axis, the North Celestial Pole gradually traces out a circle on the sky with an angular diameter of 47° and a period of 25 800 years.

The additional element of the Earth's rotational motion that causes the gradual change in axial orientation is an example of **precession**, a phenomenon that affects a wide range of rotating bodies. In the case of the Earth, the precession is mainly due to the gravitational pull of the Sun and the Moon on the Earth's bulging Equator — there would be no precession if the Earth was perfectly spherical.

Clearly, if we are to understand phenomena such as the approximate constancy of the Earth's axial orientation, or the departure from constancy represented by precession, then we need to be able to predict rotational motion, just as we have already learned to predict translational motion. It is with this aim in mind that we devote this chapter to the concepts, principles and applications of *rotational dynamics*. By the end of the chapter you will have been introduced to *torque*, the rotational analogue of force, and to *rotational kinetic energy* and *angular momentum*, the rotational analogues of translational kinetic energy and linear momentum, respectively. As you will see, each of these quantities plays an important part in the prediction of rotational motion and is vital to our understanding of our own bodies, our technology and the physical world in general, including the nature of the Earth's rotation.

2 Torque

We noted in Chapter 1 of this book that the motion of a rigid body is, in general, a combination of *translational motion* and *rotational motion*. In the case of a rigid body, a translational displacement implies that each part of the body moves the same distance and in the same direction. In contrast, a rotational displacement may always be associated with an **axis of rotation**, which may be thought of as remaining stationary while every other part of the body moves around that axis. It was shown in 1760, by Leonhard Euler (Figure 4.4), that any displacement of a rigid body can be regarded as the combination of a translation and a rotation about some appropriately chosen axis.

Euler's general analysis of bodies undergoing translation *and* rotation, which includes cases where the axis of rotation changes from moment to moment, is very complicated and well beyond the scope of this chapter. In order to keep the discussion as simple as possible, we shall often restrict our attention to systems that have only a *single* axis of rotation, such as a wheel turning on a fixed axle. It is with such a system that we begin — a door turning about its hinges.

2.1 The turning effect of a force

It is tempting to say that changes in rotational motion are caused by forces, since forces certainly cause changes in translational motion, and are definitely involved in changing rotational motion as well. However, though force is relevant to any attempt to predict rotational motion, it is far from being the whole story.

Figure 4.4 Leonhard Euler (1707–1783), was a Swiss mathematician who wrote over 600 books and papers, many of them very influential. ('Euler' is pronounced 'Oyler'.) His work encompassed physics and astronomy as well as pure and applied mathematics and originated several of the notations that are still in use today. It was Euler who introduced the Σ symbol for summations, $f(x)$ for functions and e for the mathematical constant 2.718... He was also responsible for much of the notation and terminology used in differential and integral calculus. His general analysis of the three-dimensional motion of rigid bodies is widely regarded as one of the high points of classical mechanics.

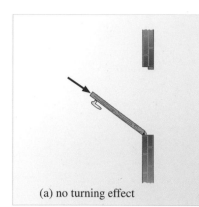

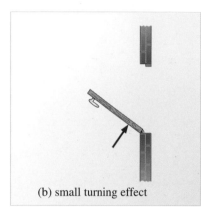

(a) no turning effect

(b) small turning effect

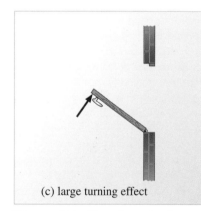

(c) large turning effect

Figure 4.5 Top view of a door. Forces of the same magnitude may have different turning effects.

To see why, consider the simple process of closing a door. To close a door, you have to make it turn about the vertical axis of rotation that passes through its hinges (see Figure 4.5). A force will certainly have to be applied to the door to make it rotate, but the effect of the force will depend on its *point of application* as well as its magnitude and direction. Pushing the outer edge of the door directly towards the hinges, for instance, will not cause the door to rotate at all (Figure 4.5a). To achieve any kind of turning effect about the axis of rotation it is essential that the *line of action* of the force does *not* go through the axis of rotation.

● The *line of action* of a force was introduced in Chapter 1. What is its definition?

○ When a force is applied to a rigid body, the force's line of action is a line that is parallel to the direction of the force and which passes through the point of application of the force. ■

The particular feature of the line of action that is significant in determining the turning effect of a force is its perpendicular distance from the axis of rotation. If you close a door by applying a given force perpendicular to its face, you will find that the turning effect of the force is increased when it is applied further from the hinges. This is why door handles are generally placed close to the outer edges of doors.

A more quantitative insight into the turning effect of a force can be gained by carrying out a simple experiment. You should be able to find the required apparatus without too much difficulty. Here's what you will need:

● a ruler of about 30 cm length (marked in mm);

● a pencil or pen (with hexagonal cross-section);

● four identical small coins;

● some sticky tape.

Place the ruler across the pen, so that it balances in the horizontal plane, as shown in Figure 4.6. Secure the pen to the ruler in this position, using sticky tape. Next, put one of the coins with its centre of mass at a point 0.120 m (i.e. 120 mm) from the central support. This will cause a rotation of the ruler about the supporting pen, which is the axis of rotation. (The ruler will stop rotating when its end hits the surface of the table.) Throughout the remainder of this experiment, you should leave this coin at the 0.120 m position and not place any other coins on this half of the ruler. You may need to tape this coin in position.

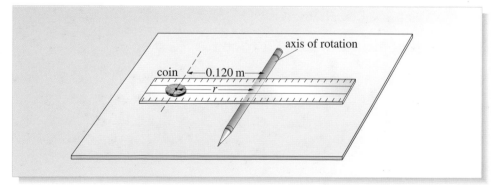

Figure 4.6 A ruler balanced on a central support (the axis of rotation).

Restore the balance by placing a single coin at a suitable point on the other side of the ruler and note its distance l from the axis in Table 4.1. Next, remove that single coin and record the distance l at which two coins, one on top of the other, will restore the balance of the ruler. Repeat this process using a stack of three coins and again note the relevant value of the distance l. You should now have three different values for l in the second column of Table 4.1. The third column, shows the magnitude F of the force due to gravity that acts on the stack of coins, expressed in terms of the mass of a single coin and the magnitude of the acceleration due to gravity (i.e. as a multiple of mg). Complete the table by writing in the fourth column the product of the entries in the second (l/m) column and the third (F) column.

Table 4.1 Results for the experiment involving coins and a ruler.

Number of coins in stack	l/m	F	$F \times l$/m
1		mg	
2		$2mg$	
3		$3mg$	

You should find that every time the ruler is balanced, the product of F and l is the same. Since, in each case, the stack of coins is being used to counterbalance the turning effect of a single fixed coin on the other side of the axis, this result indicates that the product Fl is a measure of the turning effect of a force of magnitude F when its line of action is at a perpendicular distance l from the axis of rotation.

The quantity Fl is sometimes called the 'moment' of the force. However, the turning effect of a force is more generally described by a vector quantity called *torque*, and in the case we have been considering, Fl represents the *magnitude* of the torque about the rotation axis due to the weight of the coin stack. The symbol generally used to represent a torque vector is the bold italic uppercase Greek letter gamma ($\boldsymbol{\Gamma}$), so the *magnitude* of a torque can be represented by an unemboldened gamma (Γ), and in this particular case we can write:

$$\Gamma = Fl. \tag{4.1}$$

● What is the SI unit of torque?

○ Since F is measured in newtons, and l in metres, the unit of torque is the newton metre, which may be abbreviated N m. (Energy may also be measured in newton metres, but it is conventional to use the joule (J) as the unit of energy, whereas it would be very unconventional to express any torque in terms of joules.) ■

A more general expression for the magnitude of a torque can be obtained by considering Figure 4.7a, which shows a vertically hinged door being pushed by a horizontal force $\boldsymbol{F}$ that is *not* perpendicular to the door. In this case, if $\boldsymbol{r}$ is a horizontal displacement vector that stretches from the axis of rotation to the point of application of the force, then the orientation of $\boldsymbol{F}$ in the horizontal plane is determined by the angle θ between $\boldsymbol{r}$ and $\boldsymbol{F}$ (θ is in the range 0° to 180°). One way of working out the turning effect of this force is to resolve it into two horizontal component forces: a parallel component $\boldsymbol{F}_{\text{para}}$ acting along the face of the door, directly away from the hinges, and a perpendicular component $\boldsymbol{F}_{\text{perp}}$ acting at right angles to the door. Of these two, only $\boldsymbol{F}_{\text{perp}}$ has any turning effect about the axis of rotation, and, since its magnitude is $F_{\text{perp}} = F \sin \theta$ (see Figure 4.7b), it follows from Equation 4.1 that the torque magnitude in this more general situation is

$$\Gamma = rF \sin \theta \tag{4.2}$$

where r is the magnitude of the displacement vector $\boldsymbol{r}$. It's worth noting that Equation 4.2 may also be interpreted as the magnitude of the force $\boldsymbol{F}$ times the perpendicular distance ($r \sin \theta$) from its line of action to the axis of rotation.

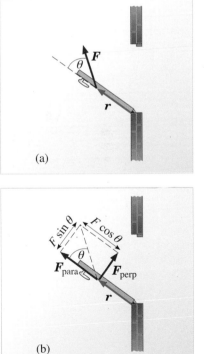

Figure 4.7 (a) Top view of a vertically hinged door subject to a horizontal force $\boldsymbol{F}$ directed at an angle θ to the horizontal vector $\boldsymbol{r}$. (b) $\boldsymbol{F}_{\text{para}}$ and $\boldsymbol{F}_{\text{perp}}$, the vector components of $\boldsymbol{F}$, parallel and perpendicular to the door.

We shall return to the vector nature of torque, including the significance of its direction, a little later. For the moment, however, let's explore, with Example 4.1, the implications of Equation 4.2.

Example 4.1

A uniform square trapdoor of side 0.80 m has a mass of 1.20 kg and is free to swing about its horizontal hinges. Find the magnitude of the torque about the hinge due to the door's own weight when the door is at an angle θ to the vertical. (b) Sketch a graph of the magnitude of this torque against the angle θ, as the door swings from being closed ($\theta = 90°$) to fully open ($\theta = 0°$).

Solution

(a) A diagram of the trapdoor is shown in Figure 4.8. The door's weight acts vertically downwards, through its centre of mass, and has magnitude $F = mg$, where m is the mass of the door and g is the magnitude of the acceleration due to gravity. The angle θ between the door and the vertical is also the angle between the weight and the displacement vector from the hinge to the centre of mass. It follows from Equation 4.2 that the magnitude of the torque about the hinge, due to the weight of the door is

$$\Gamma = rmg \sin \theta,$$

where $r = 0.80$ m/2 $= 0.40$ m, since the centre of mass of the uniform door is at its geometric centre. The required torque magnitude is therefore

$$\Gamma = (0.40 \times 1.20 \times 9.81 \sin \theta)\,\text{N m}$$
$$= (4.71 \sin \theta)\,\text{N m}.$$

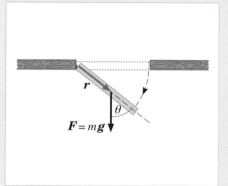

Figure 4.8 A hinged trapdoor turning under its own weight.

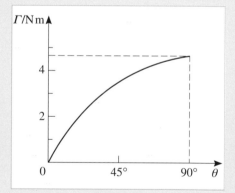

Figure 4.9 Graph of the torque magnitude in Example 4.1.

(b) The graph of Γ against θ is a sine curve as shown in Figure 4.9. The turning effect of the door's weight is reduced as its line of action gets closer to the hinge.

2.2 Torque and levers

The coin experiment described above neatly illustrates the principle of the *lever*, a simple device of such importance in everyday life that it often goes unnoticed. A **lever** is essentially a rigid beam that is supported at a single point, generally called the **fulcrum**. In practice, the fulcrum is often a **pivot**, about which the lever can rotate. As in the case of the coin experiment, the turning effect of a force applied to the lever at one point can be used to balance the turning effect of a different force acting at some other point. (This idea is treated more formally in Section 3, when we discuss static equilibrium; for the moment we treat it as a fact of common experience.)

Figure 4.10 shows an example — a simple weighing device. A load of unknown mass, *m*, is attached to one arm of a lever at a fixed distance, *L*, from a pivot. A known mass *M* is then moved along the other arm until balance is achieved when it is at a distance *l* from the pivot. Balance ensures that the torques about the pivot due to the two weights are of equal magnitude, and this condition allows the unknown mass to be determined. You can work out the details for yourself by answering the following question.

Question 4.1   A mass of 0.50 kg at 0.30 m from the pivot of a simple balance is able to counteract the turning effect arising from an unknown mass *m* suspended at a point 0.05 m on the other side of the pivot. Find the value of *m*. ■

Another kind of lever is shown in Figure 4.11. In this case the lever is the kind that you carry around with you — it's your forearm. This differs from the weighing balance in that the pivot (the elbow joint) is at one end of the lever, so both the load and the balancing force act on the same side of the pivot, but in opposite directions. A simplified model of this system is shown in Figure 4.12.

The downward force $W = m\boldsymbol{g}$, due to gravity, produces a torque about the elbow joint. This is counterbalanced by a torque due to the force $\boldsymbol{F}$ provided by the biceps muscle. (We are ignoring the weight of the forearm itself.) Note that there is also a reaction $\boldsymbol{F}'$ at the elbow, but this has no torque about the elbow. The requirement that the two balanced torques should have the same magnitude implies that

$$lF = LW,$$

so $$F = \frac{L}{l}W.$$

We can use this relation to get some idea of the magnitude of the muscular force $\boldsymbol{F}$ by substituting some typical values. If we suppose that $L = 0.400$ m, $l = 0.050$ m and $m = 10.0$ kg, then we find

$$F = \frac{(0.400\,\text{m})}{(0.050\,\text{m})} \times (10.0\,\text{kg}) \times (9.81\,\text{m}\,\text{s}^{-2}) = 785\,\text{N}.$$

This is a large force; my own mass is about 65 kg, which corresponds to a downward force of magnitude 638 N. So, holding a fairly modest weight with my forearm is equivalent to suspending more than my own body weight directly from a single biceps muscle!

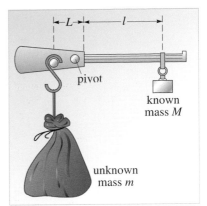

Figure 4.10   A simple weighing device.

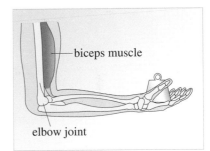

Figure 4.11   Forearm and biceps muscle holding a weight in static equilibrium.

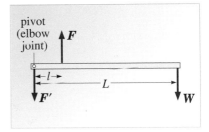

Figure 4.12   Forces acting on a forearm supporting a mass *m*.

Question 4.2   Suppose that the arm supporting the weight shown in Figure 4.11 is raised, so that it is at 45° to the horizontal. Assuming that the force applied by the biceps is still directed vertically upwards, would the magnitude of that muscular force have to be increased or decreased to maintain the weight in the new position?   ■

A variety of lever based tools is shown in Figure 4.13. In some the applied force and the load lie on the same side of the pivot (nutcrackers, wheelbarrow). In others, the applied force and the load are on opposite sides of the pivot (scissors, crowbar). Note that spanners and wrenches, used to tighten nuts on bolts, also employ the lever principle. The longer the spanner, the smaller the force that must be applied to achieve a given turning effect. It is possible to use a spanner to apply too much torque, which can result in a broken thread or a snapped bolt. This is why, in good garages, wheel nuts are normally tightened using a **torque wrench** (see Figure 4.14) that accurately measures the magnitude of the applied torque.

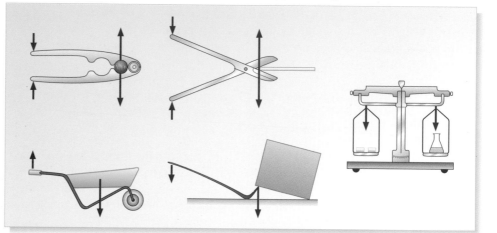

Figure 4.13   A selection of lever based tools.

Figure 4.14   A torque wrench.

2.3   A vector definition of torque

So far, we have restricted our attention to rigid systems that turn about a fixed axis, and we have only considered the turning effect of forces that act in planes perpendicular to that axis. Also, we have only dealt with torque magnitudes, not directions, even though it is quite clear that the turning effect about an axis can act in a clockwise or an anticlockwise sense. In this subsection we shall use vector notation to write down a general expression for the turning effect of any force that will overcome all of these limitations. The price we must pay for this increase in generality is a certain degree of complexity. To combat this we shall shortly introduce a powerful new mathematical concept, the *vector product*, that will enable us to rewrite the definition much more compactly.

Here then, is the general definition of the torque, or turning effect of a force.

General definition of the torque vector Γ

Given a point O, and a force F that acts at some other point whose displacement vector from O is r (see Figure 4.15a), then, if the angle between r and F is θ, the **torque** about O due to F is a vector Γ whose magnitude is $rF \sin \theta$, and which points in the direction perpendicular to the plane containing r and F, in the sense specified by the **right-hand rule**, shown in Figure 4.15b.

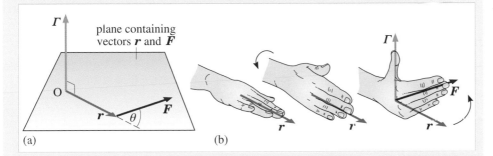

Figure 4.15 (a) The torque Γ about O, due to a force F acting at a point with displacement r from O, is a vector perpendicular to the plane defined by the vectors r and F. The magnitude of Γ is $rF \sin \theta$. (b) The direction of Γ is determined by the right-hand rule. First, point the straight palm and fingers of your right hand in the direction of vector r. Then, keeping your palm and fingers in this direction, turn your wrist (if necessary) until you can bend your fingers to point in the direction of vector F. Your extended thumb will then point in the direction of the torque vector Γ.

This is a complicated definition and you will need to study it carefully to understand it. However, there are some points to which you should pay particular attention.

1   Note that what has been defined is the torque about a *point*, not the torque about an *axis*. If you imagine an axis passing through O and pointing in some particular direction, as shown in Figure 4.16, then the torque due to F about that directed axis is the *component* of Γ along that axis. (In the fixed axis cases considered earlier, the axis was always perpendicular to the plane containing r and F, so the torque was always parallel to the axis. This is not necessarily the case in general.)

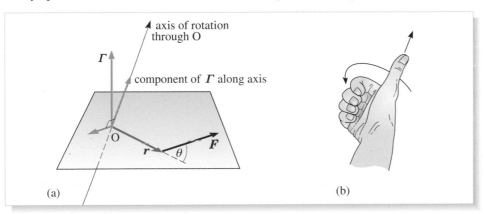

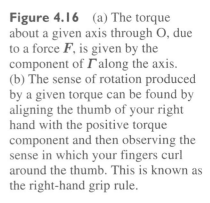

Figure 4.16 (a) The torque about a given axis through O, due to a force F, is given by the component of Γ along the axis. (b) The sense of rotation produced by a given torque can be found by aligning the thumb of your right hand with the positive torque component and then observing the sense in which your fingers curl around the thumb. This is known as the right-hand grip rule.

Do not confuse the *right-hand grip rule* with the *right-hand rule*. They are different.

2   Given a directed axis through O, the component of Γ along that axis may be positive or negative. If it is positive, it can be seen from Figure 4.16 that the torque will tend to cause anticlockwise rotation when viewed from the arrow end of the rotation axis. If the component of Γ along the rotation axis is negative, then the torque will tend to cause clockwise rotation when viewed from the arrow end of the axis. If the torque has no component along a given axis, then the force has no turning effect about that axis. (A useful way of remembering this relationship is shown in Figure 4.16b. Point the thumb of your right hand in the direction of a positive torque component, you will then find that the sense in which your fingers curl around your thumb is the same as the sense of the rotation that the torque tends to produce about the axis. This is known as the right-hand grip rule.)

3   Simply stating that Γ is perpendicular to the plane containing r and F is not sufficient to fully determine the direction of Γ; it might be pointing upwards or downwards relative to that plane. The reference to the right-hand rule is included in the general definition to remove this ambiguity.

4   To ensure that the magnitude of the torque, $\Gamma = rF \sin \theta$, is positive, as any magnitude must be, it is conventional to restrict the value of θ to be in the range $0°$ to $180°$; it certainly shouldn't be in the ranges $-180° < \theta < 0°$, or $180° < \theta < 360°$.

The general definition of a torque given above is terribly long-winded. It is just the kind of statement that cries out to be simplified by the introduction of some new piece of mathematical notation — especially if that notation can be used elsewhere. In this case, all the general definition really tells us is how to combine two given vectors (r and F) to produce a new vector (Γ) with a specific magnitude and direction. The particular method of combination described in the general definition produces what is usually referred to as the **vector product** of r and F. This is a special kind of product, so it is normally indicated by writing it as $r \times F$, with a bold cross between the vectors r and F. Using this notation we can rewrite the definition of torque as follows:

The vector product is sometimes called the *cross product*. Note that it is quite different from the *scalar product* that was introduced in Chapter 2.

> The torque about a point O due to a force F is
> $$\Gamma = r \times F \qquad\qquad (4.3)$$
> where r is the displacement vector from O to the point of application of F.

It is worth emphasizing again that Equation 4.3 is only a shorthand for the words we used in the general definition of torque. Namely, Γ is a vector of magnitude $rF \sin \theta$, perpendicular to the plane containing r and F, with its direction specified by the right-hand rule.

Equation 4.3 is completely general. It is valid for any possible orientation of the vectors r and F, and can be used to analyse all the cases we considered earlier. For instance, in the case of the trapdoor that was shown in Figure 4.8, the torque was actually directed into the page — hence its tendency to cause a clockwise rotation of the trapdoor.

●   Was the direction of the torque about the vertical hinge in Figure 4.7 upwards or downwards?

○   The torque $r \times F$ points downwards in this case and will therefore tend to cause a clockwise rotation of the trapdoor. ■

In the case of the weighing device (Figure 4.10) and the coin experiment (Figure 4.6) we didn't actually specify any particular displacement vectors, but in both cases, if we had done so, we could have drawn a diagram similar to Figure 4.17. In each case there would have been two torques to consider, $\boldsymbol{\Gamma}_1 = \boldsymbol{r}_1 \times \boldsymbol{F}_1$ and $\boldsymbol{\Gamma}_2 = \boldsymbol{r}_2 \times \boldsymbol{F}_2$, acting in opposite directions along the rotation axis. We said earlier that when balance was achieved the two torques would have the same magnitude, which is true, but you can now see that due to their opposed directions, the vector sum of the two torques would actually have been zero. Hence the lack of any net turning effect.

A general point to be drawn from this discussion is the following:

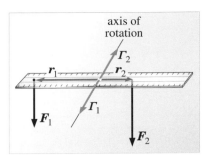

Figure 4.17 The vector sum of balanced torques about the rotation axis is zero, $\boldsymbol{\Gamma}_1 + \boldsymbol{\Gamma}_2 = 0$.

> When several different torques act at a point, the total torque at that point is given by the vector sum of the individual torques.

Note that torques are vectors, so they must be added like vectors.

Question 4.3   Given the orientations of $\boldsymbol{r}_1$, $\boldsymbol{F}_1$, $\boldsymbol{r}_2$ and $\boldsymbol{F}_2$ in Figure 4.17, describe how you would confirm that $\boldsymbol{\Gamma}_1$ and $\boldsymbol{\Gamma}_2$ have the correct orientations? Have they, in fact, been drawn with the correct orientations?

Question 4.4   A rectangular sheet of metal, with sides of length 0.400 m and 0.300 m, rests on a horizontal surface and is pivoted at one corner (point O in Figure 4.18) so that it can rotate about the z-axis. Two horizontal forces $\boldsymbol{F}_1$, and $\boldsymbol{F}_2$ of the same magnitude, 5 N, are applied simultaneously at the points indicated in Figure 4.18. Work out the total torque at O due to these forces and hence determine the direction in which the metal sheet will rotate.   ■

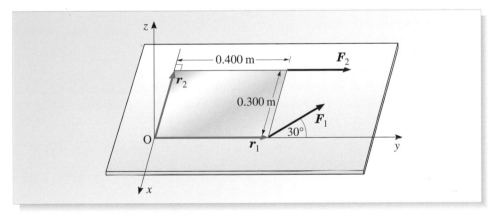

Figure 4.18   Which way will the object turn? See Question 4.4.

2.4 Torque and angular motion

Broadly speaking, as you saw in Chapter 1, forces cause translational acceleration. In similarly broad terms, it is also true that torques cause *angular acceleration*. However, the relationship between torque and angular acceleration is much more subtle than that between force and translational acceleration. In this subsection we start by generalizing our earlier discussions of angular motion in order to clarify what we mean by angular acceleration, and then make some general comments about its relationship to torque.

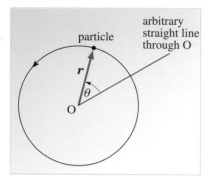

Figure 4.19 Particle moving in a circle.

It is usual to express angular speeds in radians per second rather than degrees per second.

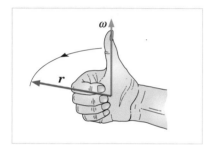

Figure 4.20 Direction of angular velocity associated with a given sense of rotation. If the fingers of the right hand are curled in the direction of rotation of the particle, then the thumb points in the direction of the angular velocity. (This is the same right-hand grip rule as was illustrated in Figure 4.16b.)

Angular position and *angular velocity* were first introduced in Chapter 3 of *Describing motion*, in the context of circular motion. Figure 4.19 shows a single particle moving in a circle about a point O. The position of the particle can be specified by the vector *r* that determines its displacement from the centre of the circle. However, because the particle is constrained to move in a circle, its position can also be specified by a single scalar parameter θ, the angle between *r* and an arbitrary fixed straight line passing through O. The angle θ therefore determines the **angular position** of the particle and will, in general, change with time. Given a point O, from which to measure *r*, and a line through O, from which to measure θ, this idea may be applied to any particle moving in a plane that contains O, irrespective of whether the motion is circular or not.

As a particle moves relative to O its angular position θ will change with time. The change can be partly described by the particle's **angular speed**, ω, which is usually measured in radians per second (rad s^{-1}). The angular speed of the particle is the *magnitude* of the rate of change of θ, so it is always a positive quantity and its definition may be expressed mathematically by the equation

$$\omega = \left| \frac{\mathrm{d}\theta}{\mathrm{d}t} \right|. \tag{4.4}$$

Being a magnitude, this quantity gives no indication of whether θ is increasing or decreasing, only that it's changing. The angular speed is therefore unable to provide any information about the sense in which the particle is moving about the point O. To overcome this shortcoming we associate with the particle a vector quantity *ω*, known as its **angular velocity** about O. The magnitude of *ω* is the particle's angular speed about O; while the direction of *ω* indicates the direction of the rotation axis through O, and is related to the sense of rotation by the same right-hand grip rule used to relate a torque to the rotation it tends to produce. This is illustrated in Figure 4.20 and has the consequence that the angular velocity associated with the particle in Figure 4.19 points out of the page.

There is an enormous difference between a particle's instantaneous velocity *v* relative to O, and its instantaneous angular velocity *ω* about O; even their units differ. Nevertheless, in the case of circular motion there is a simple relationship between these two quantities. In *Describing motion* we showed that, for a particle moving in a circle,

$$v = \omega r. \quad \text{(circular motion)} \tag{4.5}$$

Now, using the vector product introduced in the last subsection, we can write down a vector relationship that automatically takes care of directions as well:

$$\boldsymbol{v} = \boldsymbol{\omega} \times \boldsymbol{r}. \quad \text{(circular motion)} \tag{4.6}$$

This relationship is illustrated in Figure 4.21. You can confirm that the directions are correct by using the right-hand rule of Figure 4.15b, and you can see that the magnitudes are correct because *ω* and *r* are at *right angles* in this case, so $v = |\boldsymbol{\omega} \times \boldsymbol{r}| = \omega r \sin 90° = \omega r$, as required.

Figure 4.21 The instantaneous velocity of a circling particle expressed as the vector product *ω* × *r*.

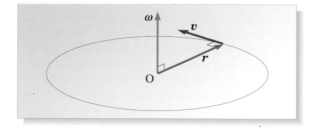

If the components of a and b are known, so that we can write $a = (a_x, a_y, a_z)$, and $b = (b_x, b_y, b_z)$, then we can determine the components of the vector product $a \times b$ using the expression

$$a \times b = (a_y b_z - a_z b_y, \, a_z b_x - a_x b_z, \, a_x b_y - a_y b_x). \tag{4.10}$$

There is no need to remember this formula, but it is worth noting that some calculators now include the evaluation of such products amongst the functions that they perform. It is also worth noting the contrast between the component expression for the vector product and that for the *scalar product* introduced in Chapter 2. The scalar product of a and b is represented by $a \cdot b$, and is defined by the relation $a \cdot b = ab \cos \theta$. In terms of components, the scalar product of two vectors may be written:

$$a \cdot b = a_x b_x + a_y b_y + a_z b_z. \tag{4.11}$$

There are many differences between Equations 4.10 and 4.11, but the most important is that Equation 4.10 defines a *vector* while Equation 4.11 defines a *scalar*. (Hence their names.)

There is one particular property of vector products that may come as a bit of a surprise. In everyday arithmetic, the result of multiplying two numbers is independent of their order, for example 2×3 is the same as 3×2. This is implicit in the general algebraic rule

$$xy = yx$$

where x and y may represent any two numbers, and we describe the result by saying that the multiplication of x and y is **commutative** or, equivalently, that x and y *commute*. The scalar product of two vectors is also commutative, in that

$$a \cdot b = b \cdot a.$$

So the result does not depend on the order in which the product is written. However, the vector product behaves quite differently — the result *does* depend on the order in which the product is written. In fact,

$$a \times b = -b \times a$$

showing that the vector product of two vectors is *not* commutative. To see why, have another look at Figure 4.24. This shows the direction of the vector product $a \times b$, using the right-hand rule and starting with vector a. Now think about what happens if you apply the same rule to the vector product $b \times a$, with vector b first and a second. If you follow the rule through, you will get a 'thumb down' result! In other words, the direction of the vector product $b \times a$ is opposite to that of $a \times b$. Example 4.2 further emphasizes the differences between scalar and vector products.

Note the minus sign in this equation.

Example 4.2

To appreciate the essential differences between the vector (or cross) product and the scalar (or dot) product of vectors, use their respective definitions to work out and compare the results $a \cdot b$ and $a \times b$ where:

(a) a and b are two parallel vectors;

(b) a and b are two identical vectors;

(c) a and b are two perpendicular vectors.

Solution

Using the definitions of the vector and scalar products given earlier:

(a) $\boldsymbol{a} \cdot \boldsymbol{b} = ab \cos 0° = ab$.

However, since $|\boldsymbol{a} \times \boldsymbol{b}| = ab \sin 0° = 0$, we have $\boldsymbol{a} \times \boldsymbol{b} = \boldsymbol{0}$.

(b) From part (a) we have $\boldsymbol{a} \cdot \boldsymbol{a} = a^2$, but $\boldsymbol{a} \times \boldsymbol{a} = \boldsymbol{0}$.

(c) $\boldsymbol{a} \cdot \boldsymbol{b} = ab \cos 90° = 0$, but $|\boldsymbol{a} \times \boldsymbol{b}| = ab \sin 90° = ab$. The direction of $\boldsymbol{a} \times \boldsymbol{b}$ is at right angles to both $\boldsymbol{a}$ and $\boldsymbol{b}$, as given by the right-hand rule.

Having performed these calculations it's also a good idea to check them. In this case we can confirm that all the scalar products are scalars and that the vector products are vectors (even though two of them are zero vectors).

Question 4.6 Compare and contrast the definition of torque, as represented by Equation 4.3, with the definition of the work done by a constant force $\boldsymbol{F}$ acting over a displacement $\boldsymbol{s}$, as discussed in Chapter 2, Section 2.5. Comment on their similarities and differences. In particular, determine the angle between the relevant vectors that will cause each quantity to attain its maximum magnitude. ■

3   Equilibrium and statics

This book is concerned with the prediction of motion. However, an important part of that process concerns predicting the *absence* of motion. This is the business of **statics**, the branch of dynamics that deals with systems at rest. Statics is of the utmost practical importance since there are a great many structures, such as bridges and buildings, that are required to remain at rest (or very nearly so) under a wide range of external conditions. Engineers often need to know the forces and torques that will act on the various parts of a static structure, and it is the principles of statics that make it possible for such information to be obtained.

3.1 Equilibrium conditions

Newton's first law tells us that if the vector sum of all forces acting on a particle is zero, then that particle has zero acceleration. Consequently, it will move with constant velocity. (This includes the possibility that the particle has zero velocity and therefore remains at rest.) As you saw in Chapter 1, Newton's laws may be extended to cover rigid bodies, in which case, if the vector sum of all the external forces is zero, the velocity of the body's centre of mass, $\boldsymbol{v}_{CM}$, remains constant and may be zero.

A body that is not undergoing translational acceleration is said to be in a state of translational equilibrium, so Newton's first law implies the following condition for **translational equilibrium**

$$\sum_i \boldsymbol{F}_i = \boldsymbol{0} \tag{4.12}$$

where the $\boldsymbol{F}_i$ are the *external* forces on the body.

We can write down a corresponding condition for a rotating rigid body by demanding that the vector sum of the external torques about the body's centre of mass should be zero. Thus we have the following condition for **rotational equilibrium**:

$$\sum_i \Gamma_i = \mathbf{0} \qquad (4.13)$$

where the Γ_i are the external torques about the centre of mass. Note though that in this case there are still situations (as pointed out in Section 2.4) where the absence of a resultant external torque is not sufficient to ensure that the angular velocity will remain constant. Nonetheless, this is, in effect, the best we can do when it comes to specifying a condition for rotational equilibrium.

To complete the terminology of equilibrium conditions, we can combine Equations 4.12 and 4.13 to provide the following condition for **mechanical equilibrium**:

$$\sum_i \mathbf{F}_i = \mathbf{0} \quad \text{and} \quad \sum_i \Gamma_i = \mathbf{0}. \qquad (4.14)$$

A static system is a special case of a system in equilibrium, so it must necessarily satisfy this condition. Remember, however, that the mechanical equilibrium condition is not *sufficient* to guarantee that a system will be static.

Question 4.7 A pair of oppositely directed forces of equal magnitude, with different lines of action is said to comprise a *couple*. Such a couple is applied to the uniform wheel of radius of 0.345 m in Figure 4.25. If the magnitude of each force is $|\mathbf{F}| = 20.2\,\text{N}$, and no other forces act on the wheel, what is the total force on the wheel? What is the torque Γ about the centre of the wheel, and will the wheel be in any sort of equilibrium?

Question 4.8 A disc of diameter 1.50 m is mounted on an axle of diameter $5.50 \times 10^{-2}\,\text{m}$, that passes through the centre of the disc and is perpendicular to its surface. The axle is supported horizontally by a worn bearing. When rotating at a constant angular speed ω, the frictional force experienced by the rotating axle is of magnitude 0.168 N. What force must be applied to the rim of the disc if it is to maintain a constant angular speed? ■

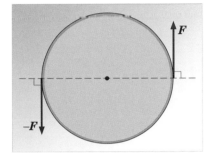

Figure 4.25 Two forces applied to the rim of a wheel. The forces are of equal magnitude but act in different directions along different lines of action. Such a pair of forces tends to cause rotational acceleration but not translational acceleration, and is sometimes referred to as a *couple*.

3.2 Static equilibrium conditions

The special case of mechanical equilibrium that occurs when a body or a system of bodies is completely motionless, i.e. when $v_{\text{CM}} = \mathbf{0}$ and $\omega = \mathbf{0}$, is known as **static equilibrium**. Under these circumstances the absence of torques around any axis *will* ensure that the angular acceleration about that axis is zero and that it therefore continues to be true that $\omega = \mathbf{0}$. Moreover, under the conditions of static equilibrium the requirement that the resultant torque about the centre of mass should be zero may be shown to be equivalent to the following more general statement:

> For a system in static equilibrium the resultant torque about any point must be zero.

The freedom to balance torques about *any* point in a static system is of enormous value when it comes to solving statics problems. A typical problem provides information about a structure and about some of the forces acting on that structure. The challenge is usually to determine one or more of the other forces that act on the structure, as efficiently as possible. By choosing to balance torques about a point where an unknown force acts, it is often possible to avoid having to calculate that force at all, since a force cannot exert any torque about its point of application (or about any point on its line of action). Example 4.3 should help to clarify this process and to give you a taste of statics problems in general.

Example 4.3

A rigid uniform beam of mass 2.50 kg and length 3.20 m, is pivoted at one end to a vertical wall. It is held horizontal by a light wire cable, attached to its other end and fixed to the wall at a point 2.40 m above the pivoted end. A sign of mass 0.85 kg is hung at 0.20 m from the outside end of the beam. Calculate the tension in the supporting cable.

Solution

To solve this problem we shall use the fact that the beam is in static equilibrium. To work out the torques involved we need a diagram indicating the forces on the beam, and the relevant displacements. A suitable diagram, (deliberately made as simple as possible) is shown in Figure 4.26. Note that the sign effectively exerts a single downward force equal to its weight. Also note that to ensure translational equilibrium there must be a force $\boldsymbol{F}_O$ acting on the pivoted end of the beam, and that the tension we seek will be the magnitude of the tension force $\boldsymbol{F}_T$ acting at the other end of the beam.

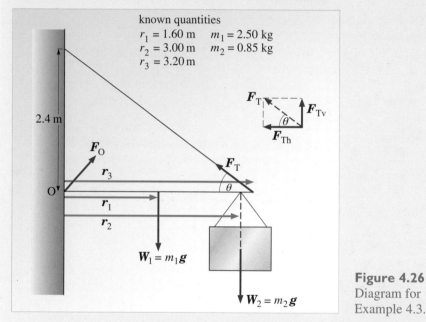

Figure 4.26
Diagram for
Example 4.3.

Since we don't know the magnitude or the direction of $\boldsymbol{F}_O$ it makes sense to consider torques about O. The condition for rotational equilibrium then implies that

$$\boldsymbol{\Gamma}_1 + \boldsymbol{\Gamma}_2 + \boldsymbol{\Gamma}_T = \boldsymbol{0},$$

where $\boldsymbol{\Gamma}_1 = \boldsymbol{r}_1 \times \boldsymbol{W}_1$, $\boldsymbol{\Gamma}_2 = \boldsymbol{r}_2 \times \boldsymbol{W}_2$ and $\boldsymbol{\Gamma}_T = \boldsymbol{r}_3 \times \boldsymbol{F}_T$. Each of these torques is directed perpendicular to the plane of the diagram, but $\boldsymbol{\Gamma}_1$ and $\boldsymbol{\Gamma}_2$ point into the page, while $\boldsymbol{\Gamma}_T$ points out of the page. It follows that, in terms of magnitudes,

$$\Gamma_T = \Gamma_1 + \Gamma_2. \tag{4.15}$$

Now, $\quad \Gamma_1 = r_1 m_1 g = 1.60 \times (2.50 \times 9.81)\,\text{N m} = 39.2\,\text{N m}$ \hfill (4.16)

and $\quad \Gamma_2 = r_2 m_2 g = 3.00 \times (0.85 \times 9.81)\,\text{N m} = 25.0\,\text{N m}.$ \hfill (4.17)

The horizontal component of the tension force $\boldsymbol{F}_{Th}$ passes through O so the torque caused by the tension force is entirely due to its vertical component vector, $\boldsymbol{F}_{Tv}$ (see Figure 4.26), and therefore has magnitude

$$\Gamma_T = r_3 F_T \sin \theta. \tag{4.18}$$

all reasonable conditions. However, they must not over-design their structures, or they will be unnecessarily expensive.

Until recently, most engineering structures were designed with very substantial safety margins, and excess cost was simply accepted as the price of safety. However, improvements in the scope and accuracy of calculations, as well as the general advance of science and technology, mean that modern engineers are better able to determine the precise physical conditions that structural members must survive, and are therefore able to design structures that meet those conditions more precisely (Figure 4.32). So, calculations based on the principles introduced in this section play an increasingly important role in helping to shape the world in which we live and work.

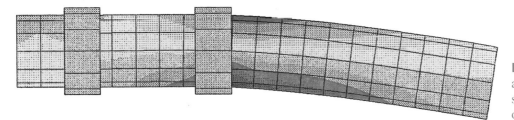

Figure 4.32 A computer-based analysis, using the principles of statics, reveals the torques acting on a loaded metal tube.

4 Rotational energy and moments of inertia

4.1 Rotational energy of particles

A particle of mass m, moving uniformly around a circle of radius r centred on a point O, will have kinetic energy by virtue of its motion. If the particle has speed v, its kinetic energy will be $\frac{1}{2}mv^2$. However, since in this case

$$v = r\omega, \tag{4.19}$$

where ω is the angular speed of the particle around O, we may regard the kinetic energy as an example of **rotational kinetic energy** and write it as

$$E_{\text{rot}} = \frac{1}{2}mr^2\omega^2. \tag{4.20}$$

We can generalize this idea to the case of a system of particles of masses m_1, m_2, m_3, … at distances r_1, r_2, r_3, … from the fixed point O. If the whole system rotates uniformly about O, with angular speed ω, so that the particles maintain their mutual separations (as indicated in Figure 4.33), then the total rotational kinetic energy of the system will be

$$E_{\text{rot}} = \frac{1}{2}m_1r_1^2\omega^2 + \frac{1}{2}m_2r_2^2\omega^2 + \frac{1}{2}m_3r_3^2\omega^2 + \cdots$$
$$= \frac{1}{2}\omega^2(m_1r_1^2 + m_2r_2^2 + m_3r_3^2 + \ldots)$$

This may be written more compactly using the summation symbol

$$E_{\text{rot}} = \frac{1}{2}\omega^2 \sum_i m_i r_i^2. \tag{4.21}$$

The sum within Equation 4.21 is said to represent the **moment of inertia** of the system of particles, about the rotation axis, and is usually given the symbol, I. We can therefore write

$$E_{\text{rot}} = \frac{1}{2}I\omega^2, \tag{4.22}$$

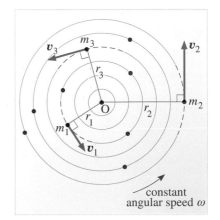

Figure 4.33 A two-dimensional system of particles in uniform rotation.

where

$$I = \sum_i m_i r_i^2. \tag{4.23}$$

Now, looking at Equation 4.22, it's pretty clear that this is the rotational analogue of $E_{trans} = \frac{1}{2}mv^2$. In place of E_{trans} we have E_{rot}, in place of v we have the angular speed ω, and in place of m we have its rotational analogue, the moment of inertia I.

We can extend these ideas to the case of a rotating rigid body, which is essentially just a continuous system of particles. However, before doing so it would be a good idea to dwell a little longer on the concept of moment of inertia since this will be of great importance throughout the rest of this chapter. Let's start by considering Example 4.4.

Example 4.4

A system consists of two very compact particles, each of mass 1.00 kg, connected by a rigid rod 0.60 m long and of negligible mass. For each of the cases shown in Figure 4.34, calculate the moment of inertia of the system about an axis perpendicular to the page and passing through the point indicated by a cross.

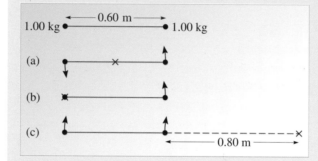

Figure 4.34 Moments of inertia of a simple two-dimensional system about different axes. In each case the axis is marked by a cross.

Solution

By way of preparation we should recall the basic definition $I = \sum_i m_i r_i^2$.

(a) For two particles rotating about their common centre of mass, as shown in Figure 4.34a

$$I = 2 \times mr^2$$

where $m = 1.00$ kg and $r = 0.30$ m from the centre of rotation. Hence

$$I = (2 \times 1.00 \times 0.09)\,\text{kg m}^2 = 1.80 \times 10^{-1}\,\text{kg m}^2.$$

(b) In this case only one particle rotates, and it is at $r = 0.60$ m from the axis, thus

$$I = mr^2 = (1.00 \times 0.36)\,\text{kg m}^2 = 3.60 \times 10^{-1}\,\text{kg m}^2.$$

(c) In this case, one particle rotates at $r_1 = 0.80$ m and the other at $r_2 = 1.40$ m from the axis of rotation. In both cases the mass is $m = 1.00$ kg. Consequently

$$I = mr_1^2 + mr_2^2 = 1.00 \times [(0.80)^2 + (1.40)^2]\,\text{kg m}^2 = 2.60\,\text{kg m}^2.$$

Example 4.4 illustrates an important general point that you should keep in mind.

> The moment of inertia of a system depends on the axis about which it is determined.

As a further illustration, consider a ring of closely spaced particles, moving around an axis that passes through the centre of the ring, and is perpendicular to the plane of the ring.

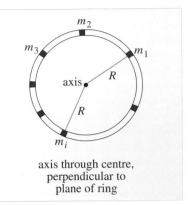

Figure 4.35 Masses in a plane and at the same distance from the centre of rotation.

Suppose that the particles have masses m_1, m_2, m_3, ..., etc., and that they are all at the same distance R from the centre, as shown in Figure 4.35.

Applying Equation 4.23, the moment of inertia of the ring about the axis through its centre will be

$$I_{ring} = m_1R^2 + m_2R^2 + m_3R^2 + \ldots = R^2(m_1 + m_2 + m_3 + \ldots)$$
$$= M_{ring}R^2$$

where M_{ring} is the total mass of the particles.

Now, although Figure 4.35 shows seven particles, our result depends only on the total mass. Therefore, our derivation applies to any number of particles, including the very large number of particles that may be thought of as composing a continuous ring. Thus, we can say that the moment of inertia of a rigid ring, of radius R and mass M, about an axis through its centre and perpendicular to its plane is

$$I = MR^2. \tag{4.25}$$

So, by imagining an appropriate system of particles and considering a suitable sum we have been able to determine the moment of inertia of a rigid body about a specific axis. As you will see in the next subsection, many other moments of inertia may be determined in a similar way.

Question 4.11 Three identical spherical marbles (small but heavy) are joined by light and rigid rods, to form a right-angled triangle of sides 3.00×10^{-2} m, 4.00×10^{-2} m and 5.00×10^{-2} m. The triangle can be made to rotate (in the same plane as the triangle) about any one of the corner marbles. Provided the angular speed is the same in each case, which centre of rotation gives the largest rotational energy? ■

The basic idea of a moment of inertia should now be clear:

> The moment of inertia of a system about a given axis characterizes the way in which the mass of the system is distributed about that axis, and thus plays an important role in determining the rotational energy of the system about that axis.

4.2 Rotational energy of rigid bodies

A three-dimensional rigid body rotating about a fixed axis may be regarded as a collection of particles rotating with a common angular velocity while maintaining fixed separations (Figure 4.36). In evaluating the moment of inertia of such a system it is convenient to imagine that the body is continuous and to replace the sum that appears in Equation 4.23 by a definite integral of the kind that was introduced in Chapter 2. We shall not pursue the mathematical details of this approach, but we will quote some of the results that emerge for rigid bodies of uniform density and specific shape, rotating about particular axes. The results are illustrated in Figure 4.37 overleaf.

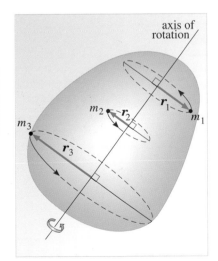

Figure 4.36 A rigid three-dimensional system of particles, rotating about a fixed axis.

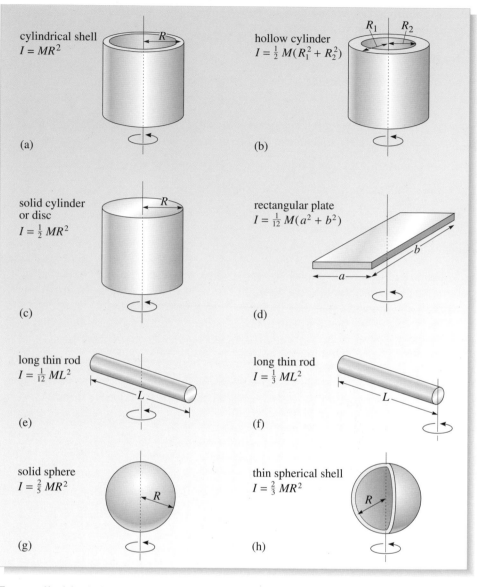

Figure 4.37 Moments of inertia for some common solids.

For a cylindrical shell or ring (Figure 4.37a) of mass M and radius R, rotating about its axis of rotational symmetry

$$I_{\text{cylindrical shell}} = MR^2. \qquad (4.26a)$$

For a hollow cylinder (Figure 4.37b) of mass M, internal radius R_2 and external radius R_1, about its axis of rotational symmetry

$$I_{\text{hollow cylinder}} = \tfrac{1}{2} M(R_1^2 + R_2^2). \qquad (4.26b)$$

For a solid cylinder or disc (Figure 4.37c) of mass M and radius R, about its axis of rotational symmetry

$$I_{\text{solid cylinder}} = \tfrac{1}{2} MR^2. \qquad (4.26c)$$

For a rectangular plate (Figure 4.37d) of mass M and sides a and b, about an axis perpendicular to its surface and through its centre

$$I_{\text{rectangular plate}} = \tfrac{1}{12} M(a^2 + b^2). \qquad (4.26d)$$

For a long thin rod (Figure 4.37e) of mass M and length L, about an axis perpendicular to the rod and through its centre

$$I_{\text{rod}} = \tfrac{1}{12} ML^2. \qquad (4.26e)$$

For a long thin rod (Figure 4.37f) of mass M and length L, about an axis perpendicular to the rod and through one of its ends

$$I_{rod} = \tfrac{1}{3} ML^2. \tag{4.26f}$$

For a solid sphere (Figure 4.37g) of mass M and radius R, about an axis through its centre

$$I_{solid\ sphere} = \tfrac{2}{5} MR^2. \tag{4.26g}$$

For a thin spherical shell (Figure 4.37h) of mass M and radius R, about an axis through its centre

$$I_{thin\ spherical\ shell} = \tfrac{2}{3} MR^2. \tag{4.26h}$$

Don't forget that the moment of inertia depends on the axis in each case.

Now, with the aid of these moments of inertia and Equation 4.22

$$E_{rot} = \tfrac{1}{2} I\omega^2, \tag{Eqn 4.22}$$

we can work out the rotational energy in a variety of situations, provided the body concerned is rotating only about the specified axis.

This last (uni-axial) restriction is a very important one. Since angular velocity is a vector quantity, it generally has three independent components, so a rigid body may simultaneously rotate about three independent axes. The methods for dealing with such cases are well known, but they are beyond the scope of this chapter.

The rotational energy that a body has due to rotation about its centre of mass is independent of any kinetic energy it has due to the motion of its centre of mass. For example, a car wheel may rotate about an axis through its centre of mass, which, at the same time, is in translational motion along the road. The total kinetic energy due to these two sources is therefore given by a simple sum of the following form

$$E_{kin} = E_{trans} + E_{rot} = \tfrac{1}{2} Mv_{CM}^2 + \tfrac{1}{2} I_{CM}\omega^2 \quad \text{(uni-axial rotation)} \tag{4.27}$$

where v_{CM} is the speed of translation of the centre of mass, and I_{CM} is the moment of inertia about the relevant axis through the centre of mass.

Question 4.12   A vehicle travels on wheels of radius 0.315 m and each wheel can be taken to be a uniform disc of mass 22.5 kg. If the vehicle travels at a speed of 40.0 km h^{-1}, find the rotational and translational kinetic energy per wheel. (Assume there is no sliding, so the point on the wheel that is in contact with the ground is instantaneously at rest.) ■

4.3 Torque, moment of inertia and angular acceleration

Now that moments of inertia have been introduced it is possible to start discussing the prediction of rotational motion.

Suppose that any one of the rigid bodies in Figure 4.37 is in uni-axial rotation with angular velocity $\boldsymbol{\omega}$ about any one of the axes shown in that figure. Then, if forces are applied to the body in such a way that they produce a net torque $\boldsymbol{\Gamma}$ that is parallel to $\boldsymbol{\omega}$, it will be found that the resulting angular acceleration of the body will be given by

$$\boldsymbol{\Gamma} = I\frac{d\boldsymbol{\omega}}{dt} \quad \text{(restricted validity)} \tag{4.28}$$

where I is the relevant moment of inertia. It should be clear at this stage that Equation 4.28 (which can also be written $\boldsymbol{\Gamma} = I\boldsymbol{\alpha}$) is the rotational analogue of $\boldsymbol{F} = m\boldsymbol{a}$. However, unlike Newton's second law, Equation 4.28 is of restricted validity and therefore of limited value. It applies to cases besides those described in Figure 4.37, but it is certainly *not* true in general. It is not the case, for example, that an arbitrary torque applied to an arbitrary rigid body will always produce an angular acceleration that is parallel to the torque.

We shall not attempt to write down the most general conditions under which Equation 4.28 can be used, preferring to leave such general discussions to Section 5. However we shall note that, despite its limitations, Equation 4.28 is an important relation since it does allow us to predict rotational motion in a variety of circumstances. It will, for example, let us predict the effect of torques on a wheel turning about a fixed axle, even though it won't let us predict the precession of the Earth, which involves a changing axis.

As an example, suppose that in some situation where Equation 4.28 does apply (a cylinder rotating about its central axis, say) we are told that Γ is actually a constant vector. (It might more generally be a function of time.) Then, in that case, we can regard Equation 4.28 as a differential equation and solve it, subject to appropriate initial conditions, to determine the angular velocity as a function of time. In such a case the result would be

Equation 4.29 could also be written $\omega = \omega_0 + \alpha t$ which further emphasizes the similarity to $v = u + at$.

$$\boldsymbol{\omega}(t) = \boldsymbol{\omega}_0 + \frac{\Gamma}{I}t, \quad \text{(restricted validity } and\ \Gamma \text{ constant)} \qquad (4.29)$$

where $\boldsymbol{\omega}_0$ is the angular velocity at time $t = 0$. (Equation 4.29 is the rotational counterpart of the uniform acceleration equation $\boldsymbol{v} = \boldsymbol{u} + \boldsymbol{a}t$.) By recalling that $\boldsymbol{\omega}$ is related to the rate of change of the angular position θ, this technique can be extended to provide a means of also predicting θ as a function of time.

Question 4.13 A disc of uniform density rotates about an axis through its centre, perpendicular to its plane. The disc has mass $M = 1.40\,\text{kg}$ and radius $R = 0.360\,\text{m}$. In a time interval $\Delta t = 5.20\,\text{s}$, the angular speed of the disc increases steadily from zero to 100 revolutions per minute, due to a constant applied torque. What is the magnitude of that torque?

Question 4.14 The rotating disc in Question 4.13 is initially at rest. At time $t = 0\,\text{s}$ a constant torque of magnitude $4.00 \times 10^{-3}\,\text{N m}$ is applied. Find the angular speed after 21.5 s. ■

4.4 The role of rotational energy in the prediction of motion

As you saw in Chapter 2, the total energy of an isolated system is always conserved, and the mechanical energy of a system is conserved as long as the only external forces that do work on the system are conservative forces. When we introduced mechanical energy earlier, it was as the sum of potential energy and translational energy. Now, we can extend its definition to include contributions arising from rotational kinetic energy. In addition, we note that a constant torque Γ acting over an angular displacement $\Delta\theta$, in a plane perpendicular to Γ, does an amount of work

$$W = \Gamma\Delta\theta \qquad (4.30)$$

and that such a torque, acting on a body rotating with angular velocity $\boldsymbol{\omega}$, is transferring energy at a rate given by the power

$$P = \boldsymbol{\Gamma} \cdot \boldsymbol{\omega}. \qquad (4.31)$$

As a consequence, we can write the mechanical energy of a rigid body as

$$E_{\text{mech}} = E_{\text{trans}} + E_{\text{rot}} + E_{\text{pot}} \qquad (4.32)$$

and we can say that this quantity will be conserved (i.e. constant) if the external forces acting on the system (including those giving rise to torques) are conservative. In the case of a body of mass M whose centre of mass has velocity v_{CM} and which has uni-axial angular speed ω about an axis through the centre of mass, the conservation of mechanical energy takes the form

$$\tfrac{1}{2}Mv_{CM}^2 + \tfrac{1}{2}I_{CM}\omega^2 + E_{pot} = \text{constant} \quad \text{(uni-axial rotation)} \quad (4.33)$$

where I_{CM} is the moment of inertia about the relevant axis through the centre of mass.

The following example (Example 4.5) will show how this relation can be used to make predictions about rotational motion.

Example 4.5

A solid cylinder rolls directly down an inclined plane without slipping. The cylinder has mass M and radius R, and the plane is inclined at an angle ϕ to the horizontal. If the cylinder starts from rest, find an expression for the speed of the cylinder when it has descended a vertical distance h.

Solution

Figure 4.38 shows the forces that act on the cylinder. The fact that the cylinder rolls without slipping means that the point of contact between the cylinder and the plane is stationary, so the frictional force that acts there does no work. Also, the normal reaction N does no work since it acts at right angles to the direction of motion. In fact, the only external force that does work on the cylinder is the gravitational force, which is conservative. Consequently, when the cylinder reaches the bottom of the incline, the decrease in its gravitational potential energy must be compensated by a corresponding increase in its kinetic energy. Thus,

$$Mgh = \tfrac{1}{2}I_{CM}\omega^2 + \tfrac{1}{2}Mv_{CM}^2 \qquad (4.34)$$

where ω is the final angular speed of the cylinder and v_{CM} is the speed of its centre of mass. Now in this case

$$I = \tfrac{1}{2}MR^2 \qquad (4.35)$$

and, since the cylinder has angular speed ω while its point of contact with the ground is instantaneously at rest,

$$v_{CM} = R\omega.$$

Hence $\quad Mgh = \tfrac{1}{2} \times \tfrac{1}{2}MR^2 \times \left(\dfrac{v_{CM}}{R}\right)^2 + \tfrac{1}{2}Mv_{CM}^2, \qquad (4.36)$

which can be written as

$$gh = \tfrac{3}{4}v_{CM}^2. \qquad (4.37)$$

Consequently $\quad v_{CM} = \sqrt{\dfrac{4gh}{3}}. \qquad (4.38)$

Notice that this result is independent of both M and R.

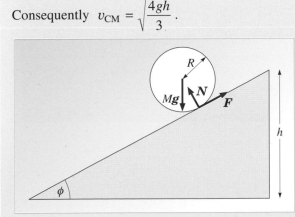

Figure 4.38 A cylinder rolling down a plane, inclined at an angle ϕ to the horizontal.

Question 4.15 Suppose that a race is arranged, in which a solid cylinder, a thin cylindrical shell, a solid sphere and a thin spherical shell, all of radius R, are rolled (without slipping) down the same slope all starting at the same time and with their respective centres at the same height. In what order will the objects arrive at the bottom of the slope? How does the value of R influence the outcome? ■

4.5 Flywheels and energy storage

A **flywheel** is a heavy rotating disc that can help to smooth the rotation of an engine. If an engine transfers energy at different rates during its cycle of operation, as most engines do, it can produce a rather jerky effect when it drives a load of some kind. A smoother effect can often be obtained by storing the energy in a flywheel and then using the flywheel to drive the load (see Figure 4.39). The reason for this can be seen by writing Equation 4.28 as

$$\frac{d\boldsymbol{\omega}}{dt} = \frac{\Gamma}{I}.$$

The flywheel is designed to have a large moment of inertia, I. Consequently, small variations in the magnitude of the torque Γ supplied by the engine will have little effect on the angular speed of the flywheel since they will only cause small angular accelerations. The flywheel is thus able to act as a sort of mechanical 'battery' that is being constantly 'recharged' (albeit at a fluctuating rate) by the engine.

Energy stored in flywheels has been, and continues to be, used in a variety of applications. Road vehicles use flywheels to smooth engine rotation, as described above, but some have employed flywheels as a major means of energy storage rather than just a smoothing mechanism. Flywheels are also found in children's 'friction powered' toys, and it is the principle of the flywheel that underlies the operation of the yo-yo (Figure 4.40). As a yo-yo descends it gains rotational energy which is exchanged for gravitational potential energy as it subsequently ascends again.

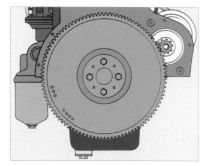

Figure 4.39 Flywheel and engine.

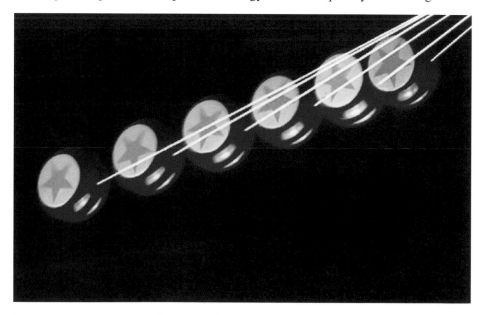

Figure 4.40 An application of the flywheel — a yo-yo.

5   Angular momentum

To complete the development of basic rotational dynamics, we need to introduce one additional quantity known as *angular momentum*. This is the rotational counterpart to linear momentum (Chapter 3). Angular momentum plays a vital role in rotational dynamics since it provides the key to dealing with general situations in which the rotation is not necessarily uni-axial.

5.1   Angular momentum of particles

The **angular momentum** l, of a particle, about a point O, is defined by the vector product of the displacement vector r of the particle from point O and the particle's linear momentum p relative to O (see Figure 4.41):

$$l = r \times p. \qquad (4.39)$$

Equation 4.39 implies that angular momentum has the following properties:

1   It is a vector.

2   Its magnitude is $l = rp \sin \theta$.

3   Its direction is perpendicular to the plane defined by the vectors r and p and in the sense specified by the right-hand rule.

Note that this definition is completely general and can be applied to any particle. The definition only involves the particle's *instantaneous* displacement and momentum; the particle does *not* need to be rotating in a circle about the point O.

● Suppose you are standing 10 m from a long straight railway track, and that a train travelling with constant velocity passes you. Does the train have angular momentum about your position? If so, does the magnitude of that angular momentum increase or decrease as the train moves away from you?

○ The train certainly has angular momentum about your position, as indicated in Figure 4.42. As the train draws away, its distance r from your position will increase, but the value of $r \sin \theta$ will remain constant since it represents the perpendicular distance to the train's line of motion, which is always 10 m. Consequently $l = rp \sin \theta$ will remain constant as long as the train's momentum may be regarded as constant. ■

Now consider the case of a particle that is moving in a circle, as shown in Figure 4.43. Suppose the particle has mass m, and that it moves in a circle of radius r, centred on the origin O of a set of Cartesian coordinates, always staying in the xy-plane. Also suppose that the particle travels in the anticlockwise sense when viewed from above, with uniform angular speed ω, so that its angular velocity ω points along the z-axis, and may be written $\omega = (0, 0, \omega)$.

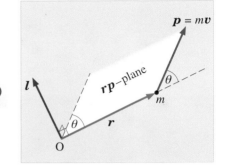

Figure 4.41   Angular momentum of a particle about point O.

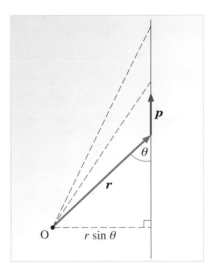

Figure 4.42   Angular momentum of a train moving along a straight track.

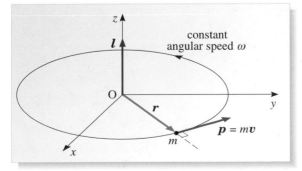

Figure 4.43   Angular momentum $l = (0, 0, l_z)$ of a particle, rotating uniformly in the xy-plane.

Using Equation 4.39, the angular momentum $\boldsymbol{l}$ of the particle will be directed along the z-axis, and may be written

$$\boldsymbol{l} = (0, 0, l_z)$$

where, in this case

$$l_z = rp \sin 90° = rmv = mr^2\omega.$$

It follows that in this particular case

$$l = |\boldsymbol{l}| = mr^2\omega$$

and we can write

$$\boldsymbol{l} = mr^2\boldsymbol{\omega}. \tag{4.40}$$

This is a useful result that we shall exploit shortly (in Example 4.6), but note that it only applies when the particle is moving in a circle about O.

Example 4.6

A geostationary satellite moves in a circular orbit about the centre of the Earth so that it remains above a fixed point on the Equator. The radius of the orbit is given by $r = 4.23 \times 10^7$ m and the satellite's angular speed ω is given by $\omega = 7.27 \times 10^{-5}$ rad s^{-1}. If the mass of the satellite is $m = 1.20 \times 10^3$ kg, what is the magnitude of the satellite's angular momentum about the centre of the Earth?

Solution

Letting the xy-plane of Figure 4.43 correspond to the Equatorial plane of the Earth, and using Equation 4.40:

$$|\boldsymbol{l}| = rmv = mr^2\omega$$

$$= (1.20 \times 10^3 \text{ kg}) \times (4.23 \times 10^7 \text{ m})^2 \times (7.27 \times 10^{-5} \text{ s}^{-1})$$

$$= 1.56 \times 10^{14} \text{ kg m}^2 \text{ s}^{-1}.$$

As a check, we note that the units of the calculated quantity are those we would expect for angular momentum, and that its sign is positive as it should be for a magnitude.

Now consider the angular momentum of a two-dimensional system of particles, all of which move in circles in the xy-plane, around O, with a common angular velocity $\boldsymbol{\omega}$. (Such a system was depicted in Figure 4.33.) The particles may be supposed to have different masses, m_i, and to be at different distances r_i from O, so the angular momentum about O of the ith particle will be $\boldsymbol{l}_i = m_i r_i^2 \boldsymbol{\omega}$. The total angular momentum about O of the whole system of particles will be given by the vector sum

$$\boldsymbol{L} = \boldsymbol{l}_1 + \boldsymbol{l}_2 + \boldsymbol{l}_3 + \ldots$$

So, $$\boldsymbol{L} = \sum_i \boldsymbol{l}_i = \sum_i m_i r_i^2 \, \boldsymbol{\omega} \tag{4.41}$$

but $\boldsymbol{\omega}$ is a constant vector that appears in every term in the sum, so we can extract it as a *common factor* and write the angular momentum about O as

$$\boldsymbol{L} = \boldsymbol{\omega} \sum_i m_i r_i^2 . \tag{4.42}$$

You should recognize the sum in this expression — it's just the moment of inertia of the system about the rotation axis. If we call it I, we can say that for this particular system the angular momentum about O is

$$\boldsymbol{L} = I\boldsymbol{\omega}. \tag{4.43}$$

It's tempting to think that this relationship between angular momentum, angular velocity and moment of inertia can be generalized from two-dimensional systems of particles, to three-dimensional rigid bodies. However, as you are about to see, such a generalization is not valid.

5.2 Angular momentum of rigid bodies

As we saw in Section 4.2, a rigid body, rotating about a fixed axis, may be regarded as a collection of particles, rotating with a common angular velocity while maintaining fixed separations. The angular momentum of such a body will be the vector sum of the angular momenta of all of those particles. In practice, it is often more convenient to regard the body as continuous and to replace the sum by a definite integral.

You might expect the outcome of this process, at least for systems in uni-axial rotation, to be a relation of the form $L = I\omega$. However, even for bodies in uni-axial rotation, this is not always true. For a body in uni-axial rotation, with angular velocity ω along a fixed axis, about which its moment of inertia is I, it will always be the case that the angular momentum has a component $L_{\text{axis}} = I\omega$, but it is quite possible for the angular momentum to also have a component at right angles to the rotation axis. This may seem rather odd, but a simple example should convince you that it is so.

Figure 4.44 shows two particles of equal mass joined by a light rod of length $2r$, inclined at a fixed angle to a vertical axis of rotation that passes through the centre of the rod. The rigid body is rotated about the axis with constant angular velocity ω. At any instant the angular momentum L about the point O will be almost entirely due to the two particles, and their angular momenta, l_1 and l_2, are always parallel and inclined at 90° to the rod, as indicated. It follows that the total angular momentum $L = l_1 + l_2$ cannot be parallel to ω. This, of course, implies that, at any instant, L has a non-zero component perpendicular to the rotation axis.

We shall not attempt to write down the general relationship between the angular momentum of a rigid body and its angular velocity. However, we shall note that the general relationship between the two quantities is well known, and that in cases where the body is one of those shown in Figure 4.37, in uni-axial rotation about one of the shown axes through the centre of mass, *then* it will be true that the angular momentum about the body's centre of mass is given by

$$L = I\omega \qquad (\textit{not a general result}). \qquad (4.44)$$

This limited amount of information about the angular momentum of rigid bodies will suffice for present purposes. Here is a reminder of the main (and rather surprising) point, in a form that is true even when the rotation is not uni-axial.

> For a rotating rigid body, the angular momentum L about a given point depends on the way the body's mass is distributed, and on the components of its angular velocity ω. It is not generally true that L is parallel to ω nor that $L = I\omega$, though there are cases in which these relations will be true.

Question 4.16 A uniform solid cylinder of total mass 5.2 kg and radius 0.125 m rotates at 3.65 revolutions per second about its axis of symmetry, which coincides with the y-axis of a coordinate system. The rotation in the xz-plane is anticlockwise, when viewed from the positive y-axis towards the origin (Figure 4.45). What is the magnitude and the direction of the angular momentum about the centre of mass?

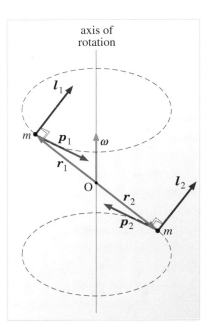

Figure 4.44 A rigid body in uni-axial rotation with angular velocity ω. In this case the angular momentum about O, $L = l_1 + l_2$, is not parallel to the angular velocity, so $L \neq I\omega$.

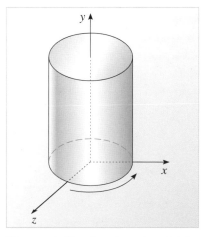

Figure 4.45 See Question 4.16.

Question 4.17   Calculate the magnitude of the Earth's orbital angular momentum L_{orb} about the Sun, and the magnitude of its spin angular momentum L_{spin} about its own centre. Use the following approximate data:

mass of the Earth:	5.98×10^{24} kg
average radius of the Earth:	6.38×10^6 m
average radius of the Earth's orbit:	1.50×10^{11} m
orbital period of the Earth:	365 days
period of spin of the Earth:	24.0 h. ■

5.3   Torque and the rate of change of angular momentum

The key to predicting the general rotational motion of rigid bodies is the equation that links the resultant external torque on a body to the rate of change of that body's angular momentum. This fundamental equation is true in general, not just for cases of uni-axial rotation, and takes the form

$$\boldsymbol{\Gamma} = \frac{\mathrm{d}\boldsymbol{L}}{\mathrm{d}t} \tag{4.45}$$

where $\boldsymbol{\Gamma}$ represents the sum of the external torques about a point and $\boldsymbol{L}$ is the angular momentum of the body about the same point. (In practice, the chosen point is usually the body's centre of mass.)

Equation 4.45 is the rotational analogue of Newton's second law expressed in the form

$$\boldsymbol{F} = \frac{\mathrm{d}\boldsymbol{p}}{\mathrm{d}t}. \tag{4.46}$$

When Equation 4.46 was first introduced, as Equation 3.2, it was stressed that it provided a more fundamental statement of Newton's second law than $\boldsymbol{F} = m\boldsymbol{a}$. We now see that in a similar way

$$\boldsymbol{\Gamma} = \frac{\mathrm{d}\boldsymbol{L}}{\mathrm{d}t} \quad \text{(a general result)} \tag{Eqn 4.45}$$

is of much wider applicability than

$$\boldsymbol{\Gamma} = I\frac{\mathrm{d}\boldsymbol{\omega}}{\mathrm{d}t} \quad \text{(a restricted result).} \tag{Eqn 4.28}$$

The difference arises because of the complicated relationship between $\boldsymbol{L}$ and $\boldsymbol{\omega}$. In cases where $\boldsymbol{L} = I\boldsymbol{\omega}$, and I is a constant, the correctness of Equation 4.28 is a direct mathematical consequence of the correctness of Equation 4.45. In other cases, only Equation 4.45 is guaranteed to be correct and its implications for the angular acceleration $\mathrm{d}\boldsymbol{\omega}/\mathrm{d}t$ must be worked out component by component.

Chapter 1 of this book described how the translational motion of a body of known mass, subject to given external forces, could be predicted. The basic approach was to substitute the given forces into Newton's second law (now recognized as Equation 4.46) and to treat the resulting *equation of motion* as a differential equation that could be solved to provide a detailed description of the motion. The process of solving the differential equation introduced various *arbitrary constants* that could only be evaluated by using additional information about the initial condition of the motion, and it was in this way that the past could be said to determine the future as far as Newtonian mechanics was concerned.

Now, thanks to Equation 4.45, we can do the same for the rotational motion of a body subjected to known torques. The torques are substituted into Equation 4.45 and, using the relationship between the body's angular momentum and its angular velocity, a set of differential equations is obtained relating the rates of change of the angular velocity components to the torque components. These differential equations are known as **Euler's equations**; they are the basic equations of motion of rigid body rotational dynamics. (Note, though, that they are essentially *consequences* of Newton's laws, not entirely new laws in their own right.) By solving Euler's equations, subject to appropriate initial conditions, the rotational motion of the body can be predicted in detail.

So, given a rigid body that is subject to known external forces and torques, it is possible to use Equation 4.46 to predict the motion of, say, its centre of mass, and Equation 4.45 to predict its rotational motion about the centre of mass. Thus, provided its initial condition is known in sufficient detail, it is possible to completely predict the subsequent motion of the body. The past does, according to classical Newtonian mechanics, determine the future.

Although we have now arrived at a general scheme for predicting rotational motion, we shall not pursue it here. Instead, we devote the rest of the chapter to topics that further illustrate the significance of Equation 4.45 and emphasize the enormous importance of angular momentum in almost all rotational phenomena.

5.4 Conservation of angular momentum

Equation 4.45 shows that in the absence of any net external torque ($\Gamma = 0$), the angular momentum of a rotating rigid body does not change with time, so L is a constant vector. That is to say it is a *conserved* quantity. This is a special case of a more general law of **conservation of angular momentum** which may be stated as follows:

> For any system, the total angular momentum about any point remains constant as long as no net external torque acts on that system.

This belongs with the conservation of energy and the conservation of linear momentum as one of the fundamental conservation principles that constrain the behaviour of the physical world.

The rest of this subsection is devoted to examples of angular momentum conservation. Most of the systems we discuss will be free of external torques, but that does not mean there are no torques at all. A system may produce *internal* torques arising from forces that one part of the system exerts on some other part. If so, a consequence of the conservation of angular momentum is that any such internally generated torque must be automatically balanced by another internal torque of the same magnitude and opposite direction. In non-rigid systems, such as the human body, this interplay of internal torques can lead to some spectacular changes in motion, as you are about to see.

Examples of conservation of angular momentum

Figure skaters

One of the best known examples of angular momentum conservation is the accelerating pirouette of a figure skater (Figure 4.46). The skater typically starts the pirouette in an 'outstretched' configuration — with both arms and possibly one leg approximately horizontal. This ensures that the skater's initial moment of inertia is relatively large and that the skater's initial angular momentum magnitude, $I_1 \omega_1$, say, is correspondingly large. By pulling the arms (and leg) towards the body, the skater's mass is redistributed and the moment of inertia reduced to some final value I_2, that is less than I_1. If the external torques due to friction, etc. are ignored, the magnitude of the angular momentum will be constant, leading to the prediction that the skaters final angular speed, ω_2, will be given by

$$I_1 \omega_1 = I_2 \omega_2. \tag{4.47}$$

This implies that ω_2 will be greater than ω_1 because I_2 is less than I_1.

Figure 4.46 Conservation of angular momentum in figure skating.

You can demonstrate a similar effect by sitting in a swivel chair with your arms outstretched, starting to turn, and then pulling your arms in. (The effect is increased if you hold some heavy objects in your hands.)

Pulsars and neutron stars

Stars, such as the Sun, spend most of their lives (typically billions of years) in a state of (almost) stable mechanical equilibrium. A star's own gravity tends to cause it to collapse, but internally generated pressures tend to make it expand, and on the whole these effects balance out. However, when some stars near the end of their lives, they develop very massive cores that grow to the point where they are simply unable to support their own weight and collapse under gravity. This is a sudden process that results in the outer parts of the star exploding to form a *supernova remnant*, while the inner part collapses to form an ultra-dense **neutron star**.

Prior to collapse, the stellar core might well have had a mass roughly equal to 1.4 times that of the Sun and a radius comparable to the Earth's. After the collapse, the mass will still be the same, but the radius will have been reduced to something like 8 to 10 km, a reduction by a factor of almost a thousand. This will have the effect of reducing the moment of inertia of the core by a factor of about 10^6 (remember that the moment of inertia of a uniform sphere is proportional to the *square* of its radius). It follows from angular momentum conservation that the angular speed of the core should undergo a million-fold increase. Now, the Sun currently has a rotation period of about 25 days; so, assuming this is typical of a stellar core, it seems that a neutron star might, very roughly, be expected to have a rotation period of about 2 s.

In 1967, in one of the most celebrated of recent astronomical discoveries, a postgraduate student in radio astronomy at the University of Cambridge, Jocelyn Bell Burnell (Figure 4.47), discovered a regularly repeating radio signal of a previously unknown type. The signal had a period of 1.34 s and came from a source known as CP1919 that would now be described as a **pulsar**. It was soon realized that what had been discovered was a highly magnetized neutron star, producing a beam of radio waves that was sweeping across the Earth once per stellar revolution, something like the beam from a lighthouse (see Figure 4.48). Over a thousand pulsars are now known, and related work on neutron stars continues to thrive at the frontiers of modern astronomy.

Figure 4.47 Jocelyn Bell Burnell, discoverer of the first pulsar, and now Head of the Department of Physics and Astronomy at the Open University, UK.

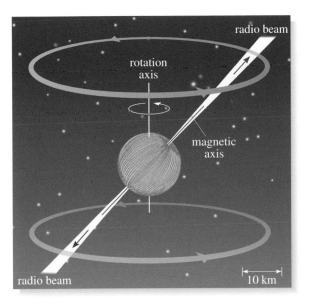

Figure 4.48 A possible model of a pulsar. The radio beam is emitted continuously, but precesses around the neutron star's rotation axis to produce the observed pulsing.

Question 4.18 A uniform spherical stellar core, collapsing to form a black hole, effectively reducing its radius from 2000 km to 1 km. Assuming that neither mass nor angular momentum is lost in this process, by what factor must the angular speed of the core increase? ■

Helicopters

When a helicopter is airborne, it experiences almost no external torque. Its total angular momentum is therefore constant. If the pilot changes the speed of the rotor blades, then their angular momentum about the rotor shaft changes. However, angular momentum must be conserved. This could be achieved by letting the entire helicopter spin, but this is obviously not very desirable. One practical way of conserving angular momentum is to have a small vertically oriented rotor at the tail

Figure 4.49   Single-rotor helicopter.

Figure 4.50   Twin-rotor helicopter.

that *can* cause an external torque about the main rotor shaft. An alternative is to have two horizontal rotor blades that rotate in opposite directions. Both these techniques are employed in helicopter design, as can be seen in Figures 4.49 and 4.50.

Platform divers and falling cats

Platform diving is all about executing difficult movements gracefully while falling towards a diving pool. Due to gravity, the diver's centre of mass inevitably accelerates during the fall in a way that cannot be controlled. The skill comes in controlling the movement of the body about the centre of mass.

The most impressive dives involve somersaults (rotations about a *transverse* axis going from left to right through the body) and often also incorporate twists (rotations about an axis from head to toe). Once the diver has left the diving board there is very little he or she can do to alter the angular momentum acquired at launch, but a great deal can be done about the body's moment of inertia and hence its angular velocity. Some examples are shown in Figure 4.51.

Figure 4.51   Somersaults and twists in platform diving.

Cats are also well known masters and mistresses of their own moment of inertia (Figure 4.52). The ability of a falling cat to land on its paws, almost irrespective of its initial orientation, is sufficiently well known that it does not need further experimental verification — fortunately for cats!

Planetary orbits

Another application of conservation of angular momentum concerns the motion of the planets about the Sun. Kepler's laws of planetary motion were discussed in Chapter 1, they can be stated as follows:

1   The orbit of each planet in the Solar System is an ellipse with the Sun at one focus.

2   A radial line from the Sun to a planet sweeps out equal areas in equal intervals of time.

3   The square of the orbital period of each planet is proportional to the cube of its semimajor axis.

These laws were originally deduced from observation, but they were subsequently shown to be a consequence of Newton's laws of motion and the gravitational attraction of the Sun, as long as perturbations due to other planets were ignored.

The gravitational attraction between a spherical planet and the Sun does not produce any net torque about the Sun. The gravitational force acts along the line connecting the centres of mass of the Sun and the planet, so that

$$\boldsymbol{\Gamma} = \boldsymbol{r} \times \boldsymbol{F} = \boldsymbol{0},$$

since the vectors $\boldsymbol{r}$ and $\boldsymbol{F}$ are in the same line (but in opposite directions) and $\sin 180° = 0$.

In the absence of torque, conservation of angular momentum must apply. The conservation of a planet's orbital angular momentum $\boldsymbol{L}$ has two immediate implications:

●   constant direction of $\boldsymbol{L}$;

●   constant magnitude $|\boldsymbol{L}| = L$.

Constant direction of $\boldsymbol{L}$ means that the plane of the planetary orbit should be constant in space. This is in good agreement with observations; the planes of planetary orbits are remarkably stable over long periods of time.

The consequences of the second point are more complicated. We can see from Figure 4.53, that when a planet moves in an elliptical orbit, its displacement vector from the Sun, $\boldsymbol{r}$, does not have a constant magnitude. Furthermore, $\boldsymbol{r}$ is not always perpendicular to the planet's instantaneous velocity $\boldsymbol{v}$. The constant magnitude of $\boldsymbol{L}$, however, implies that

$$rp \sin \theta = L = \text{constant}, \tag{4.48}$$

Figure 4.53   A planet moving in an ellipse with the Sun at one focus.

Figure 4.52   Stroboscopic photographs of a falling cat.

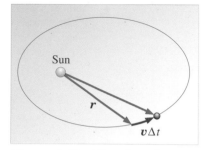

Figure 4.54 The triangular area swept out by the line joining a planet to the Sun during a short time Δt.

where θ is the angle between $\boldsymbol{v}$ and $\boldsymbol{r}$. Since the mass of the planet is constant and $p = mv$, we therefore have the condition

$$rv \sin\theta = \frac{L}{m} \tag{4.49}$$

where L/m is a constant, at all points along a given planetary orbit.

Now, during a short time interval, Δt, a planet will travel an approximate distance $v\,\Delta t$ as indicated in Figure 4.54 and its displacement vector will sweep out an area (ΔS) that is approximately equal to that of a triangle with height r and base $v\Delta t \sin\theta$, as shown in the figure

$$\Delta S \approx \tfrac{1}{2}\, r \times v\Delta t \sin\theta. \tag{4.50}$$

It follows that the *rate* at which the line joining the planet to the Sun sweeps out area is approximately

$$\frac{\Delta S}{\Delta t} \approx \tfrac{1}{2}\, rv \sin\theta. \tag{4.51}$$

If we imagine the time interval Δt getting smaller and smaller, this approximation will become increasingly accurate, and, in the limit as Δt approaches zero, the rate at which area is swept out will be exactly given by

$$\frac{dS}{dt} = \tfrac{1}{2}\, rv \sin\theta. \tag{4.52}$$

Combining this with Equation 4.49, we see that

$$\frac{dS}{dt} = \frac{L}{2m}. \tag{4.53}$$

Thus the conservation of angular momentum (the constancy of L) implies the constancy of the rate at which area is swept out. It follows that any given planet will sweep out equal areas in equal amounts of time. In other words, Kepler's second law is a consequence of conservation of angular momentum.

5.5 Tops, gyroscopes and precession

The phenomenon of precession was introduced in Section 1, in the context of the Earth's rotation. However, a more familiar demonstration is provided by the children's toy known as a **spinning top**. Such tops come in various forms, but basically consist of a heavy disc fixed to an axle, as shown in Figure 4.55. When the lower (pointed) end of the axle is placed on a horizontal surface and the top is set into a fast spin (usually by pinching the axle between two fingers and twisting), two remarkable things can be observed:

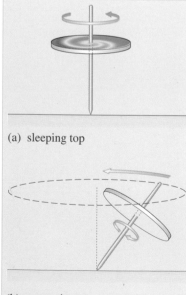

(a) sleeping top

(b) precessing top

Figure 4.55 A spinning top; (a) sleeping, (b) precessing.

1 If the axle is vertical, the top spins stably in an upright position, balancing on the point of its axle in a way that it would not do if it were not spinning (see Figure 4.55a). Such a top is said to be 'sleeping'.

2 If the axle is not vertical, the spin axis slowly rotates about the vertical (see Figure 4.55b). This motion is another example of precession; the spin axis is said to be 'precessing' about the vertical.

Both of these phenomena can be understood as consequences of angular momentum conservation and Equation 4.45 ($d\boldsymbol{L}/dt = \boldsymbol{\Gamma}$). In the first case, the top's angular momentum $\boldsymbol{L}$ is parallel to its angular velocity $\boldsymbol{\omega}$, and $\boldsymbol{L} = I\boldsymbol{\omega}$. The symmetry of the upright top ensures that there is no gravitational torque about the vertical spin axis, so if frictional forces are ignored, angular momentum will be conserved. As a result, the top is unable to tip over since that would involve a substantial change in angular momentum. The sleeping top is 'spin stabilized' against tipping over.

In the second case, the inclination of the top means that its weight will produce a torque about the point of contact, O, as indicated in Figure 4.56. This torque will be perpendicular to the spin axis and to the instantaneous angular momentum L. The torque will cause the angular momentum to change at the rate $dL/dt = \Gamma$, so that during a short time Δt the angular momentum will change by an amount

$$\Delta L = \frac{dL}{dt} \Delta t = \Gamma \Delta t. \tag{4.54}$$

Since ΔL is perpendicular to L, the effect of adding this increment to L will be to change its direction, not its magnitude, as indicated in the figure. Repeating this argument over and over, leads to the conclusion that the gravitational torque will cause the angular momentum to precess about the vertical as observed. By considering the effect of ΔL in the limit as Δt becomes very small, it is possible to work out the rate of the precession, which may be shown to be

$$\Omega = \frac{mgl}{I\omega} \tag{4.55}$$

provided ω is sufficiently large.

The effects of rotation are best demonstrated not with a child's spinning top but with a more elaborate device called a **gyroscope**. The simplest gyroscope is similar to a spinning top, except that the spinning disc is mounted on very low-friction bearings that permit it to turn freely about its centre of mass, as indicated in Figure 4.57. Conservation of angular momentum causes the axle of the spinning disc to maintain a constant direction relative to an inertial frame of reference. This constancy of orientation allows a gyroscope to be used as a navigational aid in aircraft, ships and even spacecraft. By mounting the gyroscope in a vehicle and recording movements of the gyroscope's bearings, as the vehicle changes its orientation, it is possible to determine the rotational motion of the vehicle. Naturally, such high-precision instruments are quite complicated, as the example in Figure 4.58 (overleaf) shows.

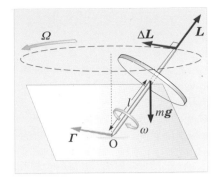

Figure 4.56 The change in the angular momentum of a top due to the gravitational torque. Note that the torque Γ at O lies in the horizontal plane and is perpendicular to the spin axis (and to the weight mg). The vector ΔL is parallel to Γ, so it is perpendicular to L and tangential to the horizontal circle.

Figure 4.57 A gyroscope — a spinning disc, mounted in such a way that it is free to rotate about three independent axes.

Figure 4.58 Gyroscopic system for use in navigation.

Finally, let us return to the case of the turning Earth, since we are now in a position to account for both the approximate constancy of its axial orientation and for the long-term departure from constancy represented by precession. The axial orientation is approximately constant for the same reason that the gyroscope's axis remains constant. The Earth has a very substantial angular momentum and it takes a good deal of work to reorient that axis. The Earth is not completely free of torques about its centre of mass, but they are sufficiently small that the reorientation is a slow process. It is that reorientation which gives rise to the 25 800-year precession. It is caused by the combined gravitational effect of the Sun and the Moon pulling on the Earth's equatorial bulge in the manner indicated in Figure 4.59.

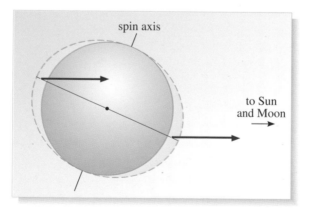

Figure 4.59 The gravitational forces on the Earth's equatorial bulges create a torque about the centre of mass. This causes precession.

The behaviour of the Earth's rotation is complicated by a number of factors including the fact that the plane of the Moon's orbit is inclined at an angle of about 5° to the plane of the Earth's orbit, and the fact that the Earth is not an ideally rigid body. However, the net effect of these complications is to make the study of the Earth's rotation even more fascinating since it can throw light on the Earth's internal structure as well as its astronomical environment and the fundamental principles of physics.

Question 4.19 Detailed observations of the Earth's rotation reveal small variations with periods of approximately 24 h and 365 days. These are attributed to changes in the Earth's moment of inertia. Suggest possible causes for variations with these periods. ■

6 Closing items

6.1 Chapter summary

1. The torque about a point O due to a force F is $\Gamma = r \times F$, where r is the displacement vector from O to the point of application of F, and the bold cross indicates that this is a vector product of magnitude $rF \sin\theta$, that points in the direction perpendicular to r and F as specified by the right-hand rule of Figure 4.15b.

2. When several different torques act about a point, the total torque about that point is given by the vector sum of the individual torques.

3. An arbitrary displacement of a rigid body may always be regarded as a combination of a translational displacement and a rotational displacement. Also, at any instant, a rigid body has a single instantaneous angular velocity, $\boldsymbol{\omega}$, and, relative to any chosen point P in the body, the instantaneous velocity of any point in the body at a displacement r from P is $v = \boldsymbol{\omega} \times r$.

4. The conditions for mechanical equilibrium are

$$\sum_i F_i = 0 \quad \text{and} \quad \sum_i \Gamma_i = 0. \tag{4.14}$$

   These conditions are necessary, but not sufficient, to ensure static equilibrium.

5. There are three types of static equilibrium: stable, unstable and neutral.

6. The moment of inertia I, about a given axis, for a system of particles is

$$I = \sum_i m_i r_i^2 \tag{4.23}$$

   where r_i is the perpendicular distance from the axis to the ith particle, and m_i is the mass of that particle. Note that the moment of inertia of a system depends on the axis about which it is determined.

7. The moment of inertia I about a given axis for a rigid body has a similar significance, though its evaluation usually involves a definite integral rather than a sum. (Some results are given in Figure 4.37.)

8. For a body rotating with angular speed ω, about a fixed axis associated with moment of inertia I, the rotational kinetic energy is given by $E_{rot} = \frac{1}{2} I\omega^2$.

9. When the centre of mass of a rigid body of mass M has speed v_{CM}, and the body rotates with angular speed ω about an axis through the centre of mass, and has a moment of inertia I_{CM} with respect to that axis then, the total kinetic energy of the body is given by

$$E_{kin} = E_{trans} + E_{rot} = \frac{1}{2} Mv_{CM}^2 + \frac{1}{2} I_{CM}\omega^2. \tag{4.27}$$

10. A constant torque Γ acting over an angular displacement $\Delta\theta$ does an amount of work $W = \Gamma \Delta\theta$ and transfers energy at a rate given by the power $P = \Gamma \cdot \boldsymbol{\omega}$.

11. The angular momentum l about a point O of a particle with linear momentum p is given by the vector product $l = r \times p$, where r is the displacement of the particle from O.

12. The total angular momentum L of a system of particles, each with angular momentum l_i, is given by the vector sum

$$L = \sum_i l_i.$$

13 For a rotating rigid body, the angular momentum L about a given point depends on the way the body's mass is distributed, and on the components of its angular velocity ω. It is not generally true that L is parallel to ω nor that $L = I\omega$, though there are cases in which these relations will be true.

14 The angular momentum L for *any* object subject to an external torque Γ satisfies the equation

$$\Gamma = \frac{dL}{dt}. \tag{4.45}$$

In situations where $L = I\omega$, and I is a constant it follows that

$$\Gamma = I\frac{d\omega}{dt}, \tag{4.28}$$

where $d\omega/dt$ is the angular acceleration. These equations can be used to predict rotational motion.

15 According to the law of conservation of angular momentum, the total angular momentum about any point remains constant for any system, as long as no net external torque acts on that system.

16 Kepler's second law states that a radial line from the Sun to a planet sweeps out equal areas in equal intervals of time. This law is a direct consequence of the conservation of angular momentum.

17 Rotating bodies, including spinning tops, gyroscopes and the Earth, may display the phenomenon of precession.

6.2 Achievements

Now that you have completed this chapter, you should be able to:

A1 Understand the meaning of all the newly defined (emboldened) terms introduced in the chapter.

A2 Solve simple problems involving torque and levers.

A3 Recognize, interpret and use vector products in a variety of contexts.

A4 Use equilibrium conditions to solve a variety of simple problems, including those that involve static equilibrium.

A5 Calculate the moment of inertia for systems of particles and sufficiently simple rigid bodies.

A6 Calculate the total kinetic energy for bodies exhibiting both translational and uni-axial rotational motion and use such calculations to solve problems concerning the motion of those bodies.

A7 Recognize, interpret and use the relation $\Gamma = dL/dt$, explain its significance in the prediction of rotational motion and both describe and exploit its relationship to the more restricted relation $\Gamma = I(d\omega/dt)$.

A8 State, use and exemplify the law of conservation of angular momentum.

A9 Describe and explain in simple terms the rotational motion of spinning tops, gyroscopes, the Earth and similar systems.

6.3 End-of-chapter questions

Question 4.20 (a) If the x-, y-, z components of a and b are given by $a = (2, 0, 0)$ and $b = (0, 3, 0)$, find $a \times (b \times a)$.

(b) If the x-, y-, z-components of a, b and c are given by $a = (2, 0, 0)$, $b = (0, 3, 0)$ and $c = (4, 0, 0)$, find $a \times b$, $a \times c$ and $b \times c$.

Question 4.21 Referring to Example 4.3, find the magnitude and direction of the force F_O provided by the pivot O. (*Hint*: Use the information about F_T found in Question 4.9.)

Question 4.22 A horizontal bridge of length L and mass M is supported at its ends. A vehicle of mass m_1 is a distance x_1 from one support and a vehicle of mass m_2 is a distance x_2 from the other support. Find expressions for the forces exerted on each end of the bridge.

Question 4.23 Consider a spherical shell (that is, a thin-walled sphere) of radius R and mass M. Explain qualitatively why its moment of inertia about a central axis is different from that of a solid sphere of the same radius and mass.

Question 4.24 A skater performing a pirouette increases the speed of his rotation by pulling in his arms and legs. Show that the kinetic energy of the skater must increase and explain where this energy comes from. ■

Chapter 5   Chaotic motion

1   Is motion always predictable?

At the beginning of this book we emphasized the power of Newton's laws as a tool for predicting motion. The success of NASA scientists in sending space probes to rendezvous with distant planets is a dramatic example of this. But is motion always so predictable?

In 1712, the watchmaker George Graham built a mechanical model of part of the Solar System. A copy of this ingenious device soon found its way into the hands of the Earl of Orrery, since when such models have been known as orreries (Figure 5.1). Orreries use gears and cogs to reproduce the motions of the planets (Figure 5.2), as prescribed by Newtonian mechanics. These can be used to predict where the planets will be observed in the sky on a particular night, and these predictions can be extended months or years ahead. Such models of the Solar System certainly appear to suggest that the motion of the planets is predictable.

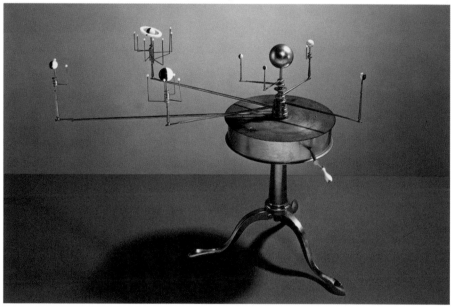

Figure 5.1   An orrery — a mechanical model of the Solar System. This one was constructed by George Adams, some time between 1789 and 1795.

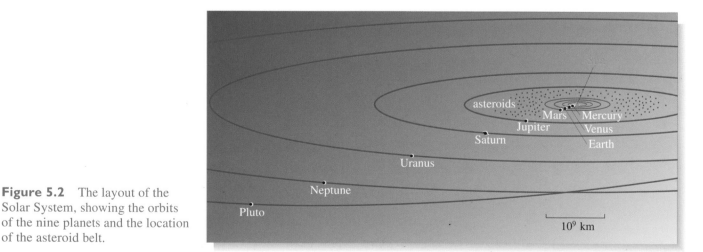

Figure 5.2   The layout of the Solar System, showing the orbits of the nine planets and the location of the asteroid belt.

The eminent eighteenth century astronomer and mathematician Pierre Simon de Laplace (1749–1827) was a great believer in the predictability of motion. As he wrote in 1776:

> The present state of nature is obviously a consequence of what it was in the preceding moment, and if we conceive of an intelligence which at a given instant comprehends all the relations of the entities of this universe, it could state the respected positions, motions and general effects of all these entities at any time in the past or future.

However, in the late 1980s, Gerald Sussman, Jack Wisdom and colleagues at the Massachusetts Institute of Technology made a *digital orrery*: a specialized computer designed to calculate the paths of the planets over extremely long periods of time. The scientists used their digital orrery to predict the position of Pluto, the planet farthest from the Sun, at various times in the future. They then repeated their calculations using a slightly different value for the current position of Pluto, a value that differed from the value used in the first calculation by less than the precision with which Pluto's position can be determined from astronomical observations.

What they found was that the predictions for Pluto's future position obtained from the second calculation differed from those obtained from the first calculation by an amount that initially increased *exponentially* with time. One consequence of this was that the two positions predicted for about 100 million years from now were on opposite sides of the Sun! Their work showed that though it is possible to make useful predictions for Pluto's position over time scales of years or centuries, predictions made for hundreds of millions of years from now are meaningless.

Exponential growth, was introduced in Chapter 2.

Subsequently, Jacques Laskar, of the Bureau des Longitudes in Paris (where Laplace once worked), investigated how a small change in the assumed position of the Earth in the Solar System affected its predicted path. He discovered that a small change in the initial position would be magnified by a factor of order 10 every 10 million years, which again indicates that the uncertainty in computed position grows *exponentially* with time.

Consequently, even if we knew the Earth's position to within $10\,\text{m}$, then, over the next 100 million years this uncertainty would grow by a factor of 10^{10} to become $10^{11}\,\text{m}$ which is roughly equal to the distance the Earth travels in a month.

So, even if the current position of the Earth is uncertain by as little as $10\,\text{m}$, we cannot reliably predict where the Earth will be in its orbit 100 million years from now. Tiny differences in initial conditions eventually produce huge differences in the subsequent position, to such an extent that the word 'predictable' becomes quite inappropriate.

This unpredictability in the motion of the planets is an example of *chaos*, or chaotic motion. The chaos here is not due to any failure of Newton's laws of motion and gravitation, nor is it due to any limitations in the precision with which computers can perform calculations. The chaotic behaviour is inherent in the equations of celestial mechanics that predict the motion of the planets. Not all systems are chaotic in this way, but similar behaviour has been observed in fields as diverse as physics, chemistry, biology, ecology, meteorology, technology, finance and social affairs!

The subject of this chapter is *deterministic chaos* (to be discussed in Section 2.6), which concerns systems that are governed by deterministic equations, yet are *not* predictable. A **deterministic system** is one for which there are well-defined rules, or equations, governing changes at all times. If these rules are known, they can be used to determine the state of the system at a later time, from information about the state of the system at an earlier time. Broadly speaking such a system is, in principle, predictable. However, if a deterministic system has the additional property of being **chaotic**, then the way in which it evolves with time will be so sensitive to the initial conditions that eventually the expected predictability will, in practice, break down. It will take a little time to explain how a system can be both deterministic *and* chaotic, but we can start by identifying two considerations that are *not* relevant to this chapter.

Here we are *not* concerned with so-called *quantum* indeterminacy. The idea that there is an element of chance built into the basic laws of nature at the atomic level was introduced in *The restless Universe* and it will be developed further in *Quantum physics: an introduction*. However, it is irrelevant here, so please set it aside for the present; we are concerned with *classical* deterministic systems in this chapter.

Nor are we concerned here with *statistical* uncertainty. When we are dealing with huge numbers of molecules in a macroscopic volume of air, for example, the position and velocity of an individual molecule very soon become unpredictable. Each molecule undergoes many collisions with other molecules, and statistical methods are needed to describe the distributions of the positions and velocities of the vast number of molecules. However, our study of chaos will deal with systems that can be described by just a *few* numbers.

Among the questions that we shall address are these: Might the flutter of a butterfly's wings in America cause changes in the world's weather that result in flooding in Bangladesh? In what ways are beating hearts like dripping taps? Could chaos in the Solar System lead to the extinction of life on Earth? How might chaos save a galaxy from extinction?

2 What is chaos?

We begin this study of chaos by looking at some deterministic systems that are much simpler than those encountered in celestial mechanics. The first of these is so simple that it is not chaotic at all, but we shall quickly move on to one that is. The advantage of these simple systems is that they allow us to study chaos while avoiding the need to solve differential equations. Their disadvantage is that they can seem rather remote from the real physical world. However, bear with us for the moment; once the basic ideas have been introduced in this simple context, we shall examine a range of real-world applications.

2.1 A linear map

The first simplification is to study changes that occur in discrete steps, rather than attempt to follow the smooth changes described by differential equations. The second simplification is to consider a system whose state is described by a single value at each step in its development.

Here is an example. Consider the bounces of a ball, focusing on the maximum height attained between bounces. Suppose we release the ball from rest at height x_0. Let x_1 be the maximum height after the first bounce, x_2 after the second bounce, and so on.

Now assume that the rule for determining the height of a bounce is that each height is an unchanging fraction k of the previous height. In symbols, we write this as $x_1 = kx_0$, $x_2 = kx_1$, $x_3 = kx_2$, etc. where $k < 1$. We can summarize this by writing

$$x_{n+1} = kx_n. \tag{5.1}$$

This is a deterministic rule that allows us to relate the height x_n of the nth bounce to the height x_0 at which the ball was released.

We know, from the above, that $x_2 = kx_1$, but we also know that $x_1 = kx_0$. Hence we can write

$$x_2 = k(kx_0), \quad \text{i.e.} \quad x_2 = k^2x_0.$$

So, in general

$$x_n = k^n x_0. \tag{5.2}$$

For example, with $k = 0.800$ we would predict $x_3 = (0.800)^3 x_0 = 0.512 x_0$.

Mathematicians refer to an equation like Equation 5.1 as a *map*. In general, a map is a well-defined rule for turning one set of values into another. Here we have a single value at each step, so the map states how the next value, x_{n+1}, depends on its predecessor, x_n. This particular map is known as a linear map, because Equation 5.1 is a linear equation: a graph of x_{n+1} against x_n is a straight line (Figure 5.3).

> A **linear map** is a linear rule for turning one set of values into another.

A map provides the rule that underlies a function. Sometimes the words 'map' and 'function' are used interchangeably.

The process of repeatedly applying a given rule, using the output of one application as the input of another is known as **iteration**. Thus, the system we are considering may be described as an *iterated linear map*. The results of successive applications of the given rule, $x_1, x_2, x_3, \ldots x_{n-1}, x_n$, are said to be the *iterates* of the initial value x_0, under the given rule.

When $k < 1$, the value of x_n given by Equation 5.2 decreases rapidly as n increases, in the same manner as the exponential function (Figure 5.4a). However, the linear map in Equation 5.2 leads to *no* chaos. If we release the ball at a slightly different initial height, the maximum heights reached between subsequent bounces will follow a similar exponential curve (Figure 5.4b). The *difference* between the predicted heights of the nth bounce in the two cases does *not* increase exponentially with n, and there is no great sensitivity to changes in initial condition. This is *not* a chaotic system.

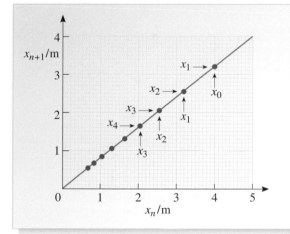

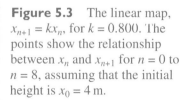

Figure 5.3 The linear map, $x_{n+1} = kx_n$, for $k = 0.800$. The points show the relationship between x_n and x_{n+1} for $n = 0$ to $n = 8$, assuming that the initial height is $x_0 = 4$ m.

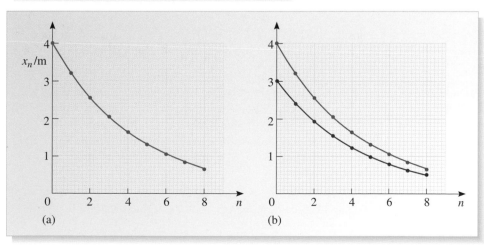

(a)

(b)

Figure 5.4 (a) The exponential dependence of the height x_n of the nth bounce of a ball. These heights are determined using the linear map shown in Figure 5.3, with $k = 0.8$ and $x_0 = 4$ m. (b) Comparison of the bounce heights when the ball is released from two different heights. Note that the difference between the curves does *not* increase as n increases.

Before going on to consider a system that is chaotic, it's worth recalling the meaning of the term 'parameter', since it will be important in what follows.

> A **parameter** is a quantity held constant in the case under consideration, but allowed to vary when one wishes to define a new case.

This precisely describes the role of k in the situation we have been considering. A particular value of k, 0.800 say, might represent a particular kind of ball bouncing on a particular kind of surface. Modelling the behaviour of a different kind of ball, or a different kind of surface, might require a different value of k. Thus, each value of k represents a particular case within a broad class of systems. In a similar way, the length of a pendulum, or the acceleration due to gravity, are considered as parameters of pendulum motion: neither quantity varies for a particular pendulum in a particular location, but you might well be interested in considering the behaviour of different pendulums, in different locations.

2.2 The logistic map

In the 1970s, ecologists studied populations of biological species, such as fish and insects, using a modification of the linear map in Equation 5.1. The map that they studied takes account of a limit to growth, and is specified by the equation

$$p_{n+1} = kp_n \left(\frac{p_{max} - p_n}{p_{max}} \right) \tag{5.3}$$

where p_n is the population in year n, and p_{max} is some maximum population, above which growth is supposed to be impossible. The factor in brackets in Equation 5.3 ensures a limit to growth. This factor is unimportant when the population p_n is small compared with p_{max}, because then it is approximately equal to one. However, when p_n is close to p_{max}, the factor in brackets is much less than one, and so it prevents p from increasing.

We can simplify this map by working with the fraction $x_n = p_n/p_{max}$, so that $x_{n+1} = p_{n+1}/p_{max}$. Dividing each side of Equation 5.3 by p_{max}, we then get

$$x_{n+1} = kx_n(1 - x_n) \tag{5.4}$$

This equation is called the **logistic map**.

The logistic map is slightly more complicated than the linear map, but it can be iterated in the same way to provide a sequence of values, $x_1, x_2, x_3, \ldots x_n$, from a given initial value x_0. For example, if $k = 1.8$ and $x_0 = 0.5$, then, from Equation 5.4 (with $n = 0$)

$$x_1 = 1.8 \times 0.5(1 - 0.5) = 0.9 \times 0.5 = 0.45$$

and applying Equation 5.4 again (with $n = 1$)

$$x_2 = 1.8 \times 0.45(1 - 0.45) = 0.81 \times 0.55 = 0.4455.$$

The sequence generated by an iterated logistic map can behave in a number of ways, depending on the value of the parameter k. For every value of k the system is deterministic, but for some values it is also chaotic while for others it is not. We will investigate some of the possibilities, starting with values of k in the range between 1 and 3.

Simple and predictable behaviour

Let's first study a case where everything is predictable, namely the logistic map with $k = 2$. Suppose we have an initial value $x_0 = 0.4$. The next value in the sequence is

$$x_1 = kx_0(1 - x_0) = 2 \times 0.4(1 - 0.4) = 0.8 \times 0.6 = 0.48.$$

- Use Equation 5.4 to calculate the values of x_2, x_3 and x_4.

○ The values that I obtained with my calculator were $x_2 = 0.4992$, $x_3 = 0.4999987$ and $x_4 = 0.5$.

(Note that this last value should be 0.500 0000 to seven significant figures, but my calculator doesn't display significant zeros on the end of a decimal number.) ■

The values appear to settle down to a final value $x_{final} = 0.5$. To understand why the final value is 0.5, we can substitute $k = 2$ into Equation 5.4 and also divide both sides by x_n. The equation then becomes

$$\frac{x_{n+1}}{x_n} = 2(1 - x_n). \tag{5.5}$$

If the sequence settles down, then successive values will be the same, which means that $x_{n+1} = x_n = x_{final}$, so $x_{n+1}/x_n = 1$ for sufficiently large values of n. Substituting $x_{n+1}/x_n = 1$ and $x_n = x_{final}$ into Equation 5.5, we get $1 = 2(1 - x_{final})$, and the solution of this is $x_{final} = 0.5$.

Note that this predicted final value does *not* depend on the starting fraction x_0 — the equation that we solved to obtain x_{final} did not contain x_0. So, whatever value we start with between 0 and 1, the logistic equation with $k = 2$ will always lead us to a final value of 0.5. (If you doubt this assertion, then choose any value of x_0 in the range $0 < x_0 < 1$ and use Equation 5.5 to calculate a sequence of values.) The steady convergence of the values to 0.5 is completely insensitive to the initial condition, and the long-term behaviour of the system is entirely predictable.

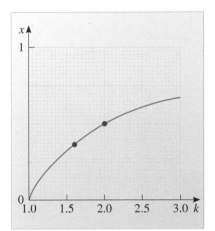

Question 5.1 Study the logistic map in Equation 5.4 for a parameter $k = 1.6$ and a starting value $x_0 = 0.6$.

(a) If the sequence settles down, what will be the value of x_{final}?

(b) Calculate the values of x_n up to $n = 5$, keeping at least five decimal places.

(c) Does the sequence appear to settle down?

(d) Would you describe this behaviour as chaotic? ■

Figure 5.5 This graph, known as a Feigenbaum plot, displays the final x-values for the logistic map on the vertical axis and the parameter k on the horizontal axis. The equation of the curve is $x_{final} = 1 - (1/k)$, and the two points that we have calculated, for k values of 1.6 and 2, are indicated.

It turns out (as you can check for yourself with a calculator) that all logistic maps that are described by Equation 5.4, and which have values of k in the range $1 < k < 3$, behave in a similar way. In all of these cases, the sequence quickly settles down to a final value that is given by $x_{final} = 1 - (1/k)$. These final values can be displayed on a rather special diagram (Figure 5.5), of a type studied by the physicist Mitchell J. Feigenbaum (Figure 5.6), working at Los Alamos, New Mexico, USA in the 1970s.

To find behaviour that is more interesting, we have to investigate the Feigenbaum plot for values of k that are greater than 3.

Limit cycles

You might have expected that the logistic map would predict that the smooth curve in Figure 5.5 continued above $k = 3$. If the sequence settles down to some value x_{final}, then we must have $1 = k(1 - x_{final})$, and hence $x_{final} = 1 - (1/k)$.

Let's try this with $k = 10/3$. Does the sequence in fact settle to the value $1 - (3/10) = 0.7$? Here are the values x_0 to x_8, with a starting value $x_0 = 0.4$:

0.400 00   0.800 00   0.533 33   0.829 63   0.471 15   0.830 56   0.469 10
0.830 15   0.470 00

Figure 5.6 Mitchell J. Feigenbaum.

There is no indication that the values are settling down to 0.7. Here's another sequence of values generated with the same constant, $k = 10/3$, but this time with a starting value of 0.5:

0.500 00   0.833 33   0.462 96   0.828 76   0.473 05   0.830 91   0.468 32
0.829 99   0.470 36

Again there is no sign of a steady value.

Now compare these two sequences.

● Can you see any similarity between the values?

○ It seems that in each case the sequence is settling down to an *oscillation* between two values, one with $x \approx 0.470$, the other with $x \approx 0.830$. ■

These sequences are not converging to a constant value in the manner that we observed with $k = 2$, but they seem fairly regular. As a check, here is how the second sequence continues:

0.470 36   0.830 40   0.469 44   0.830 22   0.469 85   0.830 30   0.469 67
0.830 27   0.469 75

We are observing what is called a *limit cycle*.

In a **limit cycle**, the sequence of values produced by an iterated map settles down to an oscillation that repeats itself after a certain number of steps.

The number of steps in the cycle is referred to as the *period*. Note that the use of the term period here is similar to its use to describe oscillations as a function of time. For a mass oscillating on a spring, the period is the time that elapses before the motion repeats itself. In the limit cycle, the period is the number of steps in the iteration that elapse before the values repeat themselves. For the case $k = 10/3$, there are two values in the limit cycle, and so the cycle has a period of 2. In fact, for all values of k between 3 and 3.4495 we obtain a limit cycle with period 2, and this type of behaviour is referred to as a *2-cycle*.

We can extend the Feigenbaum plot that we introduced in Figure 5.5 so that it includes the limit cycles with period 2, and this is done in Figure 5.7. The difference between the two values generated by the 2-cycle is very small when k is just over 3, but the difference increases steadily as k is increased.

The appearance of 2-cycles is the first indication that there is something quite different about the behaviour of systems governed by logistic maps from those governed by linear ones, and you will shortly meet even more dramatic differences. The most important observation to make here, and to carry forward to the subsequent physics, is that the reason for the different behaviour is that the logistic map is an example of a *non-linear map*. If you expand the right-hand side of Equation 5.4 you obtain $kx_n(1 - x_n) = kx_n - kx_n^2$, which is not a linear function of x_n. For the linear map (Equation 5.1) we can easily obtain a solution, such as Equation 5.2. However, for the non-linear logistic map, there is no simple way of obtaining a solution; we have to compute it step by step. This explains why the ready availability of computing power in recent decades has transformed scientists' understanding of the non-linear systems that are described by non-linear maps, or non-linear equations of motion. (An equation of motion is linear if the acceleration depends linearly on the displacement and linearly on the velocity; otherwise it is non-linear.)

A **linear system** is described by a linear map, or a linear equation of motion, whereas a **non-linear system** is described by a non-linear map, or a non-linear equation of motion.

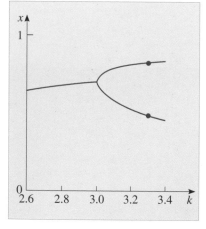

Figure 5.7   A Feigenbaum plot for the logistic map. Note that the horizontal scale here is different from that in Figure 5.5. When k has a value in the range $3 < k < 3.4495$, the final x-value oscillates between two values. The two x-values for $k = 10/3$, which we calculated in this section, are marked on the diagram. This is a 2-cycle.

Question 5.2 (a) Is the equation of simple harmonic motion linear or non-linear? (b) Consider a pendulum released from rest at an angle of 30° to the downwards vertical. Is the equation of motion linear in this case? ■

Period doubling

The logistic map provides even more interesting results for larger values of k. Above $k = 1 + \sqrt{6} = 3.4495$, something new happens. Here is a sequence of 16 results for the logistic map with $k = 3.5$:

0.400 00 0.840 00 0.470 40 0.871 93 0.390 83 0.833 29 0.486 22
0.874 34 0.384 56 0.828 35 0.497 64 0.874 98 0.382 86 0.826 98
0.500 80 0.875 00

Question 5.3 (a) Are these data settling down to a limit cycle, and, if so, what is the period of the cycle? (b) Calculate the next value in the sequence, and reassess your answer to part (a). ■

The period of the limit cycle doubles when k increases beyond 3.4495 — it changes from a cycle with period 2 to a cycle with period 4. This is shown in the Feigenbaum plot in Figure 5.8. Then at $k = 3.5441$, an 8-cycle takes over, and at 3.5644 a 16-cycle appears.

The doubling of the period of the limit cycle when the parameter k is increased is known as **period doubling**.

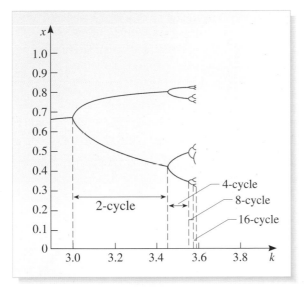

Figure 5.8 A Feigenbaum plot that shows period doubling. Note that as the period of the limit cycle increases, the range of k-values over which that period occurs gets progressively smaller. For clarity, we haven't extended the dashed lines to the top of all the branches.

Figure 5.8 now indicates the long-term behaviour of a variety of non-linear dynamical systems, each corresponding to a different value of the parameter k. The systems tend towards a single steady value when $1 < k < 3$, changing to a 2-cycle at $k = 3$, to a 4-cycle at $k = 3.4495$, to an 8-cycle at $k = 3.5441$, and to a 16-cycle at 3.5644.

Slide into chaos

It is clear from Figure 5.8 that limit cycles of progressively higher periods occur in narrower ranges of k. The parameter range for 4-cycles is quite small: $\Delta k_4 = 3.5441 - 3.4495 = 0.0946$. The range for 8-cycles is $\Delta k_8 = 3.5644 - 3.5441 = 0.0203$, which is smaller, by a factor 0.0946/0.0203 = 4.66. The range for 16-cycles is yet smaller, by approximately the same factor. Feigenbaum discovered that this progressive reduction in the range continues, and that the ranges for limit cycles of increasing periods get exponentially smaller. In the limit of large periods, each limit cycle occupies a range of k-values that is smaller by a factor of about 4.6692 than for the preceding limit cycle. The precise value of this factor is known as the *Feigenbaum constant*.

> In the limit where the periods of the limit cycles are large, the intervals between period doubling get exponentially smaller as k increases, each interval being smaller than the previous interval by a factor known as the **Feigenbaum constant** δ:
>
> $\delta = 4.669\,201\,609\,102\,990\,671\,853\,203\,820\,466\,201\,617\,258\,185\,577\,475\,768\,632\,745\,651\ldots$

Of course, when the period of the limit cycle is long, the cycle can be quite hard to identify. Look at the following 16 successive iterations of a logistic map:

0 550 21   0.883 01   0.368 60   0.830 39   0.502 52   0.891 98   0.343 79
0.804 94   0.560 23   0.879 06   0.379 33   0.840 05   0.479 42   0.890 49
0.347 95   0.809 51

It is not clear what is happening here. Is the system in the process of settling into a 4-cycle or into an 8-cycle? In fact these data were obtained after 1000 iterations of the logistic map with $k = 3.5680$. They are firmly locked into a 16-cycle, as was confirmed by computing the next 16 iterations, which agree with those above, to five decimal places.

There is, however, another snag to observing limit cycles with long periods. The Feigenbaum constant is rather large, which means that the range of k over which the 16-cycle occurs is 4.6692 times smaller than the range that we quoted for the 8-cycle above, $\Delta k_8 = 0.0203$. So $\Delta k_{16} = 0.0043$, $\Delta k_{32} = 0.0009$, and $\Delta k_{64} = 0.0002$. This explains why only the first few limit cycles are visible in the Feigenbaum plot in Figure 5.8.

Period doubling continues until the value $k = 3.5700$ is reached, at which no limit cycle is possible. For $k > 3.5700$ the system enters the chaotic regime. In this regime there are values of k that do not correspond to any limit cycle at all. For these values of k, whatever the starting value x_0, none of the values in the sequence is repeated.

Figure 5.9 shows the continuation of Figure 5.8 into the chaotic regime. Since the results obtained in this regime generally depend sensitively on the starting value, x_0, what has been plotted, at each value of k above 3.5700, is the average of many

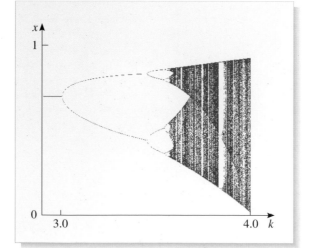

Figure 5.9 Continuation of the Feigenbaum plot into the chaotic regime, $k > 3.5700$.

results obtained with different initial values x_0. The density of dots therefore represents the probability with which x-values occur in iterated logistic maps, at given values of the parameter k.

Beyond $k = 4$ the distribution becomes uniform, but between $k = 3.5700$ and $k = 4$ there is still some structure in the Feigenbaum plot. Indeed, for some ranges of k (e.g. close to $k = 3.84$) the outcome is quite orderly. There are 'windows' of cyclic behaviour amidst the chaos.

Note that limit cycles become harder and harder to spot as their period doubles with smaller and smaller increments of k. There is no blinding flash of light as we pass $k = 3.5700$. Just below that value, we would have to wait a long time to spot a limit cycle with a long period such as $2^{20} = 1048\,576$. Just above $k - 3.5700$, we cannot find a limit cycle, however long we study the iterations. This is the way that limit cycles end and chaos takes over: not with a bang, but a whimper.

Note for Open University students: For interest we have supplied you with a simple multimedia package designed to calculate and draw Feigenbaum plots.

2.3 Some features of chaos

There is an important general point about chaos that can be seen in Figure 5.9. Even in the chaotic regime, the values of x are not spread over the whole range from 0 to 1. For any value of k, the values are all within a more restricted range. So, even if we were to start with an initial value x_0 that was close to 0, or close to 1, the sequence would rapidly move into the restricted range. There appears to be something attractive about this restricted range. We shall return to this phenomenon in Sections 2.4 and 2.5 where you will see that such attractive regions are an important characteristic of chaos. Meanwhile, let's examine some other features of chaos.

Exponential sensitivity to input

Outside the chaotic regime, for $k < 3.5700$, the eventual behaviour of the iterated logistic map is insensitive to the initial value x_0. Even in the period-doubling regime ($k = 3.0000$ to $k = 3.5700$), sequences with very different starting values get locked into the same limit cycle and converge to the same value of x.

In the chaotic regime ($k > 3.5700$), on the other hand, there is generally nowhere special to go. A wide range of x-values is possible, and the system wanders aimlessly from one value of x to another. More significantly, sequences with very similar starting values eventually lose all relation to each other. An example will help to show this (Table 5.1).

Table 5.1 shows the development of three logistic sequences, all with $k = 4$. All three start with values of x_0 that are rather close. At $n = 0$, the first two sequences differ by 0.000 10; at $n = 8$ they differ by 0.025 23, so the difference has increased by a factor of about 250. Comparing the second and third sequences, we see that the initial

Table 5.1 Three logistic sequences, all with $k = 4$, but with different starting values.

x_0	x_1	x_2	x_3	x_4	x_5	x_6	x_7	x_8
0.399 90	0.959 92	0.153 89	0.520 84	0.998 26	0.006 94	0.027 56	0.107 22	0.382 90
0.400 00	0.960 00	0.153 60	0.520 03	0.998 40	0.006 41	0.025 47	0.099 27	0.357 67
0.400 01	0.960 01	0.153 57	0.519 95	0.998 41	0.006 36	0.025 26	0.098 49	0.355 17

Table 5.2 Continuation of the three sequences of Table 5.1.

x_8	x_9	x_{10}	x_{11}	x_{12}	x_{13}	x_{14}	x_{15}	x_{16}
0.382 90	0.945 15	0.207 36	0.657 45	0.900 83	0.357 33	0.918 58	0.299 17	0.838 67
0.357 67	0.918 97	0.297 86	0.836 56	0.546 92	0.991 20	0.034 92	0.134 76	0.466 40
0.355 17	0.916 09	0.307 46	0.851 72	0.505 18	0.999 89	0.000 43	0.001 72	0.006 86

difference of 0.000 01 is likewise magnified by a factor of 250 to 0.002 50 at $n = 8$. This magnification of the difference is not due to errors in the calculations: the three sequences were computed to 100 decimals places, and were rounded to five decimal places to display them in the table.

In Table 5.2, the sequences are continued, again using 100-digit accuracy in the calculations. After $n = 10$, there is no hint that the first two sequences started with very similar values of x_0 and had similar histories until x_8; after $n = 13$, the second and third sequences have lost all trace of their even closer origins.

This — and neither more nor less than this — is what is meant by deterministic chaos. The system is completely deterministic: the rule for each step is clear and is never broken. Yet we cannot predict the eventual outcome without carrying out the detailed calculations, even a tiny difference in the initial value will eventually lead to completely different behaviour. However small the change that we make to the starting point of a logistic sequence, for this and many other values of k, the difference will *grow exponentially* with the number of steps in the sequence. (Refer back to Section 5.4 in Chapter 2 if you want to remind yourself what is meant by exponential growth.) The sensitivity to initial conditions makes long-term prediction a practical impossibility for chaotic systems, even though they are deterministic.

Limits to predictability

Laplace, as we saw in Section 1, proclaimed the predictability of the Universe, but another French scientist, Henri Poincaré (Figure 5.10), may be counted among the parents of chaos theory, though it took 70 years for the 'child' to grow up. He took seriously the possibility that small initial variations may produce much larger ones later on. In 1903, he remarked, that 'it may happen that small differences in initial conditions produce very great ones in the final phenomena'. The behaviour of the logistic map — one of the simplest non-linear models of change — bears out his suspicions.

On a modern computer, it is a simple matter to work with initial data specified to 1000 significant figures, and to compute the logistic map to this ferocious accuracy. After 3000 iterations with $k = 4$, an unimaginably small change of 10^{-1000} in the input, produces a change of order 10^{-100} in the output. The initial deviation has been enormously magnified — by 900 orders of magnitude — yet is still far below the level of practical detection. However, after 3300 iterations predictability is lost.

The number of iterations of a logistic map that occur before a change in the initial value leads to significant differences in the predicted sequence is known as the **horizon of predictability**.

The horizon of predictability will depend on the magnitude of the change that is made to the initial value x_0. With a change of 10^{-1000} in the initial value, the horizon of predictability for the $k = 4$ logistic map is 3300 iterations. With a change of 10^{-10} in

Figure 5.10 Jules Henri Poincaré (1854–1912) is regarded as one of the last universalists in both mathematics and physics. His work encompassed, extended, and in places still challenges, large areas of both subjects, notably in geometry, topology and their relation to physics. Among Einstein's predecessors, he came closest to an appreciation of relativity. His popular essays on science combine depth with accessibility.

the initial value, the horizon is much closer, with predictability lost beyond $n = 33$. In the real world, where we might have data to only three-digit precision, the horizon is closer still, at around $n = 10$.

An armchair philosopher might say: 'Oh, what I really mean by 'predictable' is that I can predict the future if you give me *totally* accurate data, and an *infinitely* powerful computer, and a *perfectly* faithful model.' Practical scientists, however, need to know whether a particular system exhibits exponential sensitivity to the initial conditions. If a system is chaotic, all is not lost; but the horizon of predictability imposes a practical limit to what can be predicted from initial data of limited accuracy.

Quadratic maps and universality

For $k = 4$, the logistic map is $x_{n+1} = 4x_n(1 - x_n)$, which may be rearranged to give

$$x_n^2 - x_n + \tfrac{1}{4} x_{n+1} = 0 \cdot$$

This is a quadratic equation for x_n. Solving it by means of the formula introduced in *Describing motion*, shows that $x_n = \tfrac{1}{2}\left(1 \pm \sqrt{1 - x_{n+1}}\right)$. So, whatever its value, x_{n+1} has *two* possible predecessors: $\tfrac{1}{2}\left(1 + \sqrt{1 - x_{n+1}}\right)$ and $\tfrac{1}{2}\left(1 - \sqrt{1 - x_{n+1}}\right)$. This is the defining characteristic of a whole class of maps, of which the logistic map is just one member. The members of this wider class are referred to as *quadratic maps*.

A map in which each value has two possible predecessors is called a **quadratic map**.

For appropriate choices of parameter value, all iterated quadratic maps exhibit chaos. But, even more remarkably:

every quadratic map approaches chaos in the *same* manner.

They all show the period-doubling slide into chaos, with intervals between period doubling getting exponentially smaller as k increases. Moreover, the intervals between successive period doublings are reduced by a factor of $\delta = 4.6692$, the Feigenbaum constant, just as they are in the logistic map.

The fact that period doubling and the slide into chaos occur in the same way for all quadratic maps tells us that we have been studying a phenomenon of great generality. The logistic map provided a route to its discovery, but what has been discovered is a general feature of quadratic maps. The fact that the Feigenbaum constant δ describes the onset of chaos for *all* quadratic maps emphasizes this generality and indicates the importance of the Feigenbaum constant. (It is as if we had discovered the value of π by studying the perimeter of a circle, and then found that the same number occurs in formulae for the area of a circle, the surface area and volume of a sphere, and a host of other results.) The finding that certain properties of iterated non-linear maps are independent of the details of the map is described by the term *universality*.

Universality refers to the observation that different systems (e.g. different iterated quadratic maps) may exhibit common behavioural features (e.g. all approaching chaos by a similar period-doubling route involving the Feigenbaum constant).

There are many examples of universality associated with chaos. Cubic maps, for example, where x_{n+1} may have three distinct predecessors, x_n, also exhibit chaos. The onset of chaos is described by a different constant in this case, but this new constant is common to all cubic maps.

Question 5.4 (a) Select a data set from Sections 2.2 or 2.3 that might have pleased Laplace.

(b) Select a data set that confirms Poincaré's misgivings.

Question 5.5 All of the following statements are incorrect. For each statement, identify what is wrong, and suggest how it should be corrected.

(a) All dynamical systems are chaotic since one can never make totally precise predictions about their behaviour.

(b) In iterations of the logistic map, the value of x moves along the curves of the Feigenbaum plot of Figure 5.8.

(c) In iterations of the logistic map in the chaotic region, the value of x jumps between curves of the Feigenbaum plot of Figure 5.8.

(d) Data from a few successive iterations of a logistic map cannot be used to determine whether the parameter k has a value of 3.5699, or is in a range where the limit cycle is very long, or has a value 3.5701 that corresponds to chaotic behaviour. ∎

2.4 Fractals

Each iteration of the linear map or the logistic map produces a single value x_n. Such systems may therefore be said to be 'one-dimensional'. However, it is quite possible to devise systems of iterated maps that are 'two-dimensional' in the sense that two values, x_n and y_n, are required to specify the state of the system at step n of the iteration. Some of the most beautiful expressions of chaos — like those in Figure 5.11 — have emerged from considering families of such maps that also require two parameters, p and q say, to specify a particular member of the family. To generate such images the parameters p and q are used to label a pair of Cartesian axes; each point in the plane then corresponds to particular values of p and q and hence to a particular two-dimensional map. One then studies some quantitative feature (call it f) of the iterations of that map, such as how long the state (x_n, y_n) remains within a specified region. If the value of f is represented by a colour, one can then plot a pixel of the appropriate colour at the relevant point in the pq-plane. By repeating the procedure for all allowed values of p and q, a two-dimensional pattern is built up. Thus each coloured pixel represents a property of a different map (characterized by specific values of p and q) belonging to the same family.

As you may deduce from this brief description, this is quite a complicated game to play, and can use up a lot of computer time. However, the results have entranced both mathematicians and artists. The coloured patterns that are produced are beautiful, as you can see from the example of Figure 5.11. Of great interest to mathematicians is the fact that such patterns are often *fractal*. This means that if you take a small region of the pattern and expand this region to the size of the original pattern, then the expanded region and the original region will look qualitatively very similar. They will both show comparable structure, even though the original scales were very different. The whole pattern is said to exhibit *self-similarity*. For example, Figure 5.11b shows an expanded version of a small region in Figure 5.11a.

Fractal patterns can be seen throughout nature — they have been called 'the real geometry of nature'.

A **fractal** pattern is a pattern for which enlarging a small part produces a pattern that is similar to the original pattern. Fractal patterns are self-similar.

What you can see in a fractal is, so to speak, a world within a world. And there is no end to this fractal structure; we can zoom in again and again (given enough computer power) and at each stage reveal detail that was not visible in the previous picture. Most people who have done this on a computer react with awe, on seeing that a simple map generates such wonderful, unending, self-similar structure.

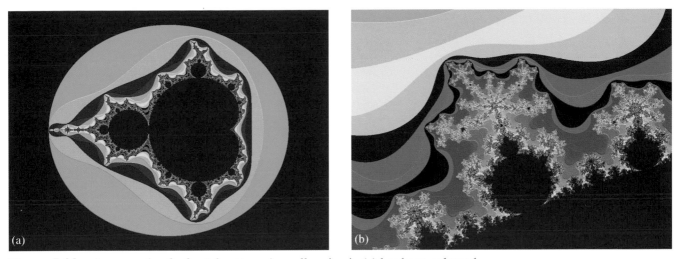

Figure 5.11 An example of a fractal pattern. A small region in (a) has been enlarged to produce the pattern in (b). The patterns exhibit self-similarity.

How does this relate to the one-dimensional logistic map? Well, the Feigenbaum plot for the logistic map in Figure 5.9 also has fractal structure. A small region of this plot with $0.9471 < x < 0.9709$ and $3.8196 < k < 3.8596$ is replotted in Figure 5.12. It shows period doubling, similar to that in the region of the parent plot between $k = 3$ and $k = 3.5700$ in Figure 5.9. Moreover, if we were to enlarge the area indicated by the arrow in Figure 5.12, then we would reveal another region of period doubling, very similar to what is shown in the whole of Figure 5.9.

Note for Open University students: For interest we have supplied you with a simple multimedia package designed to draw and enlarge the fractal pattern of Figure 5.11.

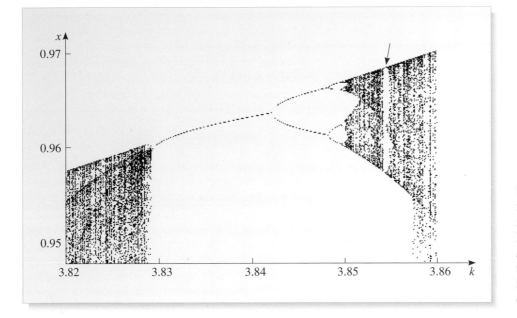

Figure 5.12 An expanded region from the Feigenbaum plot in Figure 5.9 demonstrates fractal structure. Note the ranges of x-values and k-values here are much smaller than in Figure 5.9. The area indicated with an arrow also exhibits period doubling.

An understanding of fractals can be of help in data compression, when it is necessary to capture the essential features of a picture, or pattern, without recording every last detail. If there is some degree of structure — as for example in a picture of a forest with many trees with many branches with many leaves — then the data can be compressed into a form that reflects the approximately fractal structure of nature itself. Thus fractals — which first emerged from practical concerns — now serve other practical concerns, as well as delighting the eye and brain.

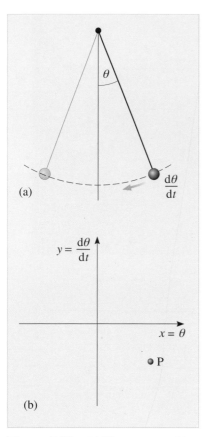

(a)

(b)

Figure 5.13 (a) The two quantities that specify the state of a simple pendulum are the angular displacement θ, and its rate of change $\mathrm{d}\theta/\mathrm{d}t$. (b) Any point plotted on a graph of $\mathrm{d}\theta/\mathrm{d}t$ against θ represents a unique state of the pendulum: the graph shows the state space of the pendulum. For point P, the angular displacement is positive, and its rate of change is negative, so P could represent the state shown in part (a).

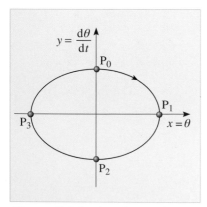

Figure 5.14 A trajectory in state space that represents simple harmonic motion of a pendulum.

2.5 State spaces and trajectories

You have seen some important features of chaos in systems described by non-linear maps, and in particular by the logistic map. In these cases, we were concerned with *discrete changes* between successive iterations. We now turn to the more realistic case of non-linear systems that *change smoothly with time*, and hence are described by differential equations. To start with, we shall introduce some of the necessary ideas by referring to a simple pendulum, which is a linear system for small oscillations and therefore *not* chaotic.

The first thing to establish about a system is the number of quantities that specify its state at any given time. For a simple pendulum, we need *two* quantities: the angular displacement θ, and its rate of change $\mathrm{d}\theta/\mathrm{d}t$ (Figure 5.13a). (These are the data that must be supplied at any time in order that the differential equation that describes the motion of the pendulum can be solved to predict the state at some subsequent time.) This means that each state of the pendulum can be represented by a point on a graph that has axes $x = \theta$ and $y = \mathrm{d}\theta/\mathrm{d}t$, as shown in Figure 5.13b. We acknowledge this by saying that the pendulum has a two-dimensional *state space*.

> A **state space** is a way of representing the state of a system. If m quantities are required to completely specify the state, then the state space is m-dimensional, and any state of the system is represented by a point in that m-dimensional space.

The periodic motion of a pendulum can be represented by a closed path in state space, as shown in Figure 5.14. Let us assume that at time $t = 0$ the state of the system is represented by the point P_0 in state space. This point corresponds to zero displacement and maximum angular velocity. After a quarter of a period, the state is represented by P_1, which corresponds to maximum displacement and zero angular velocity. After another quarter period, it reaches P_2, indicating that $\theta = 0$ again. At P_3, the pendulum has swung back as far as possible, and has zero angular velocity again. The state then returns to P_0, to complete a full cycle. This path traced out in state space provides a complete description of the motion of the pendulum, and is referred to as a *trajectory*. For a pendulum undergoing simple harmonic motion the trajectory can be shown to be elliptical.

> The path that a system traces out in state space is known as a **trajectory**.

It is important to keep in mind that a trajectory in state space is completely different from the sort of trajectories you met earlier in the course. In *Describing motion* we discussed the parabolic trajectory of a projectile: this is the path that a projectile follows in real three-dimensional position space, and it only tells us about the positions of the projectile. The trajectory in state space tells us about the velocity as well as the position.

● Suppose that a pendulum oscillates with an amplitude that is half of that for the motion indicated in Figure 5.14. Describe the shape and semimajor axis of the new trajectory for this smaller oscillation.

○ The trajectory will still be elliptical, but the semimajor axis will be half the size of the original. The points where the trajectory crosses the x-axis correspond to the maximum and minimum angular displacement, so these crossing points must be at half the distance from the origin of the original ellipse. ■

The behaviour of a simple pendulum is determined by its equation of motion,

$$\frac{\mathrm{d}^2\theta}{\mathrm{d}t^2} = -\omega^2\theta \tag{5.6}$$

Solving this second-order differential equation, allows us to express θ as a function of t. Once this has been done we can determine $d\theta/dt$ and hence the pendulum's trajectory in state space. This will lead to the elliptical trajectory of Figure 5.14.

However, the pendulum's trajectory can also be determined by solving a pair of first-order differential equations for the time dependence of x ($= \theta$) and y ($= d\theta/dt$).

The first of these equations is found by noting that since $x = \theta$, it must be the case that $dx/dt = d\theta/dt$. Since it is also the case that $y = d\theta/dt$, it follows that

$$\frac{dx}{dt} = y. \tag{5.7}$$

The second equation, that for dy/dt, can be obtained from the equation of motion (Equation 5.6). First note that

$$\frac{d^2\theta}{dt^2} = \frac{d}{dt}\left(\frac{d\theta}{dt}\right) = \frac{dy}{dt}.$$

Substituting this into Equation 5.6, and using $x = \theta$ to eliminate θ from the right-hand side, we get

$$\frac{dy}{dt} = -\omega^2 x. \tag{5.8}$$

You are not expected to be able to solve these two equations (5.7 and 5.8), but it's important for what follows to realize that such a pair of first-order differential equations can be used to determine the trajectory in state space.

Solving Equations 5.7 and 5.8 provides a useful alternative to solving Equation 5.6. Even if we didn't know how to solve any of these equations exactly (as is often the case in more complicated situations) we could still use Equations 5.7 and 5.8 to provide a numerical procedure that would enable us to approximate the trajectory of the system. This procedure is similar to that introduced in Chapter 1 when we discussed 'stepping through Newton's laws', though there we were concerned with approximating motion in position space, and now our concern is with approximating a trajectory in state space.

In the case of the pendulum, the approximate trajectory can be determined by considering a sequence of equally-spaced times, t_1, t_2, t_3, ... t_n, separated by a constant interval τ that is much smaller than the period of the pendulum. The state of the system (x_{n+1}, y_{n+1}) at time t_{n+1} can then be calculated from its state (x_n, y_n) at the slightly earlier time, t_n, as follows. The derivative dx/dt will be approximately constant throughout this short interval and can therefore be approximated by $\Delta x/\Delta t = (x_{n+1} - x_n)/(t_{n+1} - t_n)$, so Equation 5.7 implies that

$$\frac{(x_{n+1} - x_n)}{(t_{n+1} - t_n)} = y_n$$

and, since $(t_{n+1} - t_n)$ can be represented by τ, this can be rearranged to give the following expression for x_{n+1}:

$$x_{n+1} = x_n + \tau y_n. \tag{5.9}$$

Similarly, from Equation 5.8

$$\frac{(y_{n+1} - y_n)}{(t_{n+1} - t_n)} = -\omega^2 x_n$$

and this can be rearranged to give the following expression for y_{n+1}

$$y_{n+1} = y_n - \tau\omega^2 x_n. \tag{5.10}$$

Equations 5.9 and 5.10 now provide a two-dimensional map that can be iterated, starting from some initial state (x_0, y_0), to produce a sequence of states (x_1, y_1), (x_2, y_2), (x_3, y_3), ... etc. Each of these states will represent a discrete point in state space that, approximately at least, is on the pendulum's elliptical trajectory. Provided we use a sufficiently small time step τ, we can use this technique to obtain a good approximation to the accurate elliptical trajectory that passes through (x_0, y_0).

A similar iterative procedure can be carried out for state spaces with any number of dimensions, and for any equation of motion, provided sufficient computing power is available.

Equations 5.9 and 5.10 provide a *linear* map, since they only involve terms with the first power of x and y, and no terms involving products of x and y. We obtained these equations from the *linear* second-order differential equation describing the motion of the simple pendulum (Equation 5.6). Since these equations are linear, the simple pendulum does *not* show chaotic behaviour. This is analogous to what we discovered with iterated maps in earlier sections of this chapter; the linear map did not result in chaos. However, carrying out such calculations using *non-linear* maps derived from *non-linear* differential equations certainly can lead to chaos, as you will shortly see.

Question 5.6 (a) Sketch a state-space trajectory for a (linearly) damped pendulum. (b) Where in state space will the system end up? ■

2.6 Strange attractors

Working at the Massachusetts Institute of Technology in the early 1960s, the meteorologist Edward Lorenz devised a computer model that represented the changing state of the weather using a 12-dimensional state space. Wishing to recreate an interesting state, seen in a previous run, he entered (what he thought to be) the initial data used last time. The result was quite different the second time. Checking, he saw that in the second run he had rounded the last few digits of the initial data, not imagining that this would matter. We now know what had happened: the model was in its chaotic regime, with exponential sensitivity to initial data. At the time, it took great boldness to imagine that this was a genuine effect, rather than an artefact of the computational method.

Lorenz went on to find simpler models for which the trajectory in state space is also extremely sensitive to the initial state. His most dramatic discovery was in a three-dimensional state space, used to model the behaviour of gas in a box with a temperature difference between the top and bottom. The effect of the temperature difference is to cause convection, with upward flow of gas on one side and a downward flow on the other. The differential equations that describe this system are

$$\frac{\mathrm{d}x}{\mathrm{d}t} = s(y - x), \quad \frac{\mathrm{d}y}{\mathrm{d}t} = rx - y - xz, \quad \frac{\mathrm{d}z}{\mathrm{d}t} = xy - bz$$

where, at any time t, the values of x, y and z specify the state of the system, while s, r and b are parameters related to the viscosity of the gas, the temperature difference and the shape of the box. The variable x measures the rate of rotation of the gas, and y and z give a rough description of the temperature distribution within the box. You don't need to remember or understand the details of these equations; the important points are that these are three first-order differential equations describing how the state changes with time, and that the second and third equations are *non-linear* (because they contain an xz and xy term, respectively).

Employing the sort of numerical procedure outlined in Section 2.5, it is possible to use a computer to determine the trajectory of this model in its three-dimensional state space for given values of s, r and b. When this was done, the model showed something never seen before: a region of state space into which trajectories are

3   Examples of chaos

We will now look at some examples of chaotic systems. In Section 3.1 we shall discuss two types of system for which the underlying deterministic rules are not clear — dripping taps and beating hearts. We shall look to see whether their complex behaviour might be illuminated by chaos theory, in certain regimes.

The final two subsections look at chaos on a much larger scale. Section 3.2 deals with chaotic motion of planets and asteroids in the Solar System, a topic that we mentioned in Section 1, and Section 3.3 sketches a way in which the onset of chaotic motion of stars in a galaxy might save the system from being gobbled up by a black hole at its centre.

These studies of (possibly) chaotic motion are subject to a considerable degree of uncertainty: the systems are complex and scientists are still learning how to model them. The tentative conclusions — briefly reported here — may change. However, you have seen that chaos theory developed through the study of complex natural phenomena. The hope is that it will continue to inform such studies.

3.1   Dripping taps and beating hearts

The sound of a slowly dripping tap can be infuriating. A water droplet grows until its weight can no longer be supported by surface tension and then we hear the 'plop' as it splatters in the sink or basin. The process begins again and we are left waiting for the next annoying drop to fall. Measurement shows that, in this slowly dripping mode, the process is as boringly predictable as it sounds: each interval between two drips is likely to be rather close to the previous one. However, there is another mode, which you may be able to observe with one of your taps. If you slowly increase the rate of flow, then before the water forms a continuous stream, you may observe a regime in which the pitter-patter noise is less regular. Could this be an example of chaos? Are the small initial differences in how drops start to grow, now producing big differences in outcome?

Robert Shaw and colleagues, of the University of California at Santa Cruz, made a study of these irregular time intervals between drips and looked for a way of displaying the results that might expose some similarity to the strange attractors discovered in chaotic systems. They started by letting δ_n be the time interval between the nth drop and it's predecessor. Then, to make a three-dimensional plot, they assigned the three numbers (δ_n, δ_{n+1}, δ_{n+2}) to drop n. This was a rather novel thing to do with an irregular sequence. It means that the 7th drop has coordinates (δ_7, δ_8, δ_9), while its predecessor has coordinates (δ_6, δ_7, δ_8).

Figure 5.18 shows the remarkable pattern that appears when the unpredictable data for a dripping tap are plotted in this novel three-dimensional manner. This pattern was taken as circumstantial evidence for chaos, i.e. deterministic unpredictability. It is what might be expected from the operation of a deterministic rule that is exponentially sensitive to small deviations, as in the case of the three-dimensional Lorenz attractor. The sequence of time intervals is unpredictable, yet there is structure in it that may have come from the operation of a simple rule, yet to be discovered.

A tap is known as a faucet in North America.

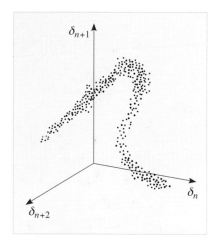

Figure 5.18   A novel way of showing the relationship between drops from an irregularly dripping tap.

Now let us turn from regularly and irregularly dripping taps to something altogether more serious: normally and abnormally beating hearts. Might it be that a healthy heart is like a steadily dripping tap, while a malfunctioning heart is chaotic? If so, perhaps we could display data for the intervals between heartbeats of patients at risk, and see something like a strange attractor, as in the case of irregularly dripping taps.

Ary Goldberger and a team working at the Harvard Medical School and Beth Israel Hospital, in Boston, took measurements from a normal heart and from one that showed a pathological variation in heart rate. Some of their results are shown in Figure 5.19a and b.

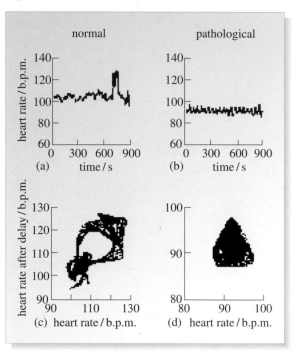

Figure 5.19 (a) A normal variation in heart rate, and (b) a pathological variation. (c) and (d) show plots of heart rate against preceding heart rate, measured at a fixed time earlier, for the normal and pathological variations shown in (a) and (b), respectively. It is the healthy heart that appears to be chaotic. Note that (c) and (d) use different scales.

Question 5.8 (a) What are the two defining characteristics of chaos? (b) What then would be the characteristics of a chaotically healthy heart? ■

In Figure 5.19c the data in part (a) are plotted as a graph of heart rate against the heart rate at a fixed time before. The dots jump around, yet form a distinctive pattern, reminiscent of those seen with the strange attractors that are associated with chaotic behaviour. By contrast, the same type of plot for a pathological variation in heart rate, in Figure 5.19d, shows no such structure. This suggests that chaos is *not* linked to malfunctioning of the heart: in fact, it appears that chaos and health go together, so far as heart rate is concerned!

Such research is still in its infancy. Yet it offers the prospect that the healthy heart is both deterministic and unpredictable, while heart disease is surprisingly associated with *loss* of chaos.

3.2 Chaos in the Solar System

The motion of planets in our Solar System is chaotic, as we indicated in Section 1. Though we can successfully predict the positions of planets on time scales of centuries, small changes of the initial data, below the level of existing observational precision, grow exponentially and lead to big differences in predicted positions 100 million years from now.

The question arises: could the chaotic motion of planets lead to instability, with one planet colliding with another, or suffering a close encounter that ejects it from the Solar System? The planets have been here for nearly 5 billion years: what is in store for them in the next few billion years?

This is clearly a difficult question to tackle, since chaos makes detailed prediction impossible. Methods have been devised for studying long-term changes, such as changes to the inclination of the plane of the orbit of a planet, and changes to the eccentricity of the orbit. The eccentricity of an elliptical orbit is a measure of the departure from circular motion, and the eccentricity of the orbit of Mercury is currently 0.206, corresponding to an ellipse in which the distance from the Sun varies between 0.8 and 1.2 times the average. If the eccentricity of Mercury's orbit were to become greater, a close encounter with Venus might wreak havoc in the Solar System. Similarly, an increase in the eccentricity of the orbit of Mars could spell danger for the Earth.

It is possible to track fluctuations in eccentricity in computer simulations over times of a few billion years. Note that these simulations do not produce firm predictions of what will actually happen, since the changes are exponentially sensitive to the initial conditions. Rather, they are indications of likely variability. For example, if the calculations show big changes in the eccentricity of the orbit of Mercury in 5 billion years, then there is a significant chance that such an effect could occur in the real Solar System in such a time, and a smaller chance that it might occur sooner. Figure 5.20 shows results of one simulation, and there are large and irregular variations in Mercury's eccentricity. The time scale of these changes is comparable to the age of the Solar System, which indicates that Mercury has been somewhat — though not extremely — fortunate in avoiding Venus in the past. However, there is a chance of Mercury being ejected from the Solar System before the Sun exhausts its fuel and swells into a planet-gobbling red giant.

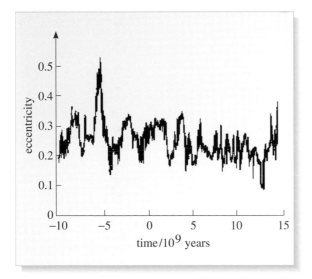

Figure 5.20 Simulation of fluctuations in the eccentricity of Mercury's orbit.

By contrast, the motion of Pluto is robustly chaotic: it is impossible to predict the precise details of its orbit beyond a horizon of order 100 million years. There seems little likelihood however, that the unpredictable fluctuations in Pluto's orbit will result in a close encounter with Neptune. As with heartbeats, this serves to underline that chaos does not necessarily imply disastrous behaviour.

The greatest potential for a disaster on Earth, due to chaotic motion in the Solar System, seems to reside in the asteroid belt, which lies between the orbits of Mars and Jupiter (Figure 5.2). Asteroids are lumps of rock, with the largest — Ceres — having a diameter of about 750 km and accounting for a third of the total mass of the entire belt. Chaotic motion of an asteroid becomes acute when the angular frequency of its orbit is simply related to that of Jupiter. Computer simulations of an asteroid orbiting with an angular frequency close to three times that of Jupiter show chaotic — and dramatic — fluctuations in eccentricity over time scales of only 100 000 years. These changes in eccentricity might be sufficient to cause an asteroid to have a close encounter with Mars, throwing the asteroid into a quite different orbit that might present a grave threat to life on Earth.

There are therefore two qualitative predictions about the asteroid belt. First, there should be few asteroids presently in the chaotic regions, since they are prone to instability. Second, any asteroids found in the chaotic regions are dangerous, since they may change their orbits unpredictably and dramatically. Figure 5.21 shows the observed distribution of asteroids. There is indeed a pronounced dip at 2.5 AU from the Sun, corresponding to a ratio of 3 : 1 between the angular frequencies of the asteroid and of Jupiter. Similar gaps occur at ratios of 5 : 2, 7 : 3 and 2 : 1. They are called *Kirkwood gaps*, after the American mathematician and astronomer Daniel Kirkwood, who noted their relation to the motion of Jupiter in 1860. However, it wasn't until the 1980s that the 3 : 1 Kirkwood gap was explained by chaos theory. Unfortunately, the gap is not quite empty: the angular frequencies of two asteroids — Alinda and Quetzalcoatl — are currently in a 3 : 1 ratio with Jupiter's angular frequency.

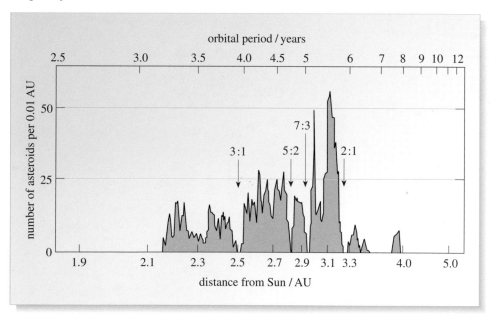

Figure 5.21 Distribution of asteroids in the asteroid belt. The ratios quoted are ratios of the asteroid angular frequency to Jupiter's angular frequency. (1 astronomical unit (1 AU) is the average distance from Sun to Earth and is equal to 1.50×10^{11} m.)

Extinction of the human species might result from chaotic motion of an asteroid currently in the region of a Kirkwood gap. This is not a cause for immediate concern; it is 65 million years since the last mass extinction of species on planet Earth, when the dinosaurs may have fallen victim to an asteroid. Chaos theory points a finger at a possible origin of such cataclysms, while also asserting their unpredictability.

3.3 Chaos in galaxies

Though chaotic motion in the Solar System may hasten the end of species on Earth, chaotic motion in galaxies may protect stars from being gobbled up by a black hole.

Galaxies can be divided into a number of different classes. A spiral galaxy (Figure 5.22a), such as our own, is disc shaped, with a thickness normally less than 10% of its diameter, and with spiral arms winding out from the centre. The structure of elliptical galaxies (Figure 5.22b) is less clear. For a while it was believed that they might differ from a spherically symmetric distribution of stars only by a flattening caused by rotation, in the same manner that the Earth's rotation makes its equatorial diameter greater than its polar diameter. Such a structure is said to be axially symmetric.

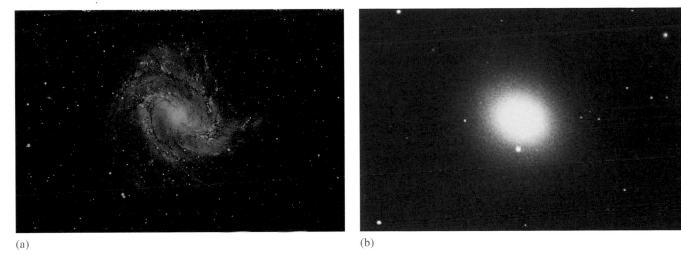

(a)   (b)

Figure 5.22   (a) Spiral galaxy. (b) Elliptical galaxy.

However, in the 1970s astronomers discovered that the rotations of elliptical galaxies are too slow to account for their departures from spherical symmetry. This raised the possibility that elliptical galaxies may lack axial symmetry and possess a more general structure, called tri-axial, since none of its three axes is a symmetry axis. There is an interesting feature of motion in a tri-axial galaxy: a star may follow an orbit in the shape of a bow tie, which confines it to a rectangular box-shaped region and allows it to pass close to the centre of the galaxy. Such orbits are called box orbits, since a bow tie fits into a rectangular box.

Along with this shift in ideas about elliptical galaxies came the realization that some may contain supermassive black holes at their centres. Such a black hole may account for a few per cent of the total mass of the galaxy. While nothing escapes from within the black hole, much heat and light may be generated as matter is sucked into it, with frictional and tidal effects converting a significant proportion of this matter into visible radiation. By such a scenario, astrophysicists seek to explain the existence of *quasars*. A quasar, or quasi-stellar object, is something that generates an energy output comparable to that of a galaxy within a region that is more like the size of a star.

These two ideas suggest a third: if a tri-axial galaxy contains a supermassive black hole, might that black hole gobble up the entire galaxy? You can see the possibility: stars in such a galaxy live dangerously, venturing close to the centre, twice in each box orbit.

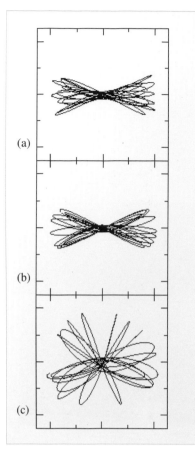

Figure 5.23 Orbit of a star in a tri-axial galaxy with (a) 0.3%, (b) 1% and (c) 3% of the galactic mass in a central black hole.

In the late 1990s, David Merritt and colleagues, at Rutgers University in New Jersey, undertook a study of motion of stars in tri-axial galaxies with a central black hole. They found that with 0.3% of the galactic mass in a black hole there is little effect on the box orbits, which maintain their general form for up to 100 orbital periods, corresponding to times of billions of years. On the other hand, in a simulation of a black hole with 3% of the galactic mass, chaotic motion became apparent within a few orbital periods. The effect of the chaos is to create an axial symmetry and thus deny the black hole its ready food supply. According to the simulations, the black hole initially increases in mass by gobbling up stars that pass close by it in the box orbits characteristic of a tri-axial galaxy. As its mass increases, it destroys the box orbits and so starves itself.

Figure 5.23 illustrates the effect: the difference in the orbits for galaxies with 1% and 3% of the mass in the black hole is rather dramatic. These findings agree well with the current record of 2.5%, set by the black hole candidate in the galaxy whose catalogue name is NGC 3115. The seeming absence of more massive galactic black holes has thus been attributed to the onset of chaos as the black hole mass increases. Here then we see chaos theory invoked to explain a control mechanism that prevents a cataclysm. As in the case of a healthily beating heart, chaos may be part of what preserves the integrity of a system, rather than signalling its doom.

4 Closing items

4.1 Chapter summary

1 Chaos is here used to mean exponential sensitivity in time to initial conditions in a deterministic system.

2 Iterated non-linear maps exhibit chaos, for certain ranges of their parameters. The logistic map $x_{n+1} = kx_n(1 - x_n)$ is chaotic for $k > 3.57$. For k between 1 and 3, by contrast, the final state is predictable: $x_{\text{final}} = 1 - (1/k)$. Between these two ranges, lies a regime of period doubling, with a 2-cycle starting at $k = 3$, a 4-cycle at $k = 1 + \sqrt{6}$, and so on. As one approaches $k = 3.57$, the ranges of successive cycles shrink by a factor of the Feigenbaum constant, $\delta \approx 4.7$, thus making the phenomenon difficult to study experimentally. Such an approach to chaos is common to all quadratic maps, i.e. those for which each x_{n+1} has two predecessors x_n. Such maps have contributed to the study of fractals: structures exhibiting self-similarity when one changes the scale on which they are examined.

3 Poincaré anticipated the possibility of chaos, at the end of the nineteenth century. The realization that deterministic chaos is common came from work in the 1960s and 1970s, in branches of science as apparently diverse as physics, ecology and meteorology.

4 Continuous change is described by differential equations, which determine how a state evolves, thus tracing out a trajectory in state space. If the equations are non-linear, one may observe chaotic behaviour, within some ranges of the parameters, as in the case of iterated non-linear maps.

5 Lorenz observed the butterfly effect in a model of weather change: a tiny change in the initial data produced a large change in the prediction of a meteorological model in a 12-dimensional state space. He was able to observe such effects in a three-dimensional state space, modelling convection. For certain ranges of the parameters, the system has a strange attractor: a region of state space into which trajectories are attracted and within which they exhibit chaos. The distinctive feature of the chaos is that the attractor has two parts, with reasonably regular motion within each part, interspersed by unpredictable jumps between the two. Other non-linear systems possess different strange attractors.

6   In the case of a dripping tap, an increased rate of flow leads to chaos, with irregular intervals between drops. When such chaotic data are suitably displayed a pattern may be observed, possibly indicating an (unknown) underlying rule.

7   In the case of a beating heart, chaos appears to be a sign of good health, whereas heart disease may be associated with loss of chaos.

8   Chaotic motion occurs in the Solar System: even with initial data of great precision, prediction becomes impossible beyond 100 million years.

9   Chaos does not necessarily imply instability: though one cannot predict the future position of Pluto in its orbit, simulations suggest that it will remain in a stable orbit, far beyond the horizon of predictability for its position.

10 In the case of Mercury, there is a chance that fluctuations in the eccentricity of its orbit may result in a close encounter with Venus, within a few billion years. This might lead to Mercury's exit from the Solar System.

11 Asteroids near to Kirkwood gaps are unpredictable over times of merely 100 000 years. A fluctuation in eccentricity might lead to a close encounter with Mars, which could be potentially hazardous to life on Earth.

12 A model of the motion of stars, in an elliptical galaxy containing a black hole, suggests that the black hole starves itself by creating chaotic motion as its mass increases. The effect is to destroy the pattern of the orbits of the stars on which it fed.

4.2 Achievements

Now that you have completed this chapter, you should be able to:

A1 Understand the meaning of all the newly defined (emboldened) terms introduced in this chapter.

A2 Recognize uses of the concept of chaos, and describes some of its occurrences.

A3 Distinguish between a limit cycle and chaos.

A4 Determine whether the onset of chaos is described by the Feigenbaum constant.

A5 Represent motion by a trajectory in state space.

A6 Recognize depictions of a chaotic attractor in state space.

A7 Suggest ways of displaying irregular data that might reveal deterministic chaos.

A8 Estimate exponential growth of uncertainties.

4.3 End-of-chapter questions

Question 5.9   Revisit the questions in the final paragraph of Section 1 and answer them in the light of what you have learnt in this chapter.

Question 5.10   The logistic map $x_{n+1} = kx_n(1 - x_n)$ produces a chaotic sequence when $k = 4$. (a) Sketch the pattern of points that would be produced on a graph of x_{n+1} against x_n, if a large number of points generated by this map were plotted. (*Hint*: You do *not* need to compute a sequence of points to do this, but you will need to think about the overall shape of the curve on which such points would have to be located.) (b) What can you say about the way the pattern of points will build up?

Question 5.11   For some people, loss of predictive power may be a matter of indifference; for others, it may be a source of concern; for yet others, it may be a cause for awe, or even joy, at the possibility of interdependence beyond the power of human computation. How might you express your own response, to a friend? Draft a few sentences of personal reflection on what you have learnt from this chapter. ■

Chapter 6   Consolidation and skills development

1   Introduction

This final chapter has two main aims: first to help you consolidate what you have learned from this book and second to help you develop your problem-solving skills.

Section 2 is mainly concerned with consolidation. It provides an overview of Chapters 1 to 5, drawing out some of the major themes that run through those chapters and linking them together. It also emphasizes the way in which *Predicting motion* has added to your mathematical skills by teaching you new techniques and allowing you to practise those you developed in *Describing motion*.

The examples and questions contained in earlier chapters have given you plenty of opportunity to acquire problem-solving skills, but Section 3 aims to further develop these skills by introducing you to a three-step problem-solving technique that will help you to approach any physics problem in a systematic way. The technique is no substitute for knowledge and experience, but it does provide a useful framework that can be used throughout the course and in your subsequent studies.

The remaining sections are devoted to consolidation activities. Section 4 contains a short test of basic skills and knowledge, Section 5 provides you with the opportunity to tackle the interactive questions for this book, and Section 6 invites you to try some of the longer questions contained in the *Physica* package. *Physica* will help you to develop some of the higher-level problem-solving skills that rely on strategy and insight rather than simple manipulation.

2   Overview of Chapters 1 to 5

This section highlights some of the major themes that link the earlier chapters of this book. It is not intended to be a summary of those chapters and does not seek to replace or repeat the material already contained in the summaries of those five chapters. Indeed, you may find it worthwhile to reread those chapter summaries before proceeding.

The subject of this book is predicting motion. In Newtonian mechanics it is Newton's laws that provide the basis for such predictions, whether they involve forces and translational motion, or torques and rotational motion. They apply to extended bodies as well as particles or systems of particles, though our considerations are largely restricted to particles and rigid bodies for simplicity.

Newton's laws justify the introduction of various ancillary concepts such as work, energy, power and momentum. Energy and momentum turn out to be important in their own right, but, within Newtonian mechanics, their main role is that of facilitating predictions when the forces acting on a system are not known in detail, as in a typical collision.

The starting point for any prediction of motion is a frame of reference that provides an observer with coordinates and clocks so that he or she can determine kinematic quantities such as displacement, velocity and acceleration. In the case of Newtonian mechanics, the most frequently encountered frames of reference belong to the class of inertial frames, since it is in these frames that Newton's laws are valid. (Fictitious forces, such as the Coriolis force and centrifugal force, arise when Newton's laws are applied in non-inertial frames.)

In situations where the forces or torques acting on a system are known, Newton's second law (or its rotational analogue) can be used to formulate the equation of motion of the system. The equation of motion is a differential equation, typically involving first and second derivatives of the relevant position coordinates. One-dimensional examples include the equation of simple harmonic motion

$$\frac{d^2 x}{dt^2} = \frac{-k}{m} x$$

and the equation of motion of the driven damped harmonic oscillator

$$\frac{d^2 x}{dt^2} = \frac{-k}{m} x - \frac{b}{m} \frac{dx}{dt} + \frac{F_0}{m} \cos(\Omega t).$$

The general solutions to such equations describe the position or configuration of the system as a function of time: $x(t) = A \sin(\omega t + \phi)$ in the case of the s.h.m. equation. The mathematical procedure for working out such solutions is beyond the scope of this course, but you should be aware that the solutions generally introduce new arbitrary constants that were not present in the equations of motion (A and ϕ in the case of s.h.m.). In order to make detailed predictions regarding any particular movement of the system, it is generally necessary to determine the values of these arbitrary constants and that is usually done by using the initial conditions of the motion itself. It is in this sense that the past may be said to determine the future in Newtonian mechanics. The Newtonian equations of motion are deterministic.

Despite their determinism, the equations of motion of Newtonian mechanics are not always able to reliably predict the future motion of a system. Some systems are known to be chaotic under certain circumstances, implying that their evolution is highly sensitive to their initial condition. As a result, any limitation in the precision with which the initial conditions can be specified inevitably leads to a breakdown in effective predictability beyond some horizon of predictability. Numerical studies indicate that even the equations governing the evolution of the Solar System are chaotic, so even the future behaviour of the planets is ultimately unpredictable, despite the deterministic nature of the equations that describe their behaviour.

Dealing with differential equations is a difficult undertaking even for experienced users of mathematics. Fortunately, in many situations, important information about the behaviour of systems may be more simply obtained by making use of conserved quantities, such as momentum (linear or angular) and energy. It should be noted that in any particular system, the conservation of energy and momentum is not guaranteed unless certain conditions are met. For instance, the linear momentum of a system is only conserved if no net force acts on the system, and the mechanical energy of a system (i.e. the sum of the kinetic and potential energies) will only be conserved if all the forces acting in the system are conservative. Still, provided the required conditions are satisfied, conservation principles often provide the quickest and simplest means of arriving at a prediction.

Einstein's relativistic modifications to the definitions of momentum and kinetic energy, and his introduction of mass energy via $E = mc^2$, show that conservation principles have a significance that extends beyond Newtonian mechanics. However, within the confines of Newtonian mechanics it is important to realize that the conservation principles are a consequence of Newton's laws, not a set of independent principles.

In addition to introducing a number of new physical principles, Chapters 1 to 5 have also shown you some of the important applications of those principles, often in a quantitative setting. In doing so, some new mathematical ideas have been introduced. Many of these have already been used more than once. Among the most important are the following.

The exponential function, $v(t) = v_0 e^{-t/\tau}$ has been introduced, extending the range of functions with which you are familiar beyond the polynomials and trigonometric functions that were introduced in *Describing motion*. This also provided the opportunity to introduce the logarithmic function, $\log_e(t)$ that undoes the effect of the exponential function in the sense that if $y(x) = e^x$ then $x = \log_e y$. Exponential processes occur throughout the natural world, arising whenever the rate of change of a quantity is proportional to the instantaneous value of that quantity.

Your knowledge of differential calculus has been augmented by seeing that

$$\frac{d e^{ax}}{dx} = a e^{ax},$$

and the vast subject of integral calculus has been opened up by introducing you to the concept of a definite integral as the area under a graph. We shall not make much use of this particular concept, but it is important to realize that it gives physicists a means of adding together continuous quantities that are not amenable to more ordinary kinds of summation. You have already seen this technique used to calculate the work done in stretching an ideal spring, and you have been told that it can also be used to determine the moment of inertia of a body about a given axis.

The unit vector, $\hat{r} = r/r$, has been introduced and used in the writing down of force laws such as Newton's law of universal gravitation

$$F_{21} = \frac{-Gm_1 m_2}{r^2} \hat{r}.$$

It is worth emphasizing yet again that a unit vector is a vector of magnitude 1, since the belief that its magnitude is 1 m or even 1 unit, though wrong, is still very common.

Another development in the use of vectors is the introduction of the scalar product
$$a \cdot b = |a||b| \cos \theta = ab \cos \theta$$
where θ is the angle between the vectors a and b.

This may also be defined in terms of the components of a and b:
$$a \cdot b = a_x b_x + a_y b_y + a_z b_z.$$

This has already been used in the expression $W = F \cdot s$ for the work done by a constant force F acting over a displacement s; in $P = F \cdot v$ for the instantaneous power delivered by a force F acting on a body moving with velocity v; and in $P = \Gamma \cdot \omega$ for the instantaneous power delivered by a torque Γ acting on a body moving with angular velocity ω.

Another way of combining vectors, the vector product, has also been introduced. This is more complicated than the scalar product but also more powerful, since it results in a vector quantity that includes directional information. The vector product of two vectors a and b is denoted $a \times b$. It is a vector of magnitude
$$|a \times b| = |a||b| \sin \theta = ab \sin \theta$$
that is directed at right angles to both a and b, in the sense specified by the right-hand rule.

Expressed in terms of components,
$$a \times b = (a_y b_z - a_z b_y, \; a_z b_x - a_x b_z, \; a_x b_y - a_y b_x).$$
Examples of vector products in *Predicting motion* include: $\Gamma = r \times F$, which describes the torque about a point O due to a force F acting at a point displaced from O by an amount r; the expression $v = \omega \times r$, for the instantaneous velocity of a particle with angular velocity ω about O; and $l = r \times p$, which describes the angular momentum about O of a particle with linear momentum p. You will meet many more vector products later in the course.

3 Problem-solving skills

Problem solving is an important skill in many subjects, but nowhere more so than in physics, where problems generally involve the formulation of a mathematical model to represent a particular physical situation, the analysis of that model, and the interpretation of the answers that it provides. This section aims to develop your problem-solving skills by providing you with an approach to problem solving that you can use in a wide variety of circumstances. The approach contains three stages.

Stage I: Preparation

This is probably the most important part of tackling a physics problem. If you don't get the preparation right, you often stand no chance of completing the problem successfully.

(i)   To begin with, write down concisely the information provided in the question. Summarize the quantities you are given, assigning symbols to each variable and constant.

(ii) It is often extremely useful to sketch a diagram of the situation. This can include defining the axes that you will use, and drawing the vectors and angles if appropriate.

(iii) Finally, before you start to solve the problem, gather together your tools for the job. These will usually be a set of equations or relationships that you think might be appropriate. Write down all the equations that you think might be relevant, using the symbols that you have already identified.

Stage II: Working

This is the heart of solving the problem. Having prepared the ground you are now ready to tackle the problem itself.

(i)   First work out your plan of attack. Using the equations you have already identified, and knowing what quantities you're given, you should be able to see which other quantities you need to calculate in order to solve the problem.

(ii) Now for the algebra. Combine the equations that you have identified, rearranging them so that you are left with the unknown quantity on the left-hand side, and only known quantities on the right.

(iii) Finally, substitute the appropriate numerical values into your algebraic equation, along with the units for each quantity. Calculate the answer, including the relevant unit.

Stage III: Checking

You may think that the problem is now complete. However, it's vital that you now check your answer, since it is here that you may identify a careless slip-up earlier in your answer.

(i)   First, check the unit of your answer. By putting in the units for the other quantities in the calculation, you should have arrived at the correct unit for the answer; if you haven't, then something has gone wrong. For instance, if you were calculating a speed, and the unit turned out to be $m^2\,s^{-2}$, it's likely that you've forgotten to take a square root somewhere along the line.

(ii) Check also that the answer you have obtained is sensible. For instance, if you are calculating a velocity your answer must be a vector and must therefore include directional information (at least implicitly), but if calculating a speed your answer must be a positive scalar. A negative speed would clearly indicate an error. If you were calculating the speed of a car, and the answer came out to be $2.7 \times 10^8\,m\,s^{-1}$, you must have made a slip — cars simply don't travel at 90% of the speed of light! All the problems you meet should be based on reality, so the values you calculate should all be realistic.

(iii) Finally, think about how your answer would alter if some of the given quantities were varied. For instance, a problem may involve a force acting on an object and producing a positive acceleration. If your answer for the final speed of the object, say, gives a *smaller* speed for a *larger* accelerating force, then something is probably wrong with your calculation. You could also look at the 'limiting case' — think what would happen if one of the input parameters were set to zero, for instance — in that case, would the final answer still make sense?

So, to summarize, the stages in solving a problem are:

Stage I: Preparation	Stage II: Working	Stage III: Checking
• Summarize the information given	• Work out a plan of attack	• Check that the units are correct
• Draw a diagram	• Do the algebra	• Check that the answer is sensible
• Write down the equations you may need	• Put in the values	• Check how the answer would vary for different input values

Examples 6.1–6.3 illustrate the problem-solving technique.

Example 6.1

A brick rests on a sloping plank. The angle α between the plank and the horizontal is gradually increased until, at an angle of 40°, the brick starts to slide. What is the coefficient of static friction between the brick and the plank?

Solution

Preparation   Let m be the mass of the brick and w the magnitude of its weight. Let f and N be the magnitudes of the frictional and normal reaction forces.

A diagram illustrating the situation is shown in Figure 6.1.

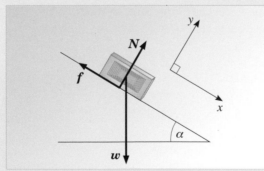

Figure 6.1   The weight of the brick, w, acts vertically downwards. The normal reaction N acts at right angles to the surface of the plank and prevents the brick from sinking into the plank. Finally, a frictional force f acts in the plane of the plank, in a direction that opposes the tendency to slide. The coordinate system has been chosen so that the normal reaction has no x-component and the frictional force no y-component.

Relevant equations are:

$$w = mg$$

and $$f_{max} = \mu_{static} N$$

where $$f_{max} \geq f.$$

Comment   *Terrestrial gravity and frictional forces can be quantified using laws described in Chapter 1. To tackle mechanics problems, you will need to have these force laws at your fingertips.*

Working   Until the brick starts to slip, it is at rest and therefore not accelerating. Consequently the total force on the brick must be zero, i.e.

$$F = w + N + f = 0.$$

This vector relation is equivalent to two scalar equations, one for each independent component of F

$$F_x = (mg \sin \alpha) - f$$
$$F_y = N - mg \cos \alpha.$$

Since each of these components is equal to zero, we have the *pair* of equations

$$f = mg \sin \alpha \tag{6.1}$$

and $N = mg \cos \alpha.$ \qquad (6.2)

The magnitude of the frictional force, f, increases as $\sin \alpha$ increases, until it reaches its maximum value $\mu_{static}N$ when α is 40° and the brick is on the point of slipping. Under these conditions, Equations 6.1 and 6.2 become

$$\mu_{static}N = mg \sin 40°$$

and $N = mg \cos 40°.$

Dividing the first of these equations by the second gives

$$\frac{\mu_{static}N}{N} = \frac{mg \sin 40°}{mg \cos 40°}$$

so, $\mu_{static} = \tan 40° = 0.84.$

The coefficient of static friction is therefore 0.84.

Checking   A coefficient of friction has no unit, so the absence of units is correct.

A value between zero and one seems reasonable for the situation described.

Referring to the final result for μ_{static}, it is clear that if the angle of slippage were smaller than 40°, the coefficient of friction would be smaller. Also, if the angle of slippage were close to 0°, the coefficient of friction would be close to zero. This behaviour seems reasonable.

In forming a plan of attack, ask yourself what is special about the brick.

Minus f appears because the frictional force acts in the opposite direction to the x-axis.

Dividing ensures that the mass will cancel.

This check considers the effect of decreasing the angle of slippage.

Example 6.2

A skier is negotiating an icy hillside (Figure 6.2). The skier — a novice ski-jumper — slides straight down the slope from rest at A and takes off from the edge of a ledge B, a vertical distance of 30 m below the start point. Assuming that the skier does not provide any kinetic energy by muscular action, calculate the speed at the moment of take-off. (Assume frictional forces are negligible and that the acceleration due to gravity has a magnitude $g = 9.8 \text{ m s}^{-2}$.)

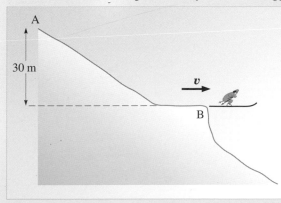

Figure 6.2   See Example 6.2.

The lack of detailed information about forces suggests that conservation of energy might be used.

Solution

Preparation Useful equations will be

$$\Delta E_{\text{grav}} = mg\Delta h \tag{6.3}$$

and, since the skier is initially at rest,

$$\Delta E_{\text{trans}} = \tfrac{1}{2} mv^2 \tag{6.4}$$

where m is the mass of skier and skis, $\Delta h = -30$ m, and v is the take-off speed at B. Since frictional forces are negligible, mechanical energy is conserved and the change in gravitational potential energy plus the change in kinetic energy is equal to zero, i.e.

$$\Delta E_{\text{grav}} + \Delta E_{\text{trans}} = 0. \tag{6.5}$$

Working   From Equation 6.5, $-\Delta E_{\text{grav}} = \Delta E_{\text{trans}}$.

So, substituting from Equations 6.3 and 6.4,

$$-mg\Delta h = \tfrac{1}{2} mv^2.$$

Dividing both sides by m, rearranging, and taking positive square roots

$$v = \sqrt{-2g\Delta h}\,.$$

Putting in the given values, we find:

$$v = \sqrt{-2 \times 9.8\,\text{m s}^{-2} \times (-30\,\text{m})} = \sqrt{588\,\text{m}^2\,\text{s}^{-2}}$$

$$= 24\,\text{m s}^{-1}.$$

Negative ΔE_{grav} makes sense as the height of the skier decreases.

Checking   The units make sense: v comes out in m s^{-1}. The speed is also plausible for a downhill skier (over twice that of a sprint athlete). If the magnitude of Δh increases, v increases as expected.

Example 6.3

In a game of snooker, the white ball strikes a stationary red ball, so as to knock it into a pocket. Before the impact, the white ball was travelling at a speed of 2.0 m s^{-1}, and after the impact it was travelling at an angle of 30° to its original direction. What are the speeds of the white ball and the red ball immediately after the impact? (Assume that the collision is elastic and both balls have the same mass.)

Solution

Preparation   The initial speed of the white ball is $u_1 = 2.0$ m s^{-1} along the x-axis. Let the speeds of the white and red balls after the impact be v_1 and v_2, respectively.

A diagram illustrating the situation is shown in Figure 6.3.

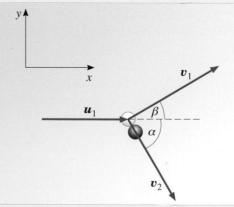

Figure 6.3   The x-axis is taken to be the original direction of travel.

Referring to Figure 6.3, $\beta = 30°$ and since this is an *elastic* collision between two balls of the same mass, one of which was initially stationary, the final velocities will be separated by 90°, hence $\alpha = 60°$.

Comment   *Collisions generally imply that conservation laws will be needed. The change in direction makes it almost certain that momentum conservation will be useful in this case.*

If we had not recalled the 90° phenomenon, we would have to use the conservation of energy, but $\beta + \alpha = 90°$ provides a short cut, so the only additional equation we will need expresses conservation of linear momentum

$$m\boldsymbol{u}_1 = m\boldsymbol{v}_1 + m\boldsymbol{v}_2.$$

We need to calculate the speeds v_1 and v_2.

Working   Applying the conservation of linear momentum in the x-direction, and cancelling the mass throughout we obtain

$$u_1 = v_1 \cos \beta + v_2 \cos \alpha \tag{6.6}$$

and similarly in the y-direction

$$0 = v_1 \sin \beta - v_2 \sin \alpha. \tag{6.7}$$

From Equation 6.7,

$$v_1 = v_2 \frac{\sin \alpha}{\sin \beta}$$

and substituting this into Equation 6.6,

$$u_1 = v_2 \frac{\sin \alpha \cos \beta}{\sin \beta} + v_2 \cos \alpha = v_2 \left(\frac{\sin \alpha \cos \beta + \cos \alpha \sin \beta}{\sin \beta} \right).$$

Since $\beta + \alpha = 90°$, it must be the case that $\cos \beta = \sin \alpha$ and $\sin \beta = \cos \alpha$. Making these substitutions on the top line, we obtain

$$u_1 = v_2 \left(\frac{\sin \alpha \sin \alpha + \cos \alpha \cos \alpha}{\sin \beta} \right) = v_2 \left(\frac{\sin^2 \alpha + \cos^2 \alpha}{\sin \beta} \right) = \frac{v_2}{\sin \beta}$$

It is a rule of trigonometry that $\sin^2 \theta + \cos^2 \theta = 1$ for any angle θ.

so $v_2 = u_1 \sin \beta$.

Putting in the given values

$$v_2 = (2.0 \, \text{m s}^{-1}) \times \sin 30° = 1.0 \, \text{m s}^{-1}$$

then from earlier

$$v_1 = \frac{v_2 \sin \alpha}{\sin \beta} = \frac{1.0 \, \text{m s}^{-1} \times \sin 60°}{\sin 30°} = 1.7 \, \text{m s}^{-1}.$$

So the speed of the white ball after the impact is $1.7 \, \text{m s}^{-1}$ and the speed of the red ball is $1.0 \, \text{m s}^{-1}$.

Checking   The units in each case come out correctly as m s^{-1}. The speeds seem reasonable in comparison with the original speed of $2.0 \, \text{m s}^{-1}$. If this original speed were larger, then the subsequent speeds of the white and red balls would each also be larger, as expected.

Now that you have seen a few examples, try Questions 6.1–6.6, following the three-stage problem-solving strategy outlined earlier.

Question 6.1   A bobsleigh slides down an icy slope that makes an angle of 20° with the horizontal. The motion is opposed by friction and air resistance. The coefficient of sliding friction between the bobsleigh and the ice is 0.10. Air resistance provides a force of magnitude Cv^2, where $C = 0.30\,\text{N}\,\text{s}^2\,\text{m}^{-2}$ and v is the speed of the bobsleigh. The combined mass of the bobsleigh and its occupants is 200 kg. At the end of a long run down the slope, the bobsleigh is travelling at a practically constant speed. Estimate this speed. (The magnitude of the acceleration due to gravity is $9.81\,\text{m}\,\text{s}^{-2}$.)

Question 6.2   A neutron of mass m travelling east at a speed of $4.0 \times 10^4\,\text{m}\,\text{s}^{-1}$ collides with a helium nucleus of mass $4m$ travelling north at a speed of $3.0 \times 10^4\,\text{m}\,\text{s}^{-1}$. If the neutron travels with unchanged speed after the collision, but moves in a direction 36.9° west of north, find the final speed and direction of the helium nucleus.

Question 6.3   A planet orbiting a star experiences a force of magnitude $4.2 \times 10^{23}\,\text{N}$ due to the gravitational attraction of the star. If the planet has a speed of $1.3 \times 10^4\,\text{m}\,\text{s}^{-1}$ and it takes $3.7 \times 10^8\,\text{s}$ to complete one orbit, calculate the radius of the planet's circular orbit and its mass.

Question 6.4   A block of mass 1.00 kg is released from rest in a vacuum and falls on to the top of a spring, which is standing vertically on a firm surface. If the total distance travelled by the block is 5.50 m when the spring is suffering maximum compression, find the maximum compression of the spring, and the speed of the block when it first hits the spring. (The spring constant is $299\,\text{N}\,\text{m}^{-1}$ and the magnitude of the acceleration due to gravity is $9.81\,\text{m}\,\text{s}^{-2}$.)

Question 6.5   The 4000 kg crane truck in Figure 6.4 has its centre of mass 1.3 m behind the front wheels. The distance between the front and rear wheels is 4.5 m. If the top of the crane is 6.0 m above ground level and 7.0 m behind the front wheels, calculate the maximum safe load which it can lift without tipping.

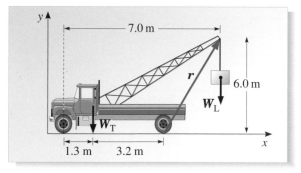

Figure 6.4   For use with Question 6.5.

Question 6.6   An arrow of mass 1.0 kg is fired into an unlatched, stationary door and strikes the door perpendicular to its face, at a distance 1.0 m from the hinges and at an impact speed of $100\,\text{km}\,\text{h}^{-1}$. The arrow embeds itself in the door, which has a width of 1.3 m and a mass of 60 kg. Find the angular velocity of the door and the translational velocity of the arrow just after impact. (The moment of inertia of a door of mass M about one edge is $I = Ma^2/3$ where a is the width of the door. You may take the moment of inertia of the embedded arrow, about the hinged edge of the door, to be $I = mb^2$ where m is the mass of the arrow and b is its distance from the edge of the door. Ignore friction at the hinges.)   ■

4   Basic skills and knowledge test

You should be able to answer these questions without referring to earlier chapters. Leave your answers in terms of π, $\sqrt{2}$, etc. where appropriate.

Where appropriate, use the three-stage problem-solving strategy.

Question 6.7   If a constant resultant force of 5 N magnitude is applied to an initially stationary body and produces an acceleration of magnitude 0.5 m s^{-2}, what is the mass of the body?

Question 6.8   A car of mass 800 kg travels on a level road with an acceleration of magnitude 1.00 m s^{-2}, caused by a driving force of magnitude 900 N and an opposing resistive force of magnitude F_R. Determine the magnitude of the resistive force.

Question 6.9   Determine the acceleration of a block of mass 2.0 kg which is pushed along a level surface by a horizontal force of magnitude 10 N, given that $\mu_{\text{slide}} = 0.5$. (Assume that the magnitude of the acceleration due to gravity is $g = 9.8$ m s^{-2}.)

Question 6.10   A spring rests horizontally on a flat, frictionless surface with one end fixed and the other attached to a particle of mass 0.5 kg. If the particle is displaced along the line of the spring such that the spring is stretched by 0.25 m, and then the particle is released, the magnitude of the initial acceleration is found to be 2.0 m s^{-2}. Find the spring constant.

Question 6.11   Three horizontal forces act on a body lying on a frictionless, horizontal surface and keep it in translational equilibrium. One force acts due north, the second due east and the third, of magnitude 10 N acts in a direction 60° west of south. Find the magnitudes of the forces in the east and north directions. (Remember, $\sin 60° = \sqrt{3}/2$.)

Question 6.12   The brakes on a car of mass 1000 kg travelling at a speed of 10 m s^{-1} are suddenly applied so that the car skids to rest in a distance of 50 m. Use energy considerations to determine the magnitude of the total frictional force acting on the tyres, assuming it to be constant throughout the braking process. What is the car's speed after the first 25 m of this skid?

Question 6.13   By how much must an ideal spring, with $k_s = 1000$ N m^{-1}, be stretched from its unextended position to give it the same potential energy as a 2 kg mass raised to a height of 1 m above the reference level at which $E_{\text{grav}} = 0$? (Assume $g = 10$ m s^{-2} for the purposes of this question.)

Question 6.14   An American football player of mass 80 kg, running east at 4.5 m s^{-1}, tackles another player whose mass is 100 kg and who is running south at a speed of 3.6 m s^{-1}. If in the tackle they cling together, what will be their common velocity immediately after the tackle? (Remember, $\cos 45° = 1/\sqrt{2}$.)

Question 6.15   Two children face each other on roller skates. The mass of the first child (including her roller skates) is 20 kg and that of the second is 25 kg, and they are initially stationary. They push against each other with a constant force of 40 N for 0.5 s and move apart. What is the subsequent momentum of each child, and what is their total momentum?

Question 6.16   The drum of a spin-drier with an internal diameter of 0.50 m rotates at a constant 3600 revolutions per minute. Find the instantaneous angular speed ω of the drum and the instantaneous speed v of a sock stuck to the inside surface of the drum of the drier.

Question 6.17   Make a rough estimate of the maximum frictional torque if, to loosen a car wheel nut, it requires your whole weight, applied at the end of a horizontal spanner of length 0.40 m.

Question 6.18   Suppose that in the situation described in Question 6.17 it is found that once the nut begins to move, the average torque over one revolution is 1/4 of its initial value. Calculate the work done in one revolution of the spanner and the average power needed to complete this operation in a time of 5 s.

Question 6.19   An aerial mast, 4.0 m high, is secured by a set of tie cords which are attached to the mast 1.0 m from the top. If each cord is secured to the ground 4.0 m from the mast and stretched to a tension of 100 N, find the magnitude of the torque exerted by a *single* cord about a fulcrum at the base of the mast.

Question 6.20   For the aerial mast of Question 6.19, calculate the magnitude of the torque exerted by a single cord about a pivot which is: (a) parallel to the ground and intersects the mast 1 m above the ground, and (b) parallel to the ground and intersects the mast at the point of attachment of the cord.

Question 6.21   (a) Write down a general expression relating the kinetic energy of a solid cylinder rolling down an inclined plane, to its translational kinetic energy and its rotational kinetic energy about its central axis.

(b) From this expression, deduce a general formula relating the angular speed to the effective vertical height fallen, the moment of inertia, the radius of the cylinder and the value of g.

Question 6.22   A particle travels at constant velocity v along a straight line. An origin of coordinates O is chosen that lies a perpendicular distance b away from this straight line. The position vector of the particle relative to O is r. Show that the vector product $r \times v$ is independent of time. (*Hint*: As a first step, split r into two vectors, one parallel to the given line, the other perpendicular to it.)

Question 6.23   A particle of mass 5.0 kg is in uniform circular motion in a plane parallel to the xy-plane, with speed 10 m s^{-1}, on a path of radius 4.0 m. The motion is analysed using an origin (O) on the axis of rotation but displaced 3.0 m below the plane of rotation as shown in Figure 6.5. Calculate (a) the magnitude of the angular momentum about O and (b) the z-component of the angular momentum about O.

Question 6.24   For the situation shown in Figure 6.5 calculate (a) the maximum value of the x-component of the angular momentum, L_x, and (b) the minimum value of L_x.

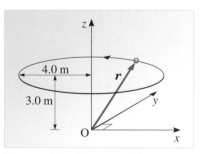

Figure 6.5   For use with Questions 6.23 and 6.24.

Question 6.25   If the vector p has components $(1, \sqrt{3}, 0)$ what is the magnitude and direction of the unit vector $\hat{p} = p/p$?

Question 6.26   If $f(x) = 6x^2 + 3x$ what is $df(x)/dx$?

Question 6.27   If $f(x) = e^{3x}$ what is $df(x)/dx$?

Question 6.28   Referring to Figure 6.6, what is the value of the scalar product $\mathbf{a} \cdot \mathbf{b}$?

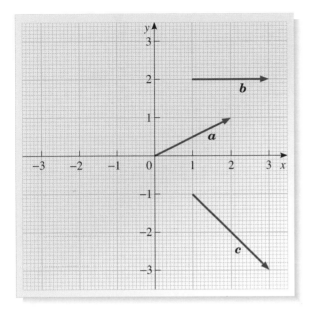

Figure 6.6   For use with Questions 6.28–6.30.

Question 6.29   Referring to Figure 6.6, what are the directions of the vectors (a) $\mathbf{a} \times \mathbf{c}$, (b) $\mathbf{a} \times \mathbf{b}$, (c) $\mathbf{b} \times \mathbf{a}$, (d) $\mathbf{b} \times (-\mathbf{a})$, (e) $\mathbf{c} \times (\mathbf{a} + \mathbf{b})$?

Question 6.30   Referring to Figure 6.6, what is the magnitude of the vector $\mathbf{c} \times \mathbf{b}$?

5   Interactive questions

Open University students should leave the text at this point and use the interactive question package for *Predicting motion*. When you have completed the questions you should return to this text. You should not spend more than 2 hours on this package.

The interactive question package includes a random number feature that alters the values used in many of the questions each time those questions are accessed. This means that if you try the questions again, as part of your end-of-course revision for instance, you will find that many of them will have changed, at least in their numerical content.

6   *Physica* problems

Open University students should leave the text at this point and tackle the *Physica* problems that relate to *Predicting motion*. You should not spend more than 2 hours on this package.

Answers and comments

Q1.1 (a) Since the painter is accelerating downwards, the train must be accelerating upwards in the painter's frame of reference. The train cannot, therefore, be moving uniformly (i.e. without acceleration) in the painter's frame. (b) The train will be revolving in the holidaymaker's frame; its changing direction of motion means it cannot be moving uniformly in that frame. (c) Since the would-be passenger is travelling at a uniform top speed, and is not accelerating, she or he would find that the train moves uniformly.

Q1.2 First, Newton's first law recognizes the existence of inertia, the tendency of a body to continue moving with uniform velocity unless disturbed in some way. Second, it provides a means of identifying inertial frames of reference, since it is only in those frames of reference that the first law (and the other laws of motion) hold true. Third, it provides a way of recognizing the action of an unbalanced force since, providing we are observing from an inertial frame, such a force will cause a body to accelerate.

Q1.3 I don't have much experience of throwing apples, but I think I could launch one with enough speed to keep up with a car travelling moderately fast, say $50\,\text{km}\,\text{h}^{-1}$, which is equivalent to about $14\,\text{m}\,\text{s}^{-1}$. I reckon I can make about ten complete throwing actions in 5 s, so allowing for the time it takes me to return my arm to the starting position after each 'throw' I estimate that it would take me about 0.15 s to accelerate the apple from rest. Assuming that the acceleration is uniform, it follows that the magnitude of the acceleration is $14\,\text{m}\,\text{s}^{-1}/0.15\,\text{s}$, which is very roughly $100\,\text{m}\,\text{s}^{-2}$ (there's no point in trying to be too precise, all these estimates are very rough). Using $F = ma$, it follows that the magnitude of the force I must have applied to the apple would be $F = 0.1\,\text{kg} \times 100\,\text{m}\,\text{s}^{-2} = 10\,\text{N}$. Having arrived at this conclusion I feel I might well have underestimated the speed or the acceleration, but I don't think I'm likely to be wrong by a factor of more than two or three, so I'd estimate that, over a short duration, I can apply a throwing force with a magnitude of about $10\,\text{N}$, but it might be as much as $20\,\text{N}$ or $30\,\text{N}$.

Q1.4 The weight of the bag is the gravitational force on the bag. This acts vertically downwards, and has magnitude

$$W = mg = 50\,\text{kg} \times 9.81\,\text{m}\,\text{s}^{-2} = 491\,\text{N}.$$

Q1.5 Modifying Equation 1.15:

$$g_{\text{Mars}} = \frac{Gm_{\text{Mars}}}{R_{\text{Mars}}^2} = \frac{6.67 \times 10^{-11} \times 6.42 \times 10^{23}}{(3.38 \times 10^6)^2}\,\text{m}\,\text{s}^{-2}$$
$$= 3.75\,\text{m}\,\text{s}^{-2}.$$

Q1.6 (a) From Equation 1.15:

$$g = \frac{Gm_{\text{E}}}{R_{\text{E}}^2} = \frac{6.67 \times 10^{-11} \times 5.98 \times 10^{24}}{(6.37 \times 10^6)^2}\,\text{m}\,\text{s}^{-2}$$
$$= 9.83\,\text{m}\,\text{s}^{-2}.$$

(b) Apart from the variations in g that arise because the Earth is not spherically symmetric, there is also an effect due to the non-inertial nature of a reference frame fixed to the (rotating) Earth. Objects observed to fall vertically in such a frame are actually subject to a fictitious 'centrifugal force' of the kind mentioned in Section 2.2. This has the effect of reducing the apparent downward force compared with its value in a truly inertial frame. The value of g predicted by Equation 1.15 is that which would apply if the Earth was not rotating.

Q1.7 From Equation 1.21, the magnitude of the frictional force is:

$$F = \mu_{\text{slide}}N.$$

But $N = mg$, and so

$$F = \mu_{\text{slide}}mg.$$

Since the frictional force is the only horizontal force acting, Newton's second law tells us that the magnitude of the horizontal acceleration, a, will be given by

$$F = ma = \mu_{\text{slide}}mg.$$

Therefore $a = \mu_{\text{slide}}g = 0.25 \times 9.81\,\text{m}\,\text{s}^{-2} = 2.45\,\text{m}\,\text{s}^{-2}$.

Note that the acceleration will be in the direction of the frictional force which is in the opposite direction to that of the motion. In other words, the car is slowing down even though we speak of its acceleration. Note also that this is independent of the mass of the car.

Q1.8 At the terminal speed, the magnitude of the drag will be equal to that of the weight of the raindrop, so

$$mg = F_{\text{air}} = k_2R^2v^2.$$

Therefore $$v^2 = \frac{mg}{k_2R^2},$$

and so $$v = \frac{1}{R}\sqrt{\frac{mg}{k_2}}$$

$$v = \frac{1}{0.50 \times 10^{-3}}\sqrt{\frac{5.00 \times 10^{-7} \times 9.81}{0.80}}\,\text{m}\,\text{s}^{-1}$$

i.e. $v = 4.95\,\text{m}\,\text{s}^{-1}$.

To check if the assumption of the v^2 dependency is valid, the value of Rv must be determined. The value is given by

$$Rv = (0.50 \times 10^{-3} \times 4.95)\,\text{m}^2\,\text{s}^{-1} = 2.48 \times 10^{-3}\,\text{m}^2\,\text{s}^{-1}.$$

This lies in the required range from $10^{-4}\,\text{m}^2\,\text{s}^{-1}$ to $1\,\text{m}^2\,\text{s}^{-1}$ for the assumption to be valid.

Q1.9 In this case it is convenient to take the downward vertical as the positive x-direction. When the monkey hangs at rest from the spring, the magnitude of the monkey's weight must equal the magnitude of the tension force applied to the monkey by the spring. Let this equilibrium position be $x = 0$. It follows from the linear nature of the force law, that the restoring force arising from any *additional* extension of the spring will be proportional to that additional extension, so even though the equilibrium position, $x = 0$, does not correspond to the unstretched length of the spring, it will still be the case that the unbalanced restoring force on the monkey will be given by $F_x = -k_s x$, where k_s is the spring constant, and x is the monkey's position. It follows from Newton's second law that

$$ma_x = -k_s x.$$

Therefore

$$a_x = \frac{-k_s x}{m} = \frac{-120 \times 0.10}{5.0}\,\mathrm{m\,s^{-2}} = -2.40\,\mathrm{m\,s^{-2}}.$$

The negative sign indicates that this initial acceleration is directed upwards, towards the equilibrium position. (Note that a negative acceleration does *not* indicate a deceleration.)

Q1.10 When at rest, we can deduce from Newton's first law that there must be no resultant force acting on the person, or on the chair. That is not to say that there is no force acting on the person, but that any forces acting must balance out (i.e. the vector sum of the forces is zero). One force that certainly acts on the person is their own weight, which arises from the gravitational attraction of the Earth. The force which balances the weight is the upward force that the chair exerts on the person, thus leaving zero resultant force on the person. This upward force is the normal reaction that the chair exerts in response to the contact force that the person exerts on the chair. The contact force on the chair together with the chair's own weight cause it to exert a downward contact force on the floor. The normal reaction to this force acts on the chair and ensures that there is no unbalanced vertical force on the chair.

When the person pushes on the floor, in addition to creating a downward thrust on the floor, a horizontal frictional force is created on the floor in the direction in which the person is facing. From Newton's third law we know that the floor must exert a frictional force of equal magnitude on the feet of the person, but in the opposite direction. This frictional force will tend to make the seated person move backwards, and will be transmitted, via a contact force to the chair also. The chair will tend to move backwards as a result of this horizontal contact force, though the motion will be opposed by air resistance and friction in the castors. However, provided the total resistive force has a smaller magnitude than that of the frictional force on the person's feet, there will be an unbalanced force acting on the chair and it will

start to accelerate backwards. Provided there is enough frictional force between the chair and the person, the two will remain in contact and the person will move with the chair, contributing to the total mass being accelerated by the unbalanced force on the chair. After a short time, the person's feet will lose contact with the floor and the frictional force on the feet will vanish, leaving only the two resistive forces acting in a direction opposite to that of the motion. For this stage of the motion, Newton's second law tells us that there will be an acceleration in the direction of these resistive forces. An acceleration in a direction opposite to that of the motion means that the chair (and its occupant) will slow down and eventually come to rest.

Q1.11 In order for a body to execute uniform circular motion, that body must be continuously accelerated by a force directed towards the centre of the circle. That force is called the centripetal force, and its magnitude is given by $F = mv^2/r$, where m is the mass of the body, v is its (uniform) instantaneous speed, and r is the radius of the circular path that the body follows. (Alternatively, $F = mr\omega^2$, where ω is the angular speed of the body.)

The centripetal force on a uniformly circling body is a real force such as the gravitational pull of a planet, or the tension force in a string (which is ultimately due to electrical attraction between atoms in the string). In contrast, centrifugal force is a fictitious force with no physical basis. It has no role in explaining uniform circular motion, though an observer in a vehicle that was executing uniform circular motion (a bus turning a corner say) would experience real phenomena which he or she might interpret as the effects of a force (the centrifugal force) if they attempted, inappropriately, to apply Newton's laws of motion in the non-inertial frame attached to the vehicle.

Q1.12 Rearranging Equation 1.65:

$$m_{\mathrm{Sun}} = \frac{4\pi^2 r^3}{GT^2}$$

$$= \frac{4\pi^2 (5.79 \times 10^{10})^3}{6.67 \times 10^{-11} \times (88.0 \times 24 \times 60 \times 60)^2}\,\mathrm{kg}$$

$$= 1.99 \times 10^{30}\,\mathrm{kg}.$$

Q1.13 Let the masses of the three bodies be $4m$, $2m$ and $3m$ and the corresponding magnitudes of their accelerations be a_1, a_2 and a_3 when the bodies are subjected to a force of magnitude F. Therefore

$$F = 4ma_1 = 2ma_2 = 3ma_3$$

and so $a_1 = F/4m$, $a_2 = F/2m$ $a_3 = F/3m$.

Multiplying each equation by 12 and expressing a_1, a_2 and a_3 as ratios gives

$$a_1 : a_2 : a_3 = 3 : 6 : 4.$$

Q1.14 The physics book is subjected to two forces; its weight, W_p, which acts vertically downwards, and the normal reaction, N_{pc}, that the chemistry book exerts vertically upwards on the physics book. The third-law pair force associated with W_p is the upward force which the physics book exerts on the Earth. The third-law pair force associated with N_{pc} is the downward contact force that the physics book exerts on the chemistry book.

The chemistry book is subjected to three forces; the weight of the chemistry book, W_c, which acts vertically downwards, the normal reaction force, N_{ct}, which the table top exerts vertically upwards on the chemistry book, and the contact force that the physics book exerts on the chemistry book (this latter force will be $-N_{pc}$). The third-law pair force associated with W_c is the upward force which the chemistry book exerts on the Earth. The third-law pair force associated with N_{ct} is the downward contact force that the chemistry book exerts on the table top. The third-law pair force associated with the contact force that the physics book exerts is the normal reaction force on the physics book, N_{pc}, that has already been discussed.

Q1.15 The block starts to slide when the magnitude of the applied force equals F_{max}, where F_{max} is given by $F_{max} = \mu_{static}N$, and N is the magnitude of the normal reaction on the block.

Therefore $\quad F_{max} = \mu_{static}mg$.

When sliding occurs, the magnitude of the friction force acting on the block drops to a value F given by $F = \mu_{slide}N = \mu_{slide}mg$.

The magnitude of the resultant force acting on the block is therefore given by

$$F_{max} - F = \mu_{static}mg - \mu_{slide}mg = (\mu_{static} - \mu_{slide})mg.$$

From Newton's second law, the magnitude a of the horizontal acceleration is given by

$$ma = (\mu_{static} - \mu_{slide})mg.$$

Therefore $\quad a = (\mu_{static} - \mu_{slide})g = (0.50 - 0.40) \times 9.81 \text{ m s}^{-2} = 0.981 \text{ m s}^{-2}$.

Q1.16 (a) Given that the positive x-direction is downwards, the person's weight produces a constant force $W_x = mg$. The force due to air resistance always opposes the velocity of the person, so it may be represented by $F_x^{air} = -C\,dx/dt$, where C is a constant of proportionality. It follows from Newton's second law, that if the mass of the person is m, then the equation of motion will be

$$\frac{d^2x}{dt^2} = g - \frac{C}{m}\frac{dx}{dt}.$$

(b) In the absence of air resistance the velocity of the falling person would increase uniformly and the velocity–time graph would be a straight line. However, due to air resistance, the falling person will eventually reach a terminal velocity after which they will cease to accelerate. As a result, the velocity–time graph will look something like that shown in Figure 1.58.

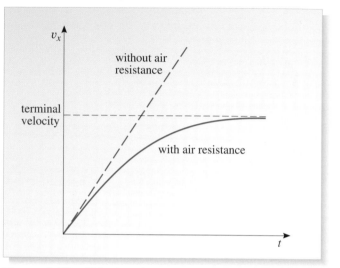

Figure 1.58 Velocity–time graph for a person falling from a plane. See Q1.16.

Q1.17 F_{ME} is given by:

$$\begin{aligned}
F_{ME} &= \frac{Gm_Mm_E}{r^2} \\
&= \frac{6.67 \times 10^{-11} \times 7.35 \times 10^{22} \times 5.98 \times 10^{24}}{(0.38 \times 10^9)^2} \text{ N} \\
&= 2.03 \times 10^{20} \text{ N}.
\end{aligned}$$

Using Newton's law of universal gravitation again, the separation d between two particles of equal mass m can be determined as follows:

$$F_{ME} = \frac{Gm^2}{d^2}.$$

Therefore $\quad d = m\sqrt{\dfrac{G}{F_{ME}}} = 1.0 \times \sqrt{\dfrac{6.67 \times 10^{-11}}{2.03 \times 10^{20}}}$ m

$$= 5.73 \times 10^{-16} \text{ m}.$$

Note that this is smaller than the diameter of an atom.

Q2.1 (a) From Equation 2.3, we know that the work done, W, on the trolley is given by $F_x s_x = (40\,\text{N}) \times (1\,\text{m})$. We can then make use of Equation 2.2. Since $u = 0$, $W = \frac{1}{2}mv^2$.

So, $\frac{1}{2}(5\,\text{kg})v^2 = (40\,\text{N}) \times (1\,\text{m}) = 40\,\text{kg m}^2\,\text{s}^{-2}$.

Thus $v^2 = 16\,\text{m}^2\,\text{s}^{-2}$ and hence $v = 4\,\text{m s}^{-1}$. This is the final speed of the trolley.

(b) Using the same argument as in (a), but with $m = 10\,\text{kg}$, leads to $\frac{1}{2}(10\,\text{kg})v^2 = 40\,\text{kg m}^2\,\text{s}^{-2}$, and hence $v = 2.8\,\text{m s}^{-1}$.

(c) Again we can use Equation 2.2, but this time with $u = 2\,\text{m s}^{-1}$, and $W = 40\,\text{N m}$ representing the *change* in kinetic energy.

With an initial kinetic energy of $\frac{1}{2}(5\,\text{kg})(2\,\text{m s}^{-1})^2 = 10\,\text{N m}$, this implies that the final kinetic energy is $\frac{1}{2}(5\,\text{kg})v^2 = (40 + 10)\,\text{N m} = 50\,\text{N m}$. So, $v = 4.5\,\text{m s}^{-1}$.

(d) In each case the change in kinetic energy is given by $W = F_x s_x = 40\,\text{N m} = 40\,\text{J}$.

Q2.2 (a) Since the slope is at $30°$ to the horizontal, the skier's weight is at $60°$ to the displacement. Hence, since $F = mg = 60 \times 9.8\,\text{N} = 588\,\text{N}$, and $s = 1000\,\text{m}$,

$$W = Fs\cos\theta = (588 \times 1000 \cos 60°)\,\text{J}$$

$$= 5.88 \times 0.5 \times 10^5\,\text{J} = 2.9 \times 10^5\,\text{J}.$$

(b) The work done is independent of the initial speed of the skier, since $W = Fs\cos\theta$, and this does not depend on the initial speed.

(c) The normal reaction is perpendicular to the displacement, so it does no work.

Q2.3 (a) If a and b are at right angles, $\theta = 90°$, so $a \cdot b = ab\cos 90° = 0$, because $\cos 90°$ is zero. You should *remember* this: the scalar product of *any* two vectors at right angles is zero.

(b) If a and b are in opposite directions, $\theta = 180°$, so $a \cdot b = ab\cos 180° = -ab$, since $\cos 180° = -1$.

(c) If a and b are equal, $a \cdot a = a^2\cos 0° = a^2$. The scalar product of a vector with itself is just the square of the magnitude of the vector.

Q2.4 Use $F \cdot s = F_x s_x + F_y s_y + F_z s_z = Fs\cos\theta$ with $F_x = 1\,\text{N}$, $F_y = 2\,\text{N}$, $F_z = 3\,\text{N}$, and $s_x = 2\,\text{m}$, $s_y = 3\,\text{m}$, $s_z = 1\,\text{m}$.

So $F^2 = F_x^2 + F_y^2 + F_z^2 = 14\,\text{N}^2$ and $s^2 = s_x^2 + s_y^2 + s_z^2 = 14\,\text{m}^2$,

implying that $(Fs)^2 = (14\,\text{N m})^2$, so that $Fs = 14\,\text{N m} = 14\,\text{J}$.

Thus, $F \cdot s = Fs\cos\theta = (14\cos\theta)\,\text{J}$,

However, $F \cdot s = F_x s_x + F_y s_y + F_z s_z = 11\,\text{J}$.

Hence, $\cos\theta = F \cdot s / Fs = 11\,\text{J}/14\,\text{J} = 0.786$, so $\theta = \arccos 0.786 = 38.2°$.

Q2.5 The frictional force on a sliding book *always opposes* motion of the book. All amounts of work done following a closed path therefore have the same negative sign; there can be no cancellation to give zero. The frictional force cannot, therefore, be a conservative force.

Q2.6 (a) We will use the constant acceleration equation, $v^2 = u^2 + 2a_x s_x$, with $a_x = -g = -9.8\,\text{m s}^{-2}$ (taking the upward direction to be positive).

We are told that $u = 200\,\text{m s}^{-1}$, and we want to determine the value of s_x for which $v = 0$, i.e. the height at which the ball comes to rest. This will be given by

$$s_x = \frac{v^2 - u^2}{2a_x},$$

hence $s_x = \dfrac{(200)^2}{2 \times 9.8}\,\text{m} = 2.0 \times 10^3\,\text{m}$.

So the cannon-ball rises to a height of $2.0\,\text{km}$.

(b) The work done by gravity on the rising cannon-ball is found by using the equation $W = F \cdot s$. We know that F is $(9.8 \times 10)\,\text{N}$ directed vertically downwards, and s is $2.0\,\text{km}$ vertically upwards. Hence $W = -98\,\text{N} \times 2000\,\text{m} = -2.0 \times 10^5\,\text{J}$.

(c) During the descent to the halfway point, the height changes by an amount $\Delta h = -1000\,\text{m}$. It follows from Equation 2.14 that the descent causes the gravitational potential energy to change by the amount $\Delta E_{\text{grav}} = -10 \times 9.8 \times 1000\,\text{J} = -9.8 \times 10^4\,\text{J}$.

(d) The work done on the cannon-ball by gravity is equal to the *reduction* in gravitational potential energy during the descent to the halfway point. It is therefore the negative of the answer to part (c), i.e. $+9.8 \times 10^4\,\text{J}$.

Q2.7 Potential energy is energy by virtue of position. In order for potential energy to be meaningfully defined it is therefore necessary that we should be able to associate a particular amount of energy with any given position. Conservative forces allow us to do this since the work done by such forces (i.e. the energy transferred by them) when a body moves from one position to another is always the same, no matter how the move is made. Non-conservative forces do not permit such a link to be made since the work they do will depend on the path followed.

Q2.8 Let the spring lie along the x-direction, with the free end of the spring at the origin. When stretched by $20\,\text{cm}$, the strain potential energy stored in the spring is

$$E_{\text{str}} = \frac{1}{2}k_s x^2 = 0.5 \times 200 \times (0.20)^2\,\text{J} = 4.0\,\text{J}.$$

If all of the stored energy is converted into kinetic energy of the projectile, then $E_{\text{trans}} = \frac{1}{2}mv^2 = E_{\text{str}} = 4.0\,\text{J}$.

Since $m = 0.005$ kg,

$$v^2 = \frac{8.0}{0.005} \, \text{m}^2 \, \text{s}^{-2}, \quad \text{so,} \quad v = 40 \, \text{m s}^{-1}.$$

Q2.9   Let the spring lie along the x-direction with the free end of the spring at the origin. The energy stored when the extension is 15 cm is

$$E_{str} = \tfrac{1}{2} k_s x^2 = 0.5 \times 200 \times (0.15)^2 \, \text{J} = 2.3 \, \text{J}.$$

When the extension is 20 cm, the energy stored is

$$E_{str} = 0.5 \times 200 \times (0.20)^2 \, \text{J} = 4.0 \, \text{J}.$$

So to stretch the spring by 5 cm from 15 cm to 20 cm requires $4.0 \, \text{J} - 2.3 \, \text{J} = 1.7 \, \text{J}$.

The energy stored when the extension is 25 cm is

$$E_{str} = 0.5 \times 200 \times (0.25)^2 \, \text{J} = 6.3 \, \text{J}.$$

So to stretch the spring by 5 cm from 20 cm to 25 cm requires $6.3 \, \text{J} - 4.0 \, \text{J} = 2.3 \, \text{J}$.

As you probably anticipated, more energy is needed to stretch the spring from 20 cm to 25 cm extension than to stretch it from 15 cm to 20 cm extension.

Q2.10   (a) My mass is about 65 kg, and my centre of mass is currently about 1.0 m above the ground, so using Equation 2.12 my gravitational potential energy is

$E_{grav} = mgh = (65 \times 9.81 \times 1.0) \, \text{J} = 6.4 \times 10^2 \, \text{J}.$

However, using Equation 2.25, my gravitational potential energy is

$$E_{grav} = \frac{-GmM_E}{R_E}$$
$$= \frac{-6.673 \times 10^{-11} \times 65 \times 5.977 \times 10^{24}}{6378 \times 10^3} \, \text{J} = -4.065 \times 10^9 \, \text{J}.$$

(b) Equation 2.12 predicts that at an altitude of 10 km my gravitational potential energy will be increased by

$$\Delta E_{grav} = mg\Delta h = (65 \times 9.81 \times 10^4) \, \text{J} = 6.4 \times 10^6 \, \text{J}.$$

Equation 2.25, on the other hand, implies that at an altitude of 10 km my gravitational potential energy is

$$E_{grav} = \frac{-GmM_E}{R_E + 10 \, \text{km}}$$
$$= \frac{-6.673 \times 10^{-11} \times 65 \times 5.977 \times 10^{24}}{6388 \times 10^3} \, \text{J} = -4.058 \times 10^9 \, \text{J}.$$

So, $\Delta E_{grav} = [(-4.058 \times 10^9) - (-4.065 \times 10^9)] \, \text{J} = 7 \times 10^6 \, \text{J}.$

So, the two methods provide consistent values for ΔE_{grav} within the precision that can be achieved.

(c) Air resistance influences the work that must be done in raising you to an altitude of 10 km, but it has no effect on the work done by the conservative gravitational force during that process. It therefore has no effect on the answer to part (b).

Q2.11   We can use Equation 2.30 to establish that the restoring force is the derivative of the potential energy

$$F_x = \frac{-dE_{pot}}{dx}.$$

If $E_{pot} = E_{str} = \tfrac{1}{2} k_s x^2$, we are concerned with distances in the x-direction so we differentiate this equation with respect to x. If we call the force exerted by the spring F_x, then

$$F_x = \frac{-d}{dx}\left(\tfrac{1}{2} k_s x^2\right) = -\tfrac{1}{2} k_s (2x) = -k_s x,$$

which is Hooke's law. The minus sign occurs because the restoring force F_x acts in the opposite direction to the extension x.

Q2.12   In one dimension, the force is given generally by Equation 2.30

$$F_x = \frac{-dE_{pot}}{dx}$$

where x may be interpreted as the displacement from some agreed origin. Since we are concerned with the case $E_{pot} = E_{grav} = mgh$, we can choose the x-direction to be vertically upwards in this case, identify x with h, and write

$$F_h = \frac{-d(mgh)}{dh} = -mg.$$

The minus sign indicates that the force is in the opposite direction to that in which h increases. (The *magnitude* of this force is mg.)

Q2.13   The (universal) gravitational potential is given by:

$$E_{grav} = \frac{-GmM_E}{r} \quad (r > R_E).$$

In one dimension, a conservative force is generally related to a potential energy by

$$F_x = \frac{-dE_{pot}}{dx}$$

so to apply this we must interpret x as r in this case. Thus

$$F_r = \frac{-dE_{grav}}{dr} = \frac{-d}{dr}\left(\frac{-GmM_E}{r}\right).$$

GmM_E does not vary with r, so may be brought outside the differentiation

$$F_r = GmM_E \frac{d}{dr}\left(\frac{1}{r}\right)$$

where the minus signs have, of course, cancelled. F_r is in the direction of increasing r.

Now $\dfrac{d}{dr}\left(\dfrac{1}{r}\right) = -\left(\dfrac{1}{r}\right)^2$. So, $F_r = \dfrac{-GmM_E}{r^2}$

which is Newton's law of gravitation. (You might also have written your final answer in vector terms, using an appropriate unit vector in the r-direction.)

Q2.14   According to Equation 2.30, the gradient of a potential energy graph such as that shown in Figure 2.26 is equal to *minus* the corresponding component of the conservative force responsible for the potential energy. In the case of Figure 2.26, the relevant force component close to the Earth's surface is $F_r = -mg$. It therefore follows that the gradient of the straight line, which approximates the potential energy curve in the boxed region near point A, will be *mg*. To put it slightly differently, the gradient of the graph of E_{grav} against r will be approximately equal to *mg* in the region close to $r = R_E$.

Q2.15   (a) Letting the initial speed be u, $E_{trans} = \frac{1}{2} mu^2 = \frac{1}{2} \times 10 \times (200)^2$ J $= 2 \times 10^5$ J.

(b) In the absence of air resistance, mechanical energy will be conserved in this case, so $E_{mech} = E_{trans} + E_{pot} = $ constant, and in any process $\Delta E_{pot} + \Delta E_{trans} = 0$.

When at its highest point the speed of the ball will be zero, so in the process of reaching that point from the ground $\Delta E_{trans} = -2 \times 10^5$ J.

However, the corresponding change in gravitational potential energy is $\Delta E_{pot} = mg\Delta h$.

So, conservation of mechanical energy implies $mg\Delta h = 2 \times 10^5$ J, from which it follows that

$$\Delta h = \frac{2 \times 10^5 \text{ J}}{10 \text{ kg} \times 9.8 \text{ m s}^{-2}} = 2.0 \times 10^3 \text{ m}.$$

So the cannon-ball rises to a height of 2 km.

(c) Similarly, in returning halfway to the ground, the change in gravitational potential energy is $\Delta E_{pot} = mg\Delta h = -1.0 \times 10^5$ J. So, if we let the speed at the halfway point be v, so that $\Delta E_{trans} = \frac{1}{2} mv^2$, then conservation of mechanical energy implies

$$v = \sqrt{\left(\frac{2 \times 1.0 \times 10^5}{10}\right)} \text{ m s}^{-1} = 1.4 \times 10^2 \text{ m s}^{-1}.$$

(d) Just before reaching the ground, the kinetic energy of the cannon-ball will be equal to the kinetic energy immediately after launch. Consequently, the speed will equal the launch speed 200 m s^{-1}.

(e) When the cannon-ball hits the ground it is brought to rest, so that it has neither translational kinetic energy nor gravitational potential energy. Thus the process of bringing the ball to rest is not one that conserves mechanical energy. We can therefore conclude that the forces involved are non-conservative.

Q2.16   If we substitute the data into Equation 2.35 we find:

$$v_{escape} = \sqrt{\frac{2GM_E}{R_E}}$$

$$= \left(\frac{2 \times 6.67 \times 10^{-11} \times 5.98 \times 10^{24}}{6.38 \times 10^6}\right)^{1/2} \text{ m s}^{-1}$$

$$= 1.1 \times 10^4 \text{ m s}^{-1}.$$

So, the required speed is 11 km s^{-1}.

Q2.17   The power is given by $P = \boldsymbol{F} \cdot \boldsymbol{v}$.

However, since the net force on the airliner is in the same direction as its velocity, this becomes

$$P = Fv = 6.72 \times 10^6 \text{ N} \times 90 \text{ m s}^{-1} = 6.0 \times 10^8 \text{ W}.$$

Q2.18   Power $= \boldsymbol{F} \cdot \boldsymbol{v} = Fv \cos \theta$
$$= 300 \text{ N} \times 2.0 \text{ m s}^{-1} \times \cos 30° = 520 \text{ W}.$$

Q2.19   Figure 2.51 shows the required energy–displacement graphs. The potential energy has the parabolic form shown in Figure 2.32, and has been drawn assuming that the potential energy is zero when the oscillator passes through its equilibrium position. The total mechanical energy is constant and is equal to the maximum value of the potential energy. The kinetic energy is equal to the difference between the total mechanical energy and the potential energy. It too, like the potential energy, is represented by a parabolic curve.

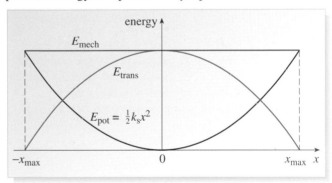

Figure 2.51   For Q2.19.

Q2.20   (a) At $t = 0$, all the exponential terms equal 1 because $e^0 = 1$. So $y_1 = 10$, $y_2 = 20$, $y_3 = 5$. Thus, y_2 has the largest value at $t = 0$.

(b) The exponential with the shortest time constant is the one which decays most quickly. Compare each exponential to the general form $v = v_0 e^{-t/\tau}$ to find τ.

$y_1 = 10e^{-12t}$ tells us that $\tau = 1/12$.

$y_2 = 20e^{-6t}$ tells us that $\tau = 1/6$.

$y_3 = 5e^{-t/6}$ tells us that $\tau = 6$.

y_1 has the shortest time constant so it decays most quickly.

Q2.21 Here is the procedure:

1 Select a starting value on the vertical axis, and read off the corresponding value on the horizontal axis. (This starting value might be v_0, in which case the corresponding value on the horizontal axis will be $t = 0$, but other choices are equally valid.)

2 Divide the chosen starting value by e (2.718 to three decimal places), locate the resulting value on the vertical axis, and read off the corresponding value on the horizontal axis.

3 Subtract the horizontal axis reading obtained in step 1 from the corresponding reading obtained in step 2. The result will be the interval along the horizontal axis that corresponds to a change by a factor of 1/e on the vertical axis. This is the value of τ.

Applying this procedure to Figure 2.37, with $v_x = 0.1\,\mathrm{m\,s^{-1}}$ as the starting value, the first reading on the horizontal axis will be $t = 0$. The required value of v_x for step 2 will be $(0.1/\mathrm{e})\,\mathrm{m\,s^{-1}} = 0.0368\,\mathrm{m\,s^{-1}}$, and the corresponding value of t (from the graph) will be $t = 2.5\,\mathrm{s}$. This will be equal to the value of τ. (In fact, the graph was plotted using the value $\tau = 2.5$.)

Q2.22 (a) The decay in Figure 2.48 is exponential, with a time constant of 3.0 s. For a given exponential process, the fractional change of amplitude in a time interval of specified length is always constant. From Figure 2.48,

$$\frac{A(2\,\mathrm{s})}{A(0\,\mathrm{s})} = 0.51, \quad \frac{A(3\,\mathrm{s})}{A(1\,\mathrm{s})} = 0.51, \quad \frac{A(4\,\mathrm{s})}{A(2\,\mathrm{s})} = 0.51.$$

In each of the 2 s intervals considered here, the amplitude drops by the same factor (to about 51% of its initial value in this example) and this indicates that the curve is exponential. The time constant τ is the time taken for the amplitude to drop to 37% (i.e. 1/e) of its original value, and for the decay in Figure 2.48 this is 3.0 s.

(b) Removing the air will reduce the damping. This will not significantly affect the period of the vibrations, but the exponential decay will be much slower, i.e. the time constant τ will be much longer than 3 s.

(c) The period of the vibrations in Figure 2.48 is 0.2 s, and so the frequency ($f = 1/T$) is 5.0 Hz, and the angular frequency ($\omega = 2\pi/T$) is $10\pi\,\mathrm{s^{-1}}$. This is the natural frequency ω_0, i.e. the angular frequency of vibration in the absence of external forces. The dissipative forces have a negligible effect on the vibration frequency except when the damping is very high.

(d) Maximum amplitude vibrations will be produced when the driving frequency is equal to the resonant frequency. This will be close to the natural frequency.

Q2.23 The vibration amplitude could be reduced by:

(i) increasing the damping (this would flatten out the resonance curve, and the response at 10 Hz would be reduced);

(ii) changing the stiffness of the springy feet;

(iii) adding extra mass to the pump.

Changes (ii) and (iii) would alter the natural frequency of the system, and therefore they would shift the resonant frequency away from 10 Hz. This would reduce the amplitude of forced vibrations at 10 Hz.

Q2.24 Work is done by a force if the force has a component in the direction of the displacement of the body on which the force acts. Positive work done tends to increase the kinetic energy of the body. Work done = $W = Fs \cos\theta$.

(a) When $\theta = 0°$, $W = 5\,\mathrm{N} \times 3\,\mathrm{m} \times \cos 0° = 15\,\mathrm{J}$.

(b) When $\theta = 53.1°$, $W = 5\,\mathrm{N} \times 3\,\mathrm{m} \times \cos 53.1° = 9\,\mathrm{J}$.

(c) When $\theta = 90°$, $W = 5\,\mathrm{N} \times 3\,\mathrm{m} \times \cos 90° = 0\,\mathrm{J}$.

Q2.25 $\frac{1}{2}k_A(2x)^2 = \frac{1}{2}k_B x^2$, so $\frac{k_A}{k_B} = \frac{1}{4}$.

Q2.26 Use $F_x = ma_x$ and the constant acceleration equation $v^2 = u^2 + 2a_x s_x$ to show that $F_x s_x = \frac{1}{2}mv^2 - \frac{1}{2}mu^2$.

This is essentially the proof of the work–energy theorem, full details may be found in Section 2.

Q2.27 From Equation 2.39, $P = \boldsymbol{F} \cdot \boldsymbol{v}$, and in this case alignment of the vectors means that $P = Fv$. Thus,

$$F = \frac{P}{v} = \frac{20\,\mathrm{W}}{8\,\mathrm{m\,s^{-1}}} = 2.5\,\mathrm{N}.$$

Q2.28 A force acting on a body is conservative if the total work done by the force on the body is zero for any round trip made by the body. (Alternatively, the work done by the force when the body moves between any two points is independent of the path followed by the body.) Non-conservative forces do not fulfil these conditions.

Potential energy is the energy associated with position or configuration, and is stored energy that can be retrieved. The concept of potential energy is meaningful only when associated with conservative forces, since it is only these forces that are guaranteed to always do the same amount of work in a given change of position or configuration, irrespective of the path followed. The change in potential energy in going from some given initial configuration of a system to some final configuration is equal to the negative of the work done by the related conservative force during that change of configuration. Thus

$$E_{\mathrm{pot}}(\text{final}) - E_{\mathrm{pot}}(\text{initial}) = \Delta E_{\mathrm{pot}} = -W_{\mathrm{cons}}(\text{initial} \to \text{final}).$$

A consequence of this, for a one-dimensional system, is that

$$F_x = \frac{-dE_{pot}}{dx}.$$

Q2.29   The force may be completely specified by its radial component F_r. It follows from Equation 2.30 that $F_r = \frac{-dE_{pot}}{dr}$, where the derivative represents the gradient of the graph of $E_{pot} = kr$ against r. Since this graph is a straight-line graph of constant gradient k, it follows that $F_r = -k$. Thus the force has the same magnitude everywhere and is directed towards O if k is positive.

Q2.30   Almost all the kinetic energy is converted into thermal energy, which causes the temperatures of the surfaces in contact to increase temporarily. The work done by the frictional force is equal to the change in the kinetic energy of the block. If the magnitude of that force is F, then $-Fd = 0 - \frac{1}{2}mu^2$.

So, $F = \dfrac{mu^2}{2d}.$

Q2.31   The escape speed for the Moon is given by an equation similar to Equation 2.35

$$v_{escape}(\text{Moon}) = \sqrt{\frac{2GM_M}{R_M}}.$$

where G is the gravitational constant $G = 6.67 \times 10^{-11}\,\text{N m}^2\,\text{kg}^{-2}$ and the parameters for the Moon are given as $M_M = 7.35 \times 10^{22}\,\text{kg}$ and $R_M = 1.74 \times 10^6\,\text{m}$. So

$$v_{escape} = \left(\frac{2 \times 6.67 \times 10^{-11} \times 7.35 \times 10^{22}}{1.74 \times 10^6} \right)^{1/2}\,\text{m s}^{-1}$$

$$= 2.4 \times 10^3\,\text{m s}^{-1}.$$

This escape speed of $2.4\,\text{km s}^{-1}$ may be compared with the escape speed of about $11\,\text{km s}^{-1}$ from the Earth.

Q2.32   The three cases are shown in the graph in Figure 2.52. An example of the use of each is the following:

(a) Heavy damping is used for the mechanism of a swing door to return but not so quickly for someone to be trapped. (b) Critical damping is used for the pointer of a measuring instrument to reach a reading as quickly as possible without overshooting more than once. (c) Light damping is used for a musical instrument to sustain the note for as long as required.

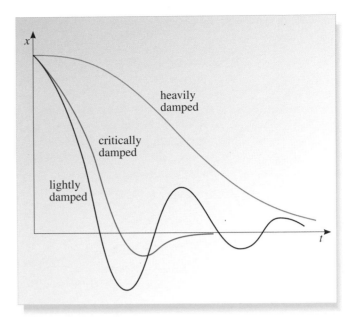

Figure 2.52   For Q2.32.

Q3.1   Start by setting up a coordinate system. Call north the y-direction and east the x-direction as in Figure 3.19.

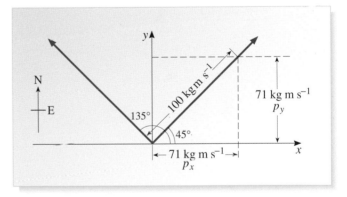

Figure 3.19   For Q3.1.

(a) From the definition of momentum, the magnitude is given by $p = mv = 5.0\,\text{kg} \times 20\,\text{m s}^{-1} = 100\,\text{kg m s}^{-1}$. The direction is 45° to the positive x- and y-axes, i.e. north-east.

Notice that multiplying the velocity *vector* by the mass, which is a scalar, gives a *vector in the same direction* as the velocity. I hope that you remembered to specify the *direction* of the momentum: as with all vector quantities, the complete specification requires *both magnitude and direction*. Notice also that the units of momentum are the units of mass × the units of velocity, i.e. kg m s^{-1}.

(b) The components in the easterly direction and in the northerly direction are p_x and p_y, respectively:

$$p_x = 1.0 \times 10^2 \, \text{kg m s}^{-1} \times \cos 45° = 71 \, \text{kg m s}^{-1};$$

$$p_y = 1.0 \times 10^2 \, \text{kg m s}^{-1} \times \sin 45° = 71 \, \text{kg m s}^{-1}.$$

(c) If the body is moving north-west at the same speed, p_y would have the same value, but p_x would be $-71 \, \text{kg m s}^{-1}$, i.e. in the opposite direction. But how does this come out of the mathematics? Well, since the components we want are p_x and p_y, the angle must be measured from the x-direction, and it is therefore 135°, as shown in Figure 3.19. Hence

$$p_x = 1.0 \times 10^2 \, \text{kg m s}^{-1} \times \cos 135° = -71 \, \text{kg m s}^{-1};$$

$$p_y = 1.0 \times 10^2 \, \text{kg m s}^{-1} \times \sin 135° = 71 \, \text{kg m s}^{-1}.$$

Q3.2 The important point to note is that the direction of the velocity vector changes after the collision (Figure 3.20).

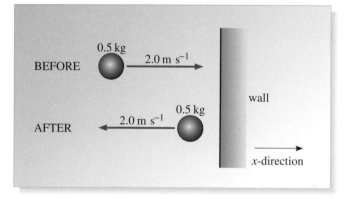

Figure 3.20 For Q3.2.

Before collision:
$$p_x = mv_x = 0.5 \, \text{kg} \times 2.0 \, \text{m s}^{-1} = 1.0 \, \text{kg m s}^{-1}.$$

After collision:

$$p_x = mv_x = 0.5 \, \text{kg} \times (-2.0 \, \text{m s}^{-1}) = -1.0 \, \text{kg m s}^{-1}.$$

The *change* in a quantity (often expressed using the symbol Δ) is given by

Δquantity = (*final* value of quantity) − (*initial* value of quantity)

and so it can be positive or negative depending on whether the value of the quantity increases or decreases.

So, Δp_x = (momentum after collision) − (momentum before collision)

i.e. $\Delta p_x = (-1.0 \, \text{kg m s}^{-1}) - (1.0 \, \text{kg m s}^{-1}) = -2.0 \, \text{kg m s}^{-1}.$

Notice that the final momentum in the x-direction, p_x(initial) + Δp_x, will be reduced because Δp_x is negative in this case.

Q3.3 You are asked for the final momentum, so you can usefully draw a vector diagram showing the momentum components after player 2 has struck the ball. Call the horizontal direction from player 1 to player 2 the x-direction, and the vertical the y-direction. The vector diagram is shown in Figure 3.21, together with the coordinate system chosen. You will need to use the equations for momentum: $p_x = mv_x$ and $p_y = mv_y$. The key to the problem is that the x-component is *not* changed by player 2, who applies only a *vertical* force.

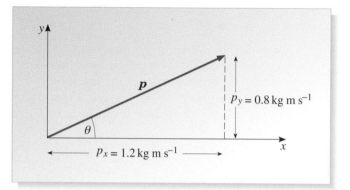

Figure 3.21 For Q3.3.

The x-component of momentum of the ball is the same before and after player 2 strikes the ball and is given by

$$p_x = mv_x = 0.40 \, \text{kg} \times 3.0 \, \text{m s}^{-1} = 1.2 \, \text{kg m s}^{-1}.$$

The question *tells* you that $p_y = 0.8 \, \text{kg m s}^{-1}$ after player 2 strikes the ball. So you now have the components of the final total momentum and can draw these as in Figure 3.21.

Using Pythagoras' theorem,

$$p^2 = p_x^2 + p_y^2$$
$$= (1.2)^2 \, \text{kg}^2 \, \text{m}^2 \, \text{s}^{-2} + (0.8)^2 \, \text{kg}^2 \, \text{m}^2 \, \text{s}^{-2}$$
$$= 2.1 \, \text{kg}^2 \, \text{m}^2 \, \text{s}^{-2}$$

Therefore $p = 1.4 \, \text{kg m s}^{-1}.$

p is at an angle θ above the original direction of motion such that $\tan \theta = 0.8/1.2$ and therefore $\theta = \arctan 0.67 = 34°$. The value for p is reasonable because it is greater than its initial magnitude ($1.2 \, \text{kg m s}^{-1}$) but not as great as the sum of its initial magnitude and the magnitude of the additional momentum supplied by player 2 (i.e. $0.8 \, \text{kg m s}^{-1} + 1.2 \, \text{kg m s}^{-1} = 2.0 \, \text{kg m s}^{-1}$). This is because the additional momentum is supplied at right angles to the initial momentum.

Q3.4   If the velocity is constant the acceleration is zero. Therefore, from Equation 3.3, with the direction of motion along the x-axis we find

$$F_x = 0 + v_x \frac{dm}{dt} = (5\,\text{m s}^{-1}) \times (6\,\text{kg s}^{-1}) = 30\,\text{N}.$$

Q3.5   (a) Call the direction of the trains the x-direction.

$$P_x = \sum_i (mv_x)_i \quad \text{where } i = 1, 2.$$

Therefore $P_x = m_1(v_x)_1 + m_2(v_x)_2$

where $(v_x)_1$ is the velocity of train 1 and $(v_x)_2$ is the velocity of train 2.

For train 1, (note the conversion from km h^{-1} to m s^{-1})

$$(p_x)_1 = m_1(v_x)_1$$

$$= 5.0 \times 10^4\,\text{kg} \times \frac{90}{3600} \times 10^3\,\text{m s}^{-1}$$

$$= 1.25 \times 10^6\,\text{kg m s}^{-1}.$$

For train 2,

$$(p_x)_2 = m_2(v_x)_2 = m_1(v_x)_1 = 1.25 \times 10^6\,\text{kg m s}^{-1}.$$

Therefore $P_x = (p_x)_1 + (p_x)_2 = 2.50 \times 10^6\,\text{kg m s}^{-1}$, which is positive, and therefore in the positive x-direction.

(b) If train 2 is travelling in the opposite direction, its velocity $(v_x)_2$ is in the opposite direction to $(v_x)_1$ and so the magnitude of its momentum is the same but the sign reverses. Consequently

$$(p_x)_2 = -1.25 \times 10^6\,\text{kg m s}^{-1}.$$

and $P_x = (p_x)_1 + (p_x)_2$

$$= 1.25 \times 10^6\,\text{kg m s}^{-1} - 1.25 \times 10^6\,\text{kg m s}^{-1} = 0.$$

Thus the *total* momentum of a system can be zero, even if constituent parts are moving *relative* to one another very quickly (in this case at 180 km h^{-1}).

Q3.6   When you are walking, you are not an isolated system since you are interacting with the surface of the Earth through friction. When you start from rest, the Earth recoils imperceptibly in the opposite direction to your motion. Similarly, when you slow down and stop, you transfer momentum to the Earth in the same direction as your motion.

Q3.7   Let the 0.3 kg block move in the $-x$-direction and the 0.2 kg block in the $+x$-direction. The total momentum of the two blocks before the impact is the vector sum of the momenta of the two individual blocks, so conservation of momentum implies

$$m_1 u_{1x} + m_2 u_{2x} = (m_1 + m_2) v_x$$

where v_x is the common final velocity. Thus

$$v_x = \frac{m_1 u_{1x} + m_2 u_{2x}}{m_1 + m_2}.$$

So, $v_x = \dfrac{0.2\,\text{kg} \times 2.0\,\text{m s}^{-1} - 0.3\,\text{kg} \times 1.0\,\text{m s}^{-1}}{0.2\,\text{kg} + 0.3\,\text{kg}}$

i.e. $v_x = 0.1\,\text{kg m s}^{-1}/0.5\,\text{kg} = 0.2\,\text{m s}^{-1}.$

Q3.8   The total momentum of the isolated system of the Earth plus the walkers is constant. On starting the walk, the walkers' total momentum *increases* in the direction of the walk by an amount given by their total mass multiplied by their speed. If we estimate the average mass per person as 50 kg, then the total increase in momentum in the direction of the walk is

$$mv = 5.6 \times 10^9 \times 50\,\text{kg} \times 1.0\,\text{m s}^{-1} = 2.8 \times 10^{11}\,\text{kg m s}^{-1}.$$

The Earth's momentum in this same direction *decreases* by this same amount. The change in the Earth's velocity, Δv, is equal to the change in the momentum, Δp, divided by the Earth's mass, so its speed changes by the amount:

$$\Delta v = \frac{\Delta p}{M_E} = \frac{2.8 \times 10^{11}\,\text{kg m s}^{-1}}{6.0 \times 10^{24}\,\text{kg}} = 4.7 \times 10^{-14}\,\text{m s}^{-1}.$$

When the walkers stop, the changes are reversed and the Earth returns to its original velocity, although the changes would have been imperceptibly small.

Q3.9   First calculate the loss of kinetic energy for the neutron, and divide by the neutron's initial kinetic energy:

The initial kinetic energy is

$$E_{\text{trans}} = \tfrac{1}{2} mu_{1x}^2$$

and the final kinetic energy is

$$E_{\text{trans}} = \tfrac{1}{2} mv_{1x}^2,$$

therefore the loss in energy is

$$\Delta E_{\text{trans}} = \tfrac{1}{2} mu_{1x}^2 - \tfrac{1}{2} mv_{1x}^2 = \tfrac{1}{2} m(u_{1x}^2 - v_{1x}^2)$$

and the fractional loss is

$$\frac{\Delta E_{\text{trans}}}{E_{\text{trans}}} = \frac{u_{1x}^2 - v_{1x}^2}{u_{1x}^2} = 1 - \left(\frac{v_{1x}}{u_{1x}}\right)^2.$$

With $u_{2x} = 0$ in Equation 3.16,

$$v_{1x} = u_{1x} \frac{m_1 - m_2}{m_1 + m_2},$$

we can write the fractional loss as

$$\frac{\Delta E_{\text{trans}}}{E_{\text{trans}}} = 1 - \left(\frac{v_{1x}}{u_{1x}}\right)^2 = 1 - \left(\frac{m_1 - m_2}{m_1 + m_2}\right)^2.$$

For gold $\dfrac{\Delta E_{\text{trans}}}{E_{\text{trans}}} = 1 - \left(\dfrac{196}{198}\right)^2 = 0.02$ (i.e. 2.0%).

For carbon $\dfrac{\Delta E_{\text{trans}}}{E_{\text{trans}}} = 1 - \left(\dfrac{11}{13}\right)^2 = 0.284$ (i.e. 28.4%).

So, a low-mass nucleus is much more effective than a more massive nucleus when it comes to slowing down fast neutrons by elastic collisions. As an aside, it is interesting to note that it is because of this fact that carbon is used in a nuclear reactor for just this purpose.

Q3.10   We designate the ball as particle 1 and the racket as particle 2, with the ball initially travelling along the positive x-direction. From Equation 3.16

$$v_{1x} = \frac{u_{1x}(m_1 - m_2) + 2m_2u_{2x}}{m_1 + m_2}$$

so $v_{1x} = [45 \text{ m s}^{-1} \times (-0.30 \text{ kg}) + 2 \times 0.35 \text{ kg} \times$
        $(-10 \text{ m s}^{-1})]/0.40 \text{ kg}$

i.e. $v_{1x} = (-13.5 \text{ kg m s}^{-1} - 7.0 \text{ kg m s}^{-1})/0.40 \text{ kg}$
        $= -51.3 \text{ m s}^{-1}$.

The server receives the ball back with interest!

Q3.11   From the triangle shown in Figure 3.8b, v_1 has a magnitude $10 \cos 20° \text{ m s}^{-1} = 9.40 \text{ m s}^{-1}$ and is at the given angle of $20°$ to the x-axis; v_2 has a magnitude $10 \sin 20° \text{ m s}^{-1} = 3.42 \text{ m s}^{-1}$ and must be at $70°$ to the x-axis.

Q3.12   Using the triangle in Figure 3.8 again,

$$v_2^2 = u_1^2 - v_1^2 = (10^2 - 6^2) \text{ m}^2 \text{ s}^{-2}$$

so that $v_2 = 8 \text{ m s}^{-1}$. Now v_1 is at an angle $\arccos(6/10) = 53.1°$, and v_2 at an angle $\arccos(8/10) = 36.9°$ to the x-axis. You will observe that the two angles add up to $90°$, as they should.

Q3.13   The meaning of the term 'high speed' obviously depends on context. However, if we simply take it to refer to speeds that are sufficiently high that there is a clear discrepancy between the Newtonian and relativistic curves in Figures 3.16 or 3.17, then we can reasonably say that high speed means greater than about $0.1c = 3 \times 10^7 \text{ m s}^{-1}$. Of the four speeds given in the question, only the last, 10^8 m s^{-1} meets this criterion. In relativistic terms, even a speed of 1000 km s^{-1} counts as a low speed.

Q3.14   The magnitudes of the momenta are as follows:
bullet $30 \times 10^{-3} \text{ kg} \times 400 \text{ m s}^{-1} = 12 \text{ kg m s}^{-1}$
brick $1 \text{ kg} \times 10 \text{ m s}^{-1} = 10 \text{ kg m s}^{-1}$.
Therefore the ratio is $12 : 10$ or $1.2 : 1.0$.

These two momenta are similar in size.

Q3.15   Since the resultant force on a body is equal to the rate of change of that body's momentum, a non-zero resultant may correspond to a changing velocity or a changing mass or a combination of the two. If a non-zero resultant force fails to produce any change in velocity, then it *must* be because the mass of the body is changing.

Q3.16   Let the velocity of the second particle after the first collision be v_{2x} and the speed of the combined second and third particles after the second collision be v_x. Using conservation of momentum along the x-axis:

first collision: $mu_x + 0 = 2mv_{2x}$ so $v_{2x} = u_x/2$

second collision: $2mv_{2x} + 0 = (2m + 3m)v_x$.

So $v_x = \frac{2v_{2x}}{5} = \frac{2(u_x/2)}{5} = \frac{u_x}{5}$.

Q3.17   The difference between the situations must be that one of the balls is more rigid than the other and so the time of contact during which the momentum of the ball changes on impact will then be less. The average force F required to effect this momentum change Δp is related to the time of contact Δt through the expression for the impulse, $\Delta p = F \Delta t$.

This implies that a much larger force is required to bring about any given change in momentum of the more rigid ball. If this force exceeds the maximum force that the window can safely sustain, then the window will break.

Q3.18   The magnitude of the momentum of the driver is $70 \text{ kg} \times 15 \text{ m s}^{-1} = 1050 \text{ kg m s}^{-1}$. Equation 3.4 gives the magnitude of the average force as

$$\frac{\Delta p}{\Delta t} = \frac{1050 \text{ kg m s}^{-1}}{0.6 \text{ s}} = 1750 \text{ kg m s}^{-2} = 1750 \text{ N}.$$

The driver's weight is of magnitude $70 \text{ kg} \times 9.8 \text{ m s}^{-2} = 686 \text{ N}$, so the average impact force is about 2.6 times the magnitude of the weight of the driver.

Q3.19   Let the masses of the spacecraft and Jupiter be m and M, respectively, the initial velocities u_{1x} and u_{2x} and the final velocities v_{1x} and v_{2x}. The law of conservation of momentum, applied along the x-direction tells us that

$$mu_{1x} + Mu_{2x} = mv_{1x} + Mv_{2x}$$

and if there is no loss of kinetic energy during the interaction

$$\tfrac{1}{2}mu_{1x}^2 + \tfrac{1}{2}Mu_{2x}^2 = \tfrac{1}{2}mv_{1x}^2 + \tfrac{1}{2}Mv_{2x}^2.$$

In these two equations there are two unknowns, v_{1x} and v_{2x}, but we are only interested in v_{1x}, so we should eliminate v_{2x}. There are many ways to do this, but the neatest is to rewrite the first of these two equations as

$$v_{2x} = \frac{mu_{1x} + Mu_{2x} - mv_{1x}}{M} = u_{2x} + \left(\frac{m}{M}\right)(u_{1x} - v_{1x})$$

and to rewrite the second equation as

$$\left(\frac{m}{M}\right)(u_{1x}^2 - v_{1x}^2) = v_{2x}^2 - u_{2x}^2.$$

Substituting for v_{2x} in the second equation gives

$$\frac{m}{M}(u_{1x}^2 - v_{1x}^2) = \left[u_{2x} + \left(\frac{m}{M}\right)(u_{1x} - v_{1x})\right]^2 - u_{2x}^2$$

$$= u_{2x}^2 + \left(\frac{m}{M}\right)^2 (u_{1x} - v_{1x})^2 + 2u_{2x}\left(\frac{m}{M}\right)(u_{1x} - v_{1x}) - u_{2x}^2.$$

Eliminating the u_{2x}^2 terms, and dividing both sides by $(m/M) \times (u_{1x} - v_{1x})$ gives

$$(u_{1x} + v_{1x}) = \left(\frac{m}{M}\right)(u_{1x} - v_{1x}) + 2u_{2x}.$$

The mass of Jupiter is so much greater than that of the spacecraft that m/M is very small and the first term on the right is negligible compared with all the other terms. (In fact, this neglected term is equal to the *change in velocity* of Jupiter.) Thus to a good approximation

$$u_{1x} + v_{1x} \approx 2u_{2x}$$

so $\quad v_{1x} \approx 2u_{2x} - u_{1x} = 2 \times 13 \text{ km s}^{-1} - (-10 \text{ km s}^{-1})$

$$= 36 \text{ km s}^{-1}.$$

The spacecraft has been speeded up very substantially! This is a simplified example of the so-called 'slingshot' method, used to give a spacecraft enough energy to reach the outermost parts of the Solar System. This additional energy comes from a reduction in the planet's kinetic energy.

Q3.20 The answer to this question may be obtained directly from Equation 3.16 or 3.17, but it is more instructive to start from first principles as we do here. If the initial velocities of the balls are u_x and $-u_x$ and the final velocity of the moving ball is v_{1x}, then the conservation of momentum along the x-axis gives:

$$m_1 u_x - m_2 u_x = m_1 v_{1x}$$

and conservation of energy gives

$$\tfrac{1}{2}m_1 u_x^2 + \tfrac{1}{2}m_2 u_x^2 = \tfrac{1}{2}m_1 v_{1x}^2.$$

Rearranging these equations gives

$$\frac{v_{1x}}{u_x} = \frac{m_1 - m_2}{m_1} \quad \text{and} \quad \left(\frac{v_{1x}}{u_x}\right)^2 = \frac{m_1 + m_2}{m_1}$$

so that $\quad \dfrac{(m_1 - m_2)^2}{m_1^2} = \dfrac{m_1 + m_2}{m_1}$

i.e. $\quad \dfrac{m_1^2 - 2m_1 m_2 + m_2^2}{m_1^2} = \dfrac{m_1 + m_2}{m_1}.$

i.e. $\quad m_1^2 - 2m_1 m_2 + m_2^2 = m_1^2 + m_1 m_2$

so, $\quad m_2^2 = 3m_1 m_2$

Which can be simplified to $m_2 = 3m_1$ and so $m_1 = 300$ g.

Q3.21 (a) If we take $m_1 = 1.00 \times 10^4$ kg, $m_2 = 8.00 \times 10^3$ kg, $m_3 = 6.00 \times 10^3$ kg, $v_{1x} = 8.00$ m s^{-1}, and $v_{2y} = -5.00$ m s^{-1}, conservation of momentum implies that

$$p_{3x} + m_1 v_{1x} = 0 \quad \text{and} \quad p_{3y} + m_2 v_{2y} = 0$$

so $\quad p_{3x} = -(1.00 \times 10^4 \text{ kg}) \times (8.00 \text{ m s}^{-1})$

$$= -8.00 \times 10^4 \text{ kg m s}^{-1}$$

and $\quad p_{3y} = -(8.00 \times 10^3 \text{ kg}) \times (-5.00 \text{ m s}^{-1})$

$$= 4.00 \times 10^4 \text{ kg m s}^{-1}.$$

Thus, the momentum of the third piece (Figure 3.22) has magnitude

$$p_3 = \sqrt{(8^2 + 4^2)} \times 10^4 \text{ kg m s}^{-1} = 8.94 \times 10^4 \text{ kg m s}^{-1}$$

and is directed at an angle to the *negative x*-axis of

$$\arctan\left(\frac{4.00 \times 10^4}{8.00 \times 10^4}\right) = 26.6°.$$

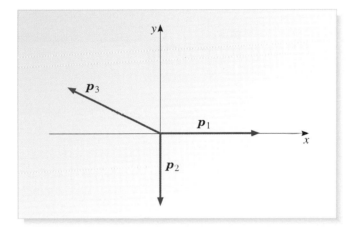

Figure 3.22 For Q3.21.

(b) The speed of the third block is

$$\frac{p_3}{m_3} = \frac{8.94 \times 10^4 \text{ kg m s}^{-1}}{6.00 \times 10^3 \text{ kg}} = 14.9 \text{ m s}^{-1}.$$

The total kinetic energy of the three blocks E_{trans} is then

$$E_{\text{trans}} = \tfrac{1}{2}[(1.00 \times 10^4 \text{ kg}) \times (8.00 \text{ m s}^{-1})^2$$

$$+ (8.00 \times 10^3 \text{ kg}) \times (5.00 \text{ m s}^{-1})^2$$

$$+ (6.00 \times 10^3 \text{ kg}) \times (14.9 \text{ m s}^{-1})^2].$$

So $E_{\text{trans}} = 1.09 \times 10^6$ J, equal to the total energy released in the explosion.

Q4.1   The gravitational force on the known mass is directed vertically downwards and has magnitude $F = (0.50\,\text{kg}) \times g$. Using Equation 4.1, the magnitude of the torque about the pivot due to this force is

$$\Gamma = Fl = (0.50\,\text{kg}) \times g \times (0.30\,\text{m}).$$

When in balance, the magnitude of the torque arising from the unknown mass, m, must be identical to this, so

$$m \times g \times (0.05\,\text{m}) = (0.50\,\text{kg}) \times g \times (0.30\,\text{m})$$

it follows that

$$m = \frac{0.50 \times 0.30}{0.05}\,\text{kg} = 3.0\,\text{kg}.$$

Q4.2   With the arm inclined at 45°, the perpendicular distance from the elbow to the line of action of the load weight will be reduced, and the magnitude of the torque about the elbow due to the load will be

$$\Gamma_{\text{L}} = LW \sin 45°.$$

However, the magnitude of the torque about the elbow due to a muscular force of magnitude F will also be affected by the inclination; it will be

$$\Gamma_{\text{M}} = lF \sin 45°.$$

Balance requires that these two magnitudes should be equal, so we require

$$LW \sin 45° = lF \sin 45°, \quad \text{i.e.} \quad F = \frac{L}{l}W.$$

This is the same result as before. Under the stated conditions there is no need for any change in the applied force, despite the change in torque. However, it should be noted that the assumption that the biceps still pulls vertically upwards is somewhat unrealistic.

Q4.3   Use the right-hand rule of Figure 4.15b. Align the fingers of your right hand with the vector r_1, and twist your wrist until you can bend your fingers into alignment with F_1, you will then find that your straightened thumb indicates that Γ_1 points out of the page as shown in Figure 4.17. A similar operation will confirm that Γ_2 is also correctly drawn.

Q4.4   Both forces act in the xy-plane, and both displacement vectors from O are also in the xy-plane, so the torques Γ_1 and Γ_2 will both point in either the positive or the negative z-direction (as determined by the right-hand rule). If we write $\Gamma_1 = (0, 0, \Gamma_{1z})$ and $\Gamma_2 = (0, 0, \Gamma_{2z})$ then

$$\Gamma_{1z} = +r_1F_1 \sin 30° = +0.400\,\text{m} \times 5\,\text{N} \times 0.5 = +1.0\,\text{N m}$$

$$\Gamma_{2z} = -r_2F_2 \sin 90° = -0.300\,\text{m} \times 5\,\text{N} = -1.5\,\text{N m}.$$

Consequently, the total torque at O is

$$\Gamma = \Gamma_1 + \Gamma_2 = (0, 0, \Gamma_{1z} + \Gamma_{2z}) = (0, 0, -0.5\,\text{N m}).$$

Since the axis of rotation of the plate is the z-axis, and since the component of the total torque along this axis is negative, it follows that the plate will rotate in the clockwise direction when viewed from above.

Q4.5   The speed v of the vehicle at time t is given by $v = at$. Now, at the time $t_{\text{f}} = 14.3$ s, the speed of the vehicle is

$$v_{\text{f}} = 100 \times 10^3\,\text{m h}^{-1} = \frac{100 \times 10^3}{3600}\,\text{m s}^{-1}$$

so,

$$a = \frac{v_{\text{f}}}{t_{\text{f}}} = \frac{100 \times 10^3\,\text{m s}^{-1}}{3600 \times 14.3\,\text{s}} = 1.94\,\text{m s}^{-2}.$$

The centre of the wheel must have the same speed as the vehicle, but the point of contact with the ground is stationary. It follows that the angular speed of the wheel is given by

$$\omega = \frac{v}{r} = \frac{at}{r} = \frac{100 \times 10^3 \times t}{3600 \times 14.3 \times 0.300}\,\text{rad s}^{-1} = 6.48t\,\text{rad s}^{-1}.$$

The magnitude of the angular acceleration is given by the magnitude of the rate of change of ω, so

$$\frac{\text{d}\omega}{\text{d}t} = 6.48\,\text{rad s}^{-2}.$$

Q4.6   The similarities are:

- Both definitions involve two vectors.
- In both cases, the result depends on the angle between the vectors.

The differences are:

- Work is a scalar quantity, but torque is a vector.
- The work done depends on the cosine of the angle between the vectors, but the magnitude of the torque depends on the sine of the angle.
- Although the two quantities have the same physical dimensions (force times distance) they are of very different natures and we distinguish between their SI units by using J for work and N m for torque.
- The work is a maximum when the two vectors are parallel, but torque has its maximum magnitude when the two vectors are perpendicular.

Q4.7   The forces comprising the couple have the same magnitude but act in opposite directions, so the total force on the wheel is zero. Each force is perpendicular to a radial displacement vector from the centre of the wheel. It follows from Equation 4.3, that the torque about the centre due to each force points vertically upwards, out of the page, and each has magnitude

$$\Gamma = rF = (0.345 \times 20.2)\,\text{N m} = 6.97\,\text{N m}.$$

The total torque Γ about the wheel's centre is the vector sum of the two individual torques, so it too points vertically upwards (out of the page) and has a magnitude of 13.94 N m. The wheel will be in translational equilibrium, but not rotational equilibrium.

Q4.8 From Equation 4.13, the total torque about the centre of the disc must be zero. So the applied torque must be equal in magnitude but opposite in direction to the frictional torque, and from Equation 4.2, the magnitude is given by $rF \sin\theta$.

If the applied force is of magnitude F, and is directed tangentially to the disc, it follows (with $\theta = 90°$) that

$$F = \left(\frac{0.0550\,\text{m}}{2}\right) \times 0.168\,\text{N} \times \left(\frac{2}{1.50\,\text{m}}\right)$$

So, $\quad F = \dfrac{0.0550}{1.50} \times 0.168\,\text{N} = 6.16 \times 10^{-3}\,\text{N}$.

Q4.9 The vertical component of $\boldsymbol{F}_\text{T}$ has magnitude

$$F_{\text{Tv}} = F_\text{T} \sin\theta = 33.4\,\text{N} \times \sin(36.870°) = 20.0\,\text{N}.$$

Since both weights act in the same direction, $|\boldsymbol{W}_1 + \boldsymbol{W}_2| = W_1 + W_2$, and

$$W_1 + W_2 = (m_1 + m_2)g = (2.50 + 0.85) \times 9.81\,\text{N} = 32.9\,\text{N}.$$

The upward force has too small a magnitude, F_{Tv}, to balance the downward force $\boldsymbol{W}_1 + \boldsymbol{W}_2$. This is to be expected because there is another force acting on the beam at the pivot O. This too will have an upward component.

Q4.10 (a) In Figure 4.60a, the cylinder is rolling on a horizontal surface. The centre of mass of the cylinder is moving with a constant velocity v. This is an example of *non-static equilibrium*. Figures 4.60b, c and d are examples of *static equilibrium*.

(b) Figure 4.60b is an example of *stable equilibrium*. The cylinder is standing on one end. If the cylinder is tilted slightly from the position shown, then it returns to this position.

In Figure 4.60c, the cylinder is balanced so that its centre of mass is exactly over one edge. This is an example of *unstable equilibrium*. The slightest increase in the tilt of the cylinder results in it moving to a new position (i.e. falling over). The cylinder does not return to its original position.

In Figure 4.60d, the cylinder is stationary on a horizontal surface. This is an example of *neutral equilibrium*. When perturbed, the cylinder does not return to its original position, but neither does it tend to move to a new position.

Q4.11 There are three possibilities: (i) rotation about A, (ii) rotation about B, and (iii) rotation about C, as shown in Figure 4.61. In all cases, there will be two particles of identical mass rotating about the third particle. However, their separations from the centre of rotation will be different. Since

$$E_\text{rot} = \tfrac{1}{2}\omega^2(mr_1^2 + mr_2^2) = \tfrac{1}{2}m\omega^2(r_1^2 + r_2^2),$$

the largest rotational energy will correspond to the largest value of $(r_1^2 + r_2^2)$. This is option (i), with $r_1 = 4.00 \times 10^{-2}\,\text{m}$ and $r_2 = 5.00 \times 10^{-2}\,\text{m}$. (The next smallest E_rot will correspond to option (iii), and the smallest to option (ii).)

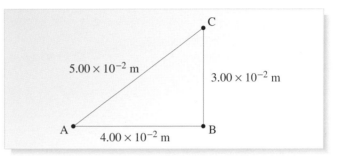

Figure 4.61 See the answer to Q4.11.

Q4.12 Using $E_\text{rot} = \tfrac{1}{2}I\omega^2$, with $I = \tfrac{1}{2}MR^2$ and $\omega = v/R$, the kinetic energy is given by

$$E_\text{rot} = \frac{1}{2}I\omega^2 = \frac{1}{2}\frac{MR^2}{2}\frac{v^2}{R^2} = \frac{Mv^2}{4}.$$

The speed, expressed in m s^{-1}, is

$$v = \frac{40.0 \times 10^3}{3600}\,\text{m s}^{-1}.$$

Therefore $\quad E_\text{rot} = \dfrac{22.5}{4} \times \left(\dfrac{40.0 \times 10^3}{3600}\right)^2\,\text{J} = 678\,\text{J}.$

Similarly, the translational kinetic energy is

$$E_\text{trans} = \tfrac{1}{2}Mv^2 = 2E_\text{rot} = 1.39 \times 10^3\,\text{J}.$$

Notice that the radius of the disc is not relevant to the solution of this question.

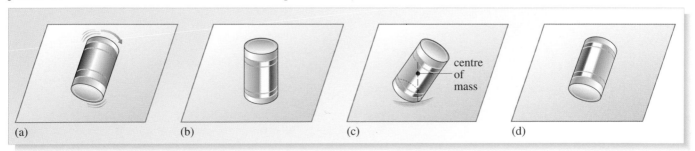

(a)   (b)   (c)   (d)

Figure 4.60 See the answer to Q4.10.

Q4.13 The increase in the angular speed is

$$\Delta\omega = \frac{100 \times 2\pi}{60} \text{ rad s}^{-1}.$$

The (steady) angular acceleration is therefore of magnitude

$$\frac{d\omega}{dt} = \frac{\Delta\omega}{\Delta t} = \frac{100 \times 2\pi}{60 \times 5.20} \text{ rad s}^{-2}.$$

Since $\boldsymbol{\Gamma} = I\dfrac{d\boldsymbol{\omega}}{dt}$ and $I = \frac{1}{2}MR^2$ in this case, it follows that

$$\Gamma = \frac{1}{2}MR^2\frac{d\omega}{dt}.$$

So,

$$\Gamma = \frac{1}{2} \times 1.40 \times (0.360)^2 \times \frac{100 \times 2\pi}{60 \times 5.20} \text{ N m} = 0.183 \text{ N m}.$$

Q4.14 Using the notation of Q4.13, $I = \frac{1}{2}MR^2$.

Setting $\omega_0 = 0$ in Equation 4.29, and taking the magnitude of both sides,

$$\omega = \frac{\Gamma}{I}t$$

giving the following expression for the magnitude of ω

$$\omega = \frac{\Gamma t}{I} = \frac{2\Gamma t}{MR^2}.$$

Hence $\omega = \dfrac{2 \times 4.00 \times 10^{-3} \times 21.5}{1.40 \times (0.360)^2} \text{ rad s}^{-1} = 0.948 \text{ rad s}^{-1}.$

Q4.15 Since it accelerates uniformly from rest, the final speed of the rolling cylinder in Example 4.5 is just twice its average speed. The same consideration applies to the four racing objects, so we need to determine the final speed in each case since a higher final speed implies a higher average speed and a shorter time of descent. Now we know that each body has a moment of inertia of the form $I_{CM} = nMR^2/m$, where, for example, $n = 1$ and $m = 2$ for the uniform cylinder. If we re-examine Example 4.5 with this in mind we see from Equation 4.36, upon dividing both sides by M, that

$$gh = \frac{n}{2m}R^2 \times \left(\frac{v_{CM}}{R}\right)^2 + \frac{1}{2}v_{CM}^2 = \left(\frac{n}{m}+1\right)\frac{v_{CM}^2}{2}$$

$$= \left(\frac{n+m}{m}\right)\frac{v_{CM}^2}{2}$$

so, $v_{CM} \propto \sqrt{\left(\dfrac{m}{n+m}\right)}$, since g and h is the same in each case.

Taking the values of n and m from Equations 4.26a, c, g and h we see that the order of arrival of the four bodies will be: solid sphere, solid cylinder, thin spherical shell

and finally thin cylindrical shell. Note that neither the masses nor the radii of the bodies have any influence on their final speed, provided their centres descend the same distance. Even if the bodies had been of different sizes, the result would have been the same!

Q4.16 This is one of the cases in which we can use $\boldsymbol{L} = I\boldsymbol{\omega}$. From Figure 4.37c, the relevant moment of inertia is $I = \frac{1}{2}MR^2$.

So, $L = \frac{1}{2}MR^2\omega.$

Thus $L = [\frac{1}{2} \times 5.20 \times (0.125)^2 \times 3.65 \times 2\pi] \text{ kg m}^2 \text{ s}^{-1}$

$\qquad = 0.932 \text{ kg m}^2 \text{ s}^{-1}.$

Using the right-hand grip rule, the direction of $\boldsymbol{\omega}$ is along the positive y-axis, which will also be the direction of $\boldsymbol{L}$ in this case.

Q4.17 The magnitude of the orbital angular momentum can be obtained from the basic definition $\boldsymbol{l} = \boldsymbol{r} \times \boldsymbol{p}$, where $\boldsymbol{p} = m\boldsymbol{v}$ and $\boldsymbol{v} = \boldsymbol{\omega} \times \boldsymbol{r}$.

For circular motion, $l = rmv = mr^2\omega$.

Replacing m by $M_{\text{Earth}} = 5.98 \times 10^{24}$ kg, r by $R_{\text{orb}} = 1.50 \times 10^{11}$ m, l by L_{orb}, and using the orbital period of the Earth to determine its orbital angular speed,

$$\omega = \frac{2\pi}{365 \times 24 \times 3600} \text{ rad s}^{-1},$$

gives

$$L_{\text{orb}} = M_{\text{Earth}} R_{\text{orb}}^2 \omega$$

$$= \frac{5.98 \times 10^{24} \times (1.50 \times 10^{11})^2 \times 2\pi}{365 \times 24 \times 3600} \text{ kg m}^2 \text{ s}^{-1}$$

$$= 2.68 \times 10^{40} \text{ kg m}^2 \text{ s}^{-1}.$$

The magnitude of the spin angular momentum can be obtained from $L = I\omega$, with $I = \frac{2}{5}MR^2$.

Replacing L by L_{spin}, M by M_{Earth}, R by $R_{\text{Earth}} = 6.38 \times 10^6$ m, and using the rotational period of the Earth to determine its rotational angular speed

$$\omega = \frac{2\pi}{24 \times 3600} \text{ rad s}^{-1},$$

gives

$$L_{\text{spin}} = \frac{2}{5}M_{\text{Earth}}R_{\text{Earth}}^2\omega$$

$$= \frac{2}{5} \times 5.98 \times 10^{24} \times (6.38 \times 10^6)^2 \times \frac{2\pi}{24 \times 3600} \text{ kg m}^2 \text{ s}^{-1}$$

$$= 7.08 \times 10^{33} \text{ kg m}^2 \text{ s}^{-1}.$$

Hence, the orbital angular momentum is very much greater than the spin angular momentum.

Q4.18 The moment of inertia of a uniform sphere is given by $I = 2MR^2/5$. Using the conservation of angular momentum, and applying Equation 4.47,

$$\frac{\omega_f}{\omega_i} = \frac{I_i}{I_f} = \frac{R_i^2}{R_f^2} = \frac{(2000)^2}{1} = 4\,000\,000$$

where i and f represent the initial and final states, respectively.

Q4.19 Any change in the distribution of matter on (or within) the Earth may change the moment of inertia.

The 24-h variation (which causes the North Pole to move in a roughly circular path of mean radius 0.5 m, each day) is due to the tidal distortion of the turning Earth by the gravitational pull of the Sun and the Moon. Because the Earth is not perfectly rigid, tides affect the solid Earth as well as the oceans and the atmosphere.

The 365-day variation is due to the annual movements of air and water associated with the global aspects of the climate and the passage of the seasons.

Q4.20 (a) From the definition of the vector product, (or from Equation 4.10 if you prefer) $b \times a$ will point in the negative z-direction and will have magnitude 6. In fact, $b \times a = (0, 0, -6)$. Using the definition again (or Equation 4.10), $a \times (b \times a) = (0, 12, 0)$.

(b) The vector b is at a right angle to both a and c. We can use the right-hand rule to get the direction. (Draw a diagram if this helps you.) So we have $a \times b = (0, 0, 6)$, $a \times c = (0, 0, 0)$ and $b \times c = (0, 0, -12)$.

Q4.21 Applying the condition of translational equilibrium, the upward directed vertical component of F_O is given by

$$F_{Ov} = W_1 + W_2 - F_{Tv} = 32.9\,\text{N} - 20.0\,\text{N} = 12.9\,\text{N}$$

and the rightward directed horizontal component is given by

$$F_{Oh} = F_{Th}\cos(36.870°) = 33.4\,\text{N} \times 0.8000 = 26.7\,\text{N}.$$

The magnitude of F_O is therefore

$$F_O = \sqrt{F_{Ov}^2 + F_{Oh}^2} = \sqrt{879.3}\,\text{N} = 29.7\,\text{N}$$

and the direction is upwards and to the right, at an angle to the horizontal given by

$$\arctan\left(\frac{F_{Ov}}{F_{Oh}}\right) = \arctan\left(\frac{12.9}{26.7}\right) = 25.8°.$$

Q4.22 The forces acting on the bridge are shown in Figure 4.62. Translational equilibrium implies that

$$F_1 + F_2 = (m_1 + m_2 + M)g. \qquad \text{(i)}$$

For rotational equilibrium, there must be no net torque. Evaluating the torque about the centre of the bridge is convenient, giving

$$F_1\frac{L}{2} + m_2 g\left(\frac{L}{2} - x_2\right) = F_2\frac{L}{2} + m_1 g\left(\frac{L}{2} - x_1\right).$$

This can be written as

$$(F_1 - F_2) = 2g\left(\frac{m_1}{2} - \frac{m_2}{2} - \frac{m_1 x_1}{L} + \frac{m_2 x_2}{L}\right). \qquad \text{(ii)}$$

Adding Equations (i) and (ii), gives

$$F_1 = \frac{g}{2}\left[2m_1\left(1 - \frac{x_1}{L}\right) + M + \frac{2m_2 x_2}{L}\right].$$

Subtracting Equations (i) and (ii), gives

$$F_2 = \frac{g}{2}\left[\frac{2m_1 x_1}{L} + M + 2m_2\left(1 - \frac{x_2}{L}\right)\right].$$

The directions of F_1 and F_2 are vertically upwards. (Don't forget that a force is a vector and therefore has a direction.)

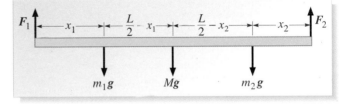

Figure 4.62 See the answer to Q4.22.

Q4.23 The moment of inertia of a hollow (thin-walled) sphere will be *greater* than that of a solid sphere of the same mass, because most of the mass will be further away from the central axis.

Q4.24 For convenience, define the following quantities:

ω_i = initial angular speed (about vertical axis through body)

ω_f = final angular speed (about vertical axis through body)

I_i = initial moment of inertia (about vertical axis through body)

I_f = final moment of inertia (about vertical axis through body)

E_i = initial rotational kinetic energy

E_f = final rotational kinetic energy.

From the conservation of angular momentum and $L = I\omega$, we have $I_i\omega_i = I_f\omega_f$ or $I_i/I_f = \omega_f/\omega_i$.

Hence $\omega_f > \omega_i$ if $I_f < I_i$.

However, from $E_{rot} = \frac{1}{2}I\omega^2$, we have $E_i = \frac{1}{2}I_i\omega_i^2$ and $E_f = \frac{1}{2}I_f\omega_f^2$.

Consequently, $\dfrac{E_i}{E_f} = \dfrac{I_i\omega_i^2}{I_f\omega_f^2} = \dfrac{\omega_f}{\omega_i} \times \dfrac{\omega_i^2}{\omega_f^2} = \dfrac{\omega_i}{\omega_f} < 1$, so $E_i < E_f$.

This increase in kinetic energy comes from the work done by the skater in pulling in his arms and legs.

Q5.1 With $k = 1.6$, the logistic map in Equation 5.4 becomes $x_{n+1} = 1.6x_n(1 - x_n)$.

(a) Dividing by x_n, we obtain $x_{n+1}/x_n = 1.6(1 - x_n)$. If $x_{n+1} = x_n = x_{final}$, at large n, then $1 = 1.6(1 - x_{final})$ and hence $x_{final} = 1 - (1/1.6) = 0.375$. So if the sequence settles down to a steady value, this value must be 0.375.

(b) The starting value is $x_0 = 0.6$. The next five values are

$$x_1 = 1.6x_0(1 - x_0) = 0.384\,00$$

$$x_2 = 1.6x_1(1 - x_1) = 0.378\,47$$

$$x_3 = 1.6x_2(1 - x_2) = 0.376\,37$$

$$x_4 = 1.6x_3(1 - x_3) = 0.375\,54$$

$$x_5 = 1.6x_4(1 - x_4) = 0.375\,22$$

to five decimal places.

(c) The differences between successive values of x_n decrease steadily, and the sequence settles down quickly to about 0.375.

(d) By no means could this steady convergence to 0.375 be called chaotic.

Q5.2 (a) The equation of simple harmonic motion is (Equation 1.70)

$$\frac{d^2x}{dt^2} = -\omega^2 x.$$

This is a linear equation: a plot of acceleration d^2x/dt^2 against displacement x gives a straight line.

(b) The equation of motion of a pendulum is (Equation 3.58, *Describing motion*)

$$\frac{d^2\theta}{dt^2} = -\frac{g}{l}\sin\theta.$$

This equation is non-linear: the angular acceleration $d^2\theta/dt^2$ is *not* proportional to the angular displacement θ, but to $\sin\theta$. A linear equation of motion is a fair *approximation* for small displacements, because then $\sin\theta \approx \theta$. At larger displacements (e.g. 30°, as in the question), the non-linear equation is needed.

Q5.3 (a) The data seem to be settling into a limit cycle of period 4.

(b) The next value is $3.5 \times 0.875\,00 \times (1 - 0.875\,00) = 0.382\,81$. This differs from the value four steps earlier by merely 0.000 05, strengthening the idea that the system is settling into a 4-cycle.

Q5.4 (a) The data for the logistic map with $k = 2$ might well have pleased Laplace, since the sequence converges quickly to 0.5, posing no predictive problem. Similarly the data you calculated in the answer to Q5.1 converge rapidly to 0.375.

(b) The data in Tables 5.1 and 5.2, for the logistic map with $k = 4$, show an exponential sensitivity on initial data, very much in the spirit of Poincaré's misgiving that 'it may happen that small differences in initial conditions produce very great ones in the final phenomena'.

Q5.5 All five statements need significant correction.

(a) Only some dynamical systems exhibit chaos. In others, such as the logistic map at $k = 2$, there is no exponential sensitivity to initial data, and hence no chaos, as defined.

(b) There is *no* sense in which a system moves *along* one of the curves in Figure 5.8, during iteration. For a particular system, the value of the parameter k is fixed. So the results for a particular system must all lie along a vertical line corresponding to the appropriate value of k. Different values of k correspond to different systems. The curve indicates the settled value(s) that emerge after sufficiently many iterations.

(c) In the period-doubling regime, the limit cycles correspond to jumps between curves in Figure 5.8. But the system is not chaotic in this regime.

(d) With just two successive values, x_n and x_{n+1}, from a logistic sequence, it is possible to measure $k = x_{n+1}/x_n(1 - x_n)$ and hence to distinguish the case with $k = 3.5699$ from that with $k = 3.5701$. Of course, it would be necessary to measure the successive values to better than five significant figures. The challenge lies not in measuring values of k from the data, but rather in determining which values of k lead to chaotic behaviour.

Q5.6 (a) Damping causes the oscillations to decrease in amplitude, so successive crossings of the x-axis must be at smaller values of x (= θ). The maximum angular speed will also decrease, so successive crossings of the y-axis will be at smaller values of y (= $d\theta/dt$). The trajectory will therefore spiral inwards, as shown in Figure 5.24.

(b) Eventually the pendulum will come to rest, at zero displacement, corresponding to the origin of state space.

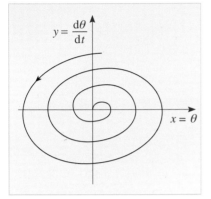

Figure 5.24 Trajectory in state space for a damped pendulum.

Q5.7 (a) The behaviour in Figure 5.17a does *not* indicate a strange attractor. The system fairly quickly settles into a regular oscillation about a positive value of x.

(b) The behaviour in Figure 5.17b *does* indicate a strange attractor. Oscillations at positive and negative values of x are punctuated by transitions from one part of state space to the other.

(c) (i) The behaviour in Figure 5.17b is chaotic, but the trajectory in (a) settles into a closed loop — a repeating oscillation — so it is not chaotic. (ii) The behaviour in both cases is governed by deterministic rules. It is important to remember that chaos is not synonymous with randomness. Deterministic chaos is governed by deterministic rules.

Q5.8 (a) The defining characteristics of chaos are: (i) a deterministic rule; (ii) unpredictable outcomes due to an exponential sensitivity to initial data.

(b) A chaotically healthy heart would be governed by a rule that maintains function while leading to changes in heart rate that cannot be predicted.

Q5.9 We do not know enough about the weather to exclude the possibility of a strange attractor, similar to that found by Lorenz in a simplified model of convection. Were such chaos to occur in the development of weather patterns, a tiny change in initial conditions could produce a large effect, beyond the horizon of predictability.

Healthily beating hearts are akin to chaotically dripping taps: changes in heart rate are not predictable, yet appear to originate from a deterministic rule.

Chaotic motion of asteroids close to Kirkwood gaps may result in large fluctuations of eccentricity, producing close encounters with Mars. A rogue asteroid would then have the possibility of doing great damage to the Earth.

Chaotic motion of stars in a tri-axial elliptical galaxy containing a black hole becomes more pronounced as the black hole increases in mass by capturing stars following box orbits. The chaos results in an axially symmetric distribution upon which the black hole cannot feed so readily. Its mass is therefore limited to a few per cent of the galactic mass.

Q5.10 (a) A plot of x_{n+1}, on the vertical (y-) axis, against x_n, on the horizontal (x-) axis, will produce a scatter of points all of which lie on the parabola $y = 4x(1 - x)$. It is clear that $x_{n+1} = 0$ when $x_n = 0$ and when $x_n = 1$, and x_{n+1} must be positive for x_n between 0 and 1. At intermediate points it is straightforward to calculate other pairs of values:

x_n	0.25	0.5	0.75
x_{n+1}	0.75	1.0	0.75

This is enough information to sketch a graph like the one shown in Figure 5.25.

(b) You can't specify how the pattern of points will build up because this depends on the starting value. But successive points will be scattered along the parabola.

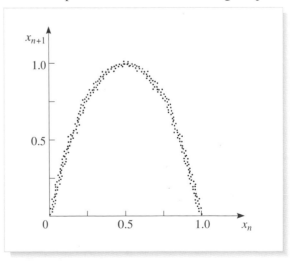

Figure 5.25 Points scattered on the parabola $x_{n+1} = 4x_n(1 - x_n)$.

Q5.11   The only correct answer to this question, as formulated, is the one that you gave.

Comment: *Here are the author's reflections; yours need not be in accord.*

I find chaos — as here defined — quite unsurprising. It seems to me that Poincaré had in mind precisely what has been discovered: exponential sensitivity to initial conditions. It is wonderful that we have learnt of places where this occurs. Yet I cannot believe that this ranks alongside the changes in world-view that were demanded by Newtonian mechanics, by relativity theory, and — most dramatically of all — by quantum physics.

When I learnt of Feigenbaum's work, in the late 1970s, there was a feature I found truly remarkable: the discovery of a new constant in many different transitions to chaos. I reflect that $\sqrt{2}$ was held in awe by Pythagoras, since it could not be expressed as the ratio of two integers; yet an approximate value was used to build pyramids. In the same vein, π is a number that enables one to do practical sums with circles and spheres; yet it cannot be expressed as a ratio of integers, nor in terms of square roots of integers. I was pleased that Feigenbaum had discovered an even subtler number, while engaged in the practice of science.

I feel that some of the interest in chaos theory, which burgeoned in the 1980s, says more about the social climate of that period than about the fabric of the Universe. Attention was focused on the idea that scientists had somehow discovered a drastic new limit to what may be learnt by existing methods and must thus adopt a radically new approach. To my mind, chaos theory constitutes solid progress in the description of natural phenomena. To regard the findings as a profound reverse seems to me to be a mistake in both magnitude and sign.

Q6.1   Preparation   Figure 6.7 shows the forces acting on the bobsleigh. Initially, the bobsleigh accelerates under the combined influence of these forces. As the speed increases, however, the air resistance increases and the acceleration becomes smaller. At the end of a long run, air resistance is large enough to cancel the effects of the other forces and the bobsleigh will move uniformly. Its terminal speed is found by evaluating the resultant force and setting it equal to zero.

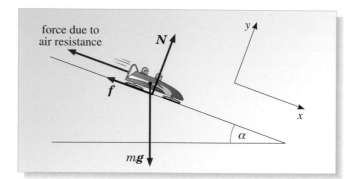

Figure 6.7   For use with Q6.1.

Working   The x- and y-components of the resultant force are

$$F_x = (mg \sin \alpha) - f - Cv^2$$

$$F_y = N - mg \cos \alpha.$$

Setting F_y equal to zero gives

$$N = mg \cos \alpha$$

and setting F_x equal to zero gives

$$Cv^2 = (mg \sin \alpha) - f.$$

Since the bobsleigh is sliding, the following equation can be used to substitute for f

$$f = \mu_{\text{slide}} N.$$

So $Cv^2 = (mg \sin \alpha) - \mu_{\text{slide}} N$

$$= (mg \sin \alpha) - \mu_{\text{slide}}(mg \cos \alpha)$$

$$= mg(\sin \alpha - \mu_{\text{slide}} \cos \alpha)$$

$$= 200 \, \text{kg} \times 9.81 \, \text{m s}^{-2} \times (\sin 20° - 0.10 \times \cos 20°)$$

$$= 486 \, \text{N}.$$

The terminal speed of the bobsleigh is therefore

$$v = \sqrt{\frac{486 \, \text{N}}{C}} = \sqrt{\frac{486 \, \text{N}}{0.3 \, \text{N s}^2 \, \text{m}^{-2}}} = 40 \, \text{m s}^{-1}$$

to the accuracy of the data given in the question.

Checking   The speed comes out in the unit of m s^{-1}, and the actual speed, about four times that of a sprint athlete, does not seem unreasonable. The formulae show that a higher terminal speed would be achieved if the bobsleigh were more massive or if gravity were stronger. These predictions seem plausible.

Q6.2 Preparation We will define east along the positive x-axis and north along the positive y-axis. The neutron travels with the same speed u_n throughout the collision. After the collision, let the helium nucleus travel at a speed v_{He} in a direction inclined at an angle θ to the positive y-axis. The pre- and post-collision situations are shown in Figure 6.8a and b.

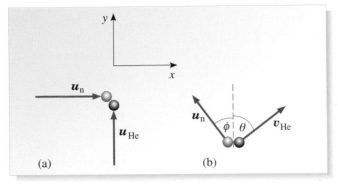

Figure 6.8 For use with Q6.2.

We will use conservation of momentum in the two directions in order to find the angle and speed of the helium nucleus after the collision.

Working Using conservation of momentum along the x-direction:

$$mu_n = 4mv_{He} \sin\theta - mu_n \sin\phi$$

the mass m cancels, so

$$\sin\theta = \frac{u_n(1 + \sin\phi)}{4v_{He}}$$

and using conservation of momentum along the y-direction:

$$4mu_{He} = 4mv_{He} \cos\theta + mu_n \cos\phi$$

the mass m again cancels, so

$$\cos\theta = \frac{4u_{He} - u_n \cos\phi}{4v_{He}}.$$

Dividing these two expressions to find $\tan\theta$ we obtain:

$$\tan\theta = \frac{\sin\theta}{\cos\theta} = \frac{u_n(1 + \sin\phi)}{4u_{He} - u_n \cos\phi}$$

and putting in the values for the two initial speeds and the value of ϕ:

$$\tan\theta = \frac{4.00 \times 10^4\,\mathrm{m\,s^{-1}} \times 1.60}{(4 \times 3.00 \times 10^4\,\mathrm{m\,s^{-1}}) - (4.00 \times 10^4\,\mathrm{m\,s^{-1}} \times 0.800)}$$

$$= 0.727.$$

This gives $\theta = 36.0°$.

The expression for $\sin\theta$ can be rearranged to give:

$$v_{He} = \frac{u_n(1 + \sin 36.9°)}{4\sin\theta} = \frac{4.00 \times 10^4\,\mathrm{m\,s^{-1}} \times 1.60}{4 \times 0.588}$$

$$= 2.72 \times 10^4\,\mathrm{m\,s^{-1}}.$$

Therefore, after the collision the helium nucleus is travelling at a speed of $2.72 \times 10^4\,\mathrm{m\,s^{-1}}$ at $36.0°$ east of north.

Checking The units for the speed come out as $\mathrm{m\,s^{-1}}$ and the tangent of the angle is dimensionless, both as required. Both values seem plausible in that they are of the same order as the other speeds and angles in the question. In this case it is difficult to check easily how the values would vary if the initial speeds were different.

Q6.3 Preparation We are given the period T, speed v and the magnitude of the centripetal force F for the planet. We can first calculate its angular speed ω from the orbital period using $\omega = 2\pi/T$, then the radius of its orbit using $v = \omega r$ and finally its mass using $F = mv^2/r$.

Working So, using $\omega = 2\pi/T$,

$$\omega = \frac{2\pi}{3.7 \times 10^8}\,\mathrm{s^{-1}} = 1.7 \times 10^{-8}\,\mathrm{s^{-1}}.$$

Rearranging $v = \omega r$ gives $r = v/\omega$

so $$r = \frac{1.3 \times 10^4\,\mathrm{m\,s^{-1}}}{1.7 \times 10^{-8}\,\mathrm{s^{-1}}} = 7.6 \times 10^{11}\,\mathrm{m}.$$

Rearranging $F = mv^2/r$, gives $m = Fr/v^2$

so $$m = \frac{(4.2 \times 10^{23}\,\mathrm{N}) \times (7.6 \times 10^{11}\,\mathrm{m})}{(1.3 \times 10^4\,\mathrm{m\,s^{-1}})^2} = 1.9 \times 10^{27}\,\mathrm{kg}.$$

Checking All units for the calculated quantities come out correctly. If the centripetal force were larger, the mass of the planet that could be accommodated in the same orbit would be larger, and this is to be expected. If the speed were larger, the radius of the orbit in which the planet would travel during one period would be larger, and this is as expected too.

If you happen to know the parameters for the Earth's orbit around the Sun, you can work out that the orbital period of this planet corresponds to about 12 (Earth) years (one year is about 3×10^7 s). So it is not unreasonable that the radius of the planet's orbit should be about five times that of the radius of the Earth's orbit about the Sun (about 150 million km). The mass of the planet is about 300 times that of the Earth (about 6×10^{24} kg), which is also reasonable.

Q6.4 **Preparation** Let the reference level for zero gravitational potential energy correspond to maximum compression, x, of the spring and suppose that h is the total distance fallen by the block at maximum compression, as shown in Figure 6.9.

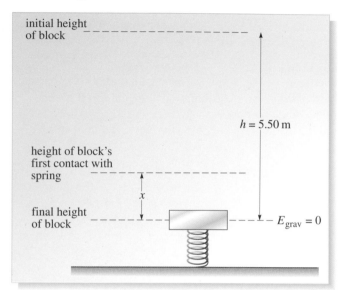

Figure 6.9 For use with Q6.4.

The law of conservation of mechanical energy means that gravitational energy is converted into strain energy of the spring, so at the instant of maximum compression

$$\tfrac{1}{2} k_s x^2 = mgh. \qquad (6.8)$$

We can use this formula to calculate the compression of the spring x.

Next we can calculate the distance fallen by the block at the instant it hits the spring as $h - x$, and finally equate the loss in gravitational potential energy to the gain in kinetic energy at this point, i.e.

$$\tfrac{1}{2} mv^2 = mg(h - x). \qquad (6.9)$$

This can be rearranged to calculate the speed v.

Working Rearranging Equation 6.8 which results from equating strain and gravitational energy:

$$x = \sqrt{\frac{2mgh}{k_s}}$$

and putting in the given values

$$x = \sqrt{\frac{2 \times 1.00\,\text{kg} \times 9.81\,\text{m s}^{-2} \times 5.50\,\text{m}}{299\,\text{N m}^{-1}}} = 0.60\,\text{m}.$$

So the maximum compression of the spring is 60 cm.

Therefore, the distance fallen when the block first hits the spring is

$$h - x = (5.50 - 0.60)\,\text{m} = 4.90\,\text{m}.$$

Rearranging Equation 6.9 which results from equating kinetic and gravitational energy

$$v = \sqrt{2g(h - x)}$$

and putting in the values

$$v = (2 \times 9.81\,\text{m s}^{-2} \times 4.90\,\text{m})^{1/2} = 9.80\,\text{m s}^{-1}.$$

So the speed of the block at the instant it lands on the spring is $9.80\,\text{m s}^{-1}$.

Checking All the units come out as expected and the value of the compression and speed seem reasonable given the other values in the question. If the block were more massive, then the compression of the spring would be larger, as expected.

Q6.5 **Preparation** When the truck is just about to tip under the load, the line of contact between the rear wheels and the ground becomes the fulcrum. We can calculate the torques about this axis due to weight of the load and the weight of the truck. The magnitude of the torque is found using $\Gamma = Wd$ in each case, where W represents the magnitude of the relevant weight, and d is the maximum distance from the fulcrum to the line of action of the weight. The direction of the torque (into or out of the page) can be indicated by a negative or a positive sign. When tipping is just about to occur, the two torques must be equal in magnitude and opposite in direction.

Working The torque Γ_L about the rear wheels, due to the load, is

$$\Gamma_L = -W_L \times (7.0 - 4.5)\,\text{m} = -W_L \times (2.5\,\text{m})$$

where the negative sign indicates that the torque is directed into the page, so the associated rotation is clockwise.

The torque Γ_T due to the weight of the truck W_T about the same fulcrum is

$$\Gamma_T = (4000\,\text{kg}) \times (9.8\,\text{m s}^{-2}) \times (3.2\,\text{m}) = 1.25 \times 10^5\,\text{N m}$$

where the positive sign indicates that the torque is directed out of the page and the associated rotation is anticlockwise.

Thus on the point of tipping, $\Gamma_T + \Gamma_L = 0$.

So $\quad 1.25 \times 10^5\,\text{N m} = W_L \times (2.5\,\text{m})$

and $\quad W_L = 1.25 \times 10^5\,\text{N m}/2.5\,\text{m} = 5.0 \times 10^4\,\text{N}.$

The maximum safe load will therefore be $m = W_L/g$

$$m = 5.0 \times 10^4\,\text{N}/9.8\,\text{m s}^{-2} = 5.1 \times 10^3\,\text{kg}$$

i.e. 5.1 tonnes.

Checking All the units work out as expected, and the load of around 5 tonnes is of the same order as the mass of the truck, which seems reasonable. If the truck were more massive, then Γ_T would be larger and so would the maximum load, which is also reasonable.

Q6.6 Preparation We can treat the door and arrow, shown in Figure 6.10, as a single system. Then, immediately before and after impact, the total angular momentum of this system is unaltered, since there are no external torques acting on it. Using the definition, $L = r \times p$, immediately before impact the angular momentum of the door is zero and the magnitude of the angular momentum of the arrow with respect to the axis of the hinges is given by

$$L_{before} = mvb$$

where m is the mass of the arrow, v is its speed and b is the distance from the hinges to the impact point.

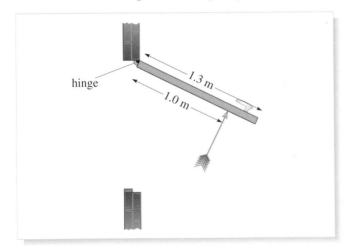

Figure 6.10 For use with Q6.6.

In this case we can use $L = I\omega$, where ω is the angular speed, so immediately after the impact, the angular momentum of the door plus arrow system has magnitude

$$L_{after} = (\tfrac{1}{3}Ma^2 + mb^2)\omega$$

where M is the mass of the door, a is its width and the quantity in brackets is the moment of inertia of the door plus arrow system about an axis through its hinges.

Working Since angular momentum is conserved, equating the angular momenta before and after the arrow strikes the door gives

$$mvb = \left(\frac{Ma^2}{3} + mb^2\right)\omega.$$

The angular speed is therefore

$$\omega = \frac{mvb}{\dfrac{Ma^2}{3} + mb^2}.$$

Noting that $100\,km\,h^{-1}$ is equal to $100 \times 10^3\,m/3600\,s = 27.8\,m\,s^{-1}$, putting in the given values we find

$$\omega = \frac{1.0\,kg \times 27.8\,m\,s^{-1} \times 1.0\,m}{\tfrac{1}{3} \times 60\,kg \times (1.3\,m)^2 + 1.0\,kg \times (1.0\ m)^2}$$

$$= 0.80\,rad\,s^{-1}.$$

The angular velocity is a vector of magnitude $0.80\,rad\,s^{-1}$ directed along the axis of rotation, in the sense specified by the right-hand grip rule (i.e. out of the page).

The magnitude of the linear velocity of the arrow after impact is

$$v = \omega r = 0.80\,m\,s^{-1}.$$

The linear velocity is the vector of magnitude $0.80\,m\,s^{-1}$ directed along the initial direction of the arrow.

Checking All the units come out as expected. The question asks for velocities, not just speeds, so the numerical answers have been supplemented by directional information as required. An angular speed of $0.80\,rad\,s^{-1}$ means that the door would swing through (say) 45° (or $\pi/4$ radians) in about 1 s, which seems reasonable. If the mass of the door were larger, the angular speed of the door would be smaller, as expected; and if the speed of the arrow were greater, the angular speed of the door would be greater, again as expected.

Q6.7 Using $F_x = ma_x$, we obtain $m = F_x/a_x = 5\,N/0.5\,m\,s^{-2} = 10\,kg$.

Q6.8 If m is the mass of the car and F_D is the magnitude of the driving force, then applying Newton's second law, we find $ma_x = F_D - F_R$, therefore $F_R = F_D - ma_x = 900\,N - (800\,kg \times 1.00\,m\,s^{-2}) = 100\,N$.

Q6.9 Let the magnitudes of the forces acting be F_P for the pushing force, f for the frictional force and N for the normal reaction force. Applying Newton's second law to the mass m, with the acceleration along the x-axis, we find the resultant force is

$$F_x = F_P - f = ma_x$$

with $f = \mu_{slide}N$ and $N = mg$

so $ma_x = F_P - \mu_{slide}mg$.

Dividing both sides of the equation by m gives

$$a_x = (F_P/m) - \mu_{slide}g$$

and substituting the given values

$$a_x = (10\,N/2.0\,kg) - (0.5 \times 9.8\,m\,s^{-2})$$

$$= (5.0 - 4.9)\,m\,s^{-2} = 0.1\,m\,s^{-2}.$$

Q6.10 A spring of spring constant k_s, stretched through a distance x, applies a tension force $F_{Tx} = -k_s x$ to an attached particle of mass m, causing the particle to accelerate with an initial acceleration a_x. Newton's second law implies that

$$ma_x = F_{Tx} = -k_s x.$$

So, $k_s = \dfrac{-ma_x}{x} = \dfrac{0.5\,\text{kg} \times 2.0\,\text{m s}^{-2}}{0.25\,\text{m}} = 4\,\text{N m}^{-1}$.

(Note that in this case the acceleration a_x is negative.)

Q6.11 Let F_N and F_E be the magnitudes of the forces in the north and east directions, respectively. If we consider the components along the north–south direction, we find

$$F_N - (10\,\text{N} \times \cos 60°) = 0.$$

So $F_N = 5.0\,\text{N}$.

Considering the components along the east–west direction

$$F_E - (10\,\text{N} \times \sin 60°) = 0$$

So $F_E = 10\,\text{N} \times \sqrt{3}/2 = 5\sqrt{3}\,\text{N}$.

Q6.12 The work done on a car, of mass m and initial velocity u_x, by a frictional force f_x, acting over a displacement s_x, is equal to the change in the translational kinetic energy, i.e. $f_x s_x = \frac{1}{2} m u_x^2$.

It follows that

$$f_x = \frac{m u_x^2}{2 s_x} = \frac{1000\,\text{kg} \times (10\,\text{m s}^{-1})^2}{2 \times 50\,\text{m}} = 1000\,\text{N}.$$

When the car has skidded half the total stopping distance, half of the original kinetic energy remains. If the velocity at this point is v_x then

$$\tfrac{1}{2} m v_x^2 = \tfrac{1}{4} m u_x^2.$$

The mass cancels, so

$$v_x = \frac{u_x}{\sqrt{2}} = \frac{10}{\sqrt{2}}\,\text{m s}^{-1} = 5\sqrt{2}\,\text{m s}^{-1}.$$

Q6.13 The strain potential energy is given by $\frac{1}{2} k_s x^2$; whereas the gravitational potential energy is given by mgh. For these to be equal, $\frac{1}{2} k_s x^2 = mgh$. So, rearranging

$$x = \sqrt{\frac{2mgh}{k_s}}$$

i.e. $x = \sqrt{\dfrac{2 \times 2\,\text{kg} \times 10\,\text{m s}^{-2} \times 1\,\text{m}}{1000\,\text{N m}^{-1}}} = 0.2\,\text{m}$.

Q6.14 Call the north–south direction the y-axis with north positive, and the east–west direction the x-axis with east positive, Let the final speed of the two players be v and the angle of their travel be θ measured anticlockwise from the positive x-axis. Momentum is conserved during the tackle, so in the x-direction

$$80\,\text{kg} \times 4.5\,\text{m s}^{-1} = (100\,\text{kg} + 80\,\text{kg}) \times v \cos \theta$$

i.e. $360\,\text{kg m s}^{-1} = 180\,\text{kg} \times v \cos \theta$

and in the y-direction

$$100\,\text{kg} \times (-3.6\,\text{m s}^{-1}) = (100\,\text{kg} + 80\,\text{kg}) \times v \sin \theta$$

i.e. $-360\,\text{kg m s}^{-1} = 180\,\text{kg} \times v \sin \theta$.

Dividing the second equation by the first

$$\sin \theta / \cos \theta = \tan \theta = -1.$$

So $\theta = -45°$, in other words, the final direction is south of east.

The speed of the two players after the tackle can be found from either of the two resolved equations as

$$v = \frac{360\,\text{kg m s}^{-1}}{180\,\text{kg} \times \cos 45°} = 2\sqrt{2}\,\text{m s}^{-1}.$$

So the final velocity of the players is $2\sqrt{2}\,\text{m s}^{-1}$ at an angle $45°$ south of east.

Q6.15 Let the force on the first child be in the positive x-direction so that she accelerates along the x-axis. Now, from Newton's third law, the force on the second child must be 40 N directed along the negative x-axis, and so he accelerates along the negative x-axis. To calculate the first child's final momentum, we note that F_x is a constant ($+40\,\text{N}$) over the time ($\Delta t = 0.5\,\text{s}$) for which it acts. The change of momentum is then the rate of change of momentum (which is equal to the force) multiplied by the time for which the force acts. Thus

$$\text{change of momentum} = \text{rate of change of momentum} \times \Delta t$$

i.e. $\Delta p_x = \mathrm{d}p_{x1}/\mathrm{d}t \times \Delta t = F_x \Delta t = 40\,\text{N} \times 0.5\,\text{s} = 20\,\text{kg m s}^{-1}$.

Thus she gains momentum of $20\,\text{kg m s}^{-1}$ in the positive x-direction. However for the second child, $F_x = -40\,\text{N}$ (by the third law), and so $\Delta p_x = -20\,\text{kg m s}^{-1}$. Since both children start from rest, their final momenta are the same as their respective changes in momenta, and these are equal in magnitude but opposite in sign:

$$p_{x1} = 20\,\text{kg m s}^{-1}$$

$$p_{x2} = -20\,\text{kg m s}^{-1}.$$

The children were initially at rest, so their total momentum was zero. Since the total momentum is conserved, their final total momentum is also equal to zero.

Q6.16 There are 3600 revolutions of the spin-drier drum in one minute which means there are 3600/60 revolutions in one second. But there are 2π radians per revolution.

So, $\omega = \dfrac{3600 \times 2\pi}{60}\,\text{rad s}^{-1} = 120\pi\,\text{rad s}^{-1}$.

The speed of the sock is given by $v = r\omega$.
So, $v = 0.25\text{ m} \times 120\pi\text{ rad s}^{-1} = 30\pi\text{ m s}^{-1}$.

Q6.17 Suppose your weight is of magnitude 700 N and that this is just sufficient to loosen the nut when you stand on a spanner of length 0.40 m, fitted horizontally. The magnitude of the torque due to your weight is:

$$\Gamma = 700\text{ N} \times 0.40\text{ m} = 280\text{ N m} \approx 300\text{ N m}.$$

This must also be the maximum frictional torque on the nut. The actual value you calculate will be different depending on the value you estimated for your weight.

Q6.18 The work done is the average torque multiplied by the angle (in radians). The angle turned through is one revolution (2π radians) and the magnitude of the average torque is, say, $\Gamma = 280\text{ N m}/4 = 70\text{ N m}$. So, $\Delta W = 70\text{ N m} \times 2\pi = 140\pi\text{ J}$. If this is completed in 5 s, the average power is given by $P = \Delta W/\Delta t = 140\pi\text{ J}/5\text{ s} = 28\pi\text{ W}$.

Q6.19 Figure 6.11 shows the mast OP, tension force $\boldsymbol{F}_\text{T}$, and one of the tie cords PQ, in side elevation. The magnitude of the torque about O is:

$$\Gamma = F_\text{T}d = F_\text{T}r \sin\theta$$

where r is the distance from O to P, which is 3.0 m in this case.

From the triangle OPQ,

$$\sin\theta = 4.0\text{ m}/(3.0^2 + 4.0^2)^{1/2}\text{ m} = 4.0/5.0.$$

So $\Gamma = (100\text{ N} \times 3.0\text{ m} \times 4.0)/5.0 = 240\text{ N m}.$

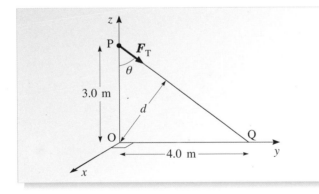

Figure 6.11   For use with Q6.19 and Q6.20.

Q6.20 We can refer again to Figure 6.11. (a) If the fulcrum is taken to be on the mast 1.0 m above the ground, then it is 2.0 m below the point of attachment of the cord at P. The torque about this fulcrum is calculated as in Q6.19, but with 2.0 m replacing the 3.0 m in that calculation (θ is unchanged):

$$\Gamma' = F_\text{T}d' = F_\text{T}r' \sin\theta$$

$$= (100\text{ N} \times 2.0\text{ m} \times 4.0)/5.0 = 160\text{ N m}.$$

(b) If the fulcrum is taken to be coincident with the point of attachment of the cord at P, then the distance of the line of action of the force from the fulcrum is zero and so the torque is zero.

Q6.21 The translational kinetic energy is $\frac{1}{2}Mv^2$ and the rotational kinetic energy about the central axis is $\frac{1}{2}I_\text{CM}\omega^2$.

(a) The total kinetic energy $E_\text{kin} = \frac{1}{2}Mv^2 + \frac{1}{2}I_\text{CM}\omega^2$.

Since $v = r\omega$ and $I_\text{CM} = \frac{1}{2}Mr^2$ we can write this total kinetic energy as:

$$E_\text{kin} = \frac{1}{2}M(r\omega)^2 + \frac{1}{4}Mr^2\omega^2 = \frac{3}{4}Mr^2\omega^2.$$

(b) The loss of gravitational potential energy in falling a height h is

$$E_\text{grav} = Mgh$$

and this is equal to the gain in total kinetic energy. So

$$Mgh = \tfrac{3}{4}Mr^2\omega^2 \quad\text{and}\quad \omega = \sqrt{\frac{4gh}{3r^2}}.$$

Q6.22 The position vector $\boldsymbol{r}$ of the particle can be resolved into a (perpendicular) displacement $\boldsymbol{r}_\perp$ from the origin O to the nearest point on the straight line and a (parallel) displacement $\boldsymbol{r}_\parallel$ along the straight line: $\boldsymbol{r} = \boldsymbol{r}_\perp + \boldsymbol{r}_\parallel$. Then,

$$\boldsymbol{r} \times \boldsymbol{v} = (\boldsymbol{r}_\perp \times \boldsymbol{v}) + (\boldsymbol{r}_\parallel \times \boldsymbol{v}).$$

Because $\boldsymbol{r}_\parallel$ is parallel to $\boldsymbol{v}$, the second vector product is zero and we are left with $\boldsymbol{r}_\perp \times \boldsymbol{v}$. This is a constant vector (because $\boldsymbol{r}_\perp$ and $\boldsymbol{v}$ are both constant vectors). It has a constant direction (perpendicular to the plane containing O and the straight line, in the sense given by the right-hand rule) and a constant magnitude equal to bv (where $\boldsymbol{r}_\perp$ has length b and is perpendicular to $\boldsymbol{v}$).

Q6.23 (a) The angular momentum vector is given by the vector product

$$\boldsymbol{L} = \boldsymbol{r} \times \boldsymbol{p} = \boldsymbol{r} \times m\boldsymbol{v}.$$

The vectors $\boldsymbol{r}$ and $\boldsymbol{p}$ are perpendicular (in spite of the fact that O does not lie in the plane of the orbit) so the magnitude of $\boldsymbol{L}$ is given by

$$|\boldsymbol{L}| = \sqrt{(3.0\text{ m})^2 + (4.0\text{ m})^2} \times (5\text{ kg} \times 10\text{ m s}^{-1}) \times \sin 90°$$
$$= 5.0\text{ m} \times (5\text{ kg} \times 10\text{ m s}^{-1}) \times 1 = 250\text{ kg m s}^{-1}.$$

(b) $\boldsymbol{L}$ is perpendicular to $\boldsymbol{r}$ and $\boldsymbol{p}$ and, for the sense of rotation shown in Figure 6.12, has a positive z-component (by the right-hand rule). In Figure 6.12, the vector $\boldsymbol{r}$ makes an angle θ with the z-axis, so the vector $\boldsymbol{L}$ makes an angle $(90 - \theta)°$ with the z-axis. Thus

$$L_z = |\boldsymbol{L}|\cos(90° - \theta) = |\boldsymbol{L}|\sin\theta$$

$$L_z = 250 \, \text{kg m s}^{-1} \times \frac{4.0 \, \text{m}}{\sqrt{(3.0 \, \text{m})^2 + (4.0 \, \text{m})^2}}$$

$$= 250 \, \text{kg m s}^{-1} \times \frac{4.0}{5.0} = 200 \, \text{kg m s}^{-1}.$$

Q6.30 The magnitude of c is $\sqrt{2^2 + 2^2} = 2\sqrt{2}$ and the magnitude of b is 2. The angle between b and c is $45°$ which has both a sine and a cosine of $2/(2\sqrt{2}) = 1/\sqrt{2}$ (as may be seen from Figure 6.6). So the magnitude of $c \times b$ is

$$bc \sin\theta = (2\sqrt{2}) \times 2 \times (1/\sqrt{2}) = 4.$$

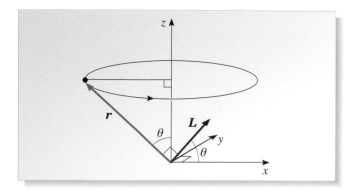

Figure 6.12   For use with Q6.23 and Q6.24.

Q6.24   (a) The maximum value of L_x occurs when the particle is at the point indicated in Figure 6.12. In terms of the angle θ,

$$(L_x)_{\text{max}} = |\mathbf{L}|\cos\theta = 250 \, \text{kg m s}^{-1} \times \frac{3.0}{5.0}$$

$$= 150 \, \text{kg m s}^{-1}.$$

(b) The minimum value of L_x occurs when the particle is at the diametrically opposite point on the circle to that shown in Figure 6.12, then

$$(L_x)_{\text{min}} = -150 \, \text{kg m s}^{-1}.$$

Q6.25   The magnitude of the unit vector is 1, and its direction is in the xy-plane at an angle θ to the x-axis such that $\tan\theta = \sqrt{3}/1$, in other words the same direction as $\mathbf{p}$, $\theta = 60°$.

Q6.26 $\mathrm{d}f(x)/\mathrm{d}x = 12x + 3$.

Q6.27 $\mathrm{d}f(x)/\mathrm{d}x = 3\mathrm{e}^{3x}$.

Q6.28   The magnitude of $\mathbf{a}$ is $\sqrt{2^2 + 1^2} = \sqrt{5}$ and the magnitude of $\mathbf{b}$ is 2, while the cosine of the angle between them is $2/\sqrt{5}$. So the value of $\mathbf{a} \cdot \mathbf{b}$ is $\sqrt{5} \times 2 \times 2/\sqrt{5} = 4$.

Alternatively, note that $\mathbf{a} \cdot \mathbf{b} = a_x b_x + a_y b_y = 2 \times 2 + 1 \times 0 = 4$.

Q6.29   The vector product is directed at right angles to the two individual vectors from which it is calculated, in the sense specified by the right-hand rule. So $\mathbf{a} \times \mathbf{c}$, $\mathbf{a} \times \mathbf{b}$ and $\mathbf{b} \times (-\mathbf{a})$ are directed *into* the plane of the paper, whilst $(\mathbf{b} \times \mathbf{a})$ and $\mathbf{c} \times (\mathbf{a} + \mathbf{b})$ are directed *out of* the plane of the paper.

Acknowledgements

Grateful acknowledgement is made to the following sources for permission to reproduce material in this book:

Front cover – Science Photo Library;

Fig. 1.1 NASA; *Fig. 1.7* NRSC Ltd. Science Photo Library; *Fig. 1.8* Crown Copyright 1996. Reproduced by permission of the controller of HMSO. National Physical Laboratory; *Fig. 1.9 particle tracks* – CERN/Science Photo Library; *Jupiter* – Science Photo Library; *galaxy* – Anglo Australian Observatory. Photograph by David Malin; *Coma cluster* – NASA & AURA/STSc; *Fig. 1.10 Sun* – Courtesy of Calvin J. Hamilton. National Solar Observatory/Sacramento Peak; *Fig. 1.31 Concorde* – Quadrant Pictures; *car* – Science Photo Library; *cyclist* – Allsport; *p. 36* Millennium Dome – Andrew Putler NMEC; *Fig. 1.57a* Toomre, A. and Toomre, J. (1994) 'Galaxies and the Universe', *Images of the Cosmos*, in Jones, B. W., *et al.* (eds), The Open University. © Toomre, A. (MIT) and Toomre, J. (University of Colorado); *Fig. 1.57b* Royal Observatory Edinburgh/Anglo Australian Telescope Board. Photograph by David Malin;

Fig.2.1 Reproduced from 'The Man in the Moone' edited by Faith K. Pizor and T. Allan Comp. Publishers Sidgwick & Jackson Ltd., 1971; *Fig. 2.2* British Film Institute; *Fig. 2.3* ESA; *Fig. 2.4* Science Picture Library; *Fig. 2.5 Earth* – NASA; *meteorite impact and supertanker* – Science Photo Library; *elephant* – Heather Angel Biofotos; *running child* – Gary Mortimore/Allsport; *Fig. 2.13 piledriver* QA Photos; *dam* – J.M. Petit/Science Photo Library; *Fig. 2.28a* Science and Society Picture Library; *Fig. 2.28b* Irish Academy; *Figure 2.29* Science and Society Picture Library; *Fig. 2.30 Sun* – Courtesy of Calvin J. Hamilton. National Solar Observatory/Sacramento Peak; *power station* – John Phillips/Photofusion; *Fig. 2.43* Smithsonian Institution; *Fig. 2.46* Heather Angel Biofotos; *Fig. 2.47* EERI Bridge Reconnaissance Team and the Federal Highway Administration;

Fig. 3.1 Science and Society Picture Library/Science Museum; *Fig. 3.10* NASA; *Fig. 3.11* Science Photo Library; *Fig. 3.12* Original artwork Don Davies/NASA; *Fig. 3.13* © Mira; *Fig. 3.14* © Mira; *Fig. 3.18* CERN;

Fig. 4.2 David Parker Science Photo Library; *Fig. 4.4* Mary Evans Picture Library; *Fig. 4.31* Courtesy of Michael Scott, Builders; *Fig. 4.32* Courtesy of Keith Martin, Technology Faculty, Open University; *Fig. 4.39* Permission granted by Readers Digest Ltd., AA Book of the Car; *Fig. 4.40* Richard Megna/Fundamental Science Photo Library; *Fig. 4.47* Courtesy of Professor Jocelyn Bell Burnell; *Fig. 4.49* Image Select; *Fig. 4.50* Quadrant Picture Library; *Fig. 4.51* Allsport; *Fig. 4.58* Sperry Marine;

Fig. 5.1 Science Museum/Science Picture Library; *Fig. 5.6* William Coupon; *Fig. 5.10* Mary Evans Picture Library; *Fig. 5.11a,b* 'Stone Soup Group' created by FRACTINT; *Fig. 5.15a,b, 5.16, 5.17a,b* By kind permission of Blair D. Fraser; *Fig. 5.18* Crutchfield, J. P. *et al.* (1986) 'Chaos', *Scientific American,* December, p. 47. Scientific American. © Crutchfield, J. P. *et al.*; *Fig. 5.19a,b* Adapted from Ary L. Goldberger, M. D., Beth Israel Deaconess Medical Center/Harvard Medical School, Boston, Massachussetts, USA; *Fig. 5.20* Laskar, J. (1994) 'Large-scale chaos in the solar system', A*stronomy and Astrophysics,* Vol. 287, L9–L12. Springer-Verlag GMBH & Co. KG; *Fig. 5.22a* Photograph by David Malin,

Index

Entries and page numbers in **bold type** refer to key words which are printed in **bold** in the text and which are defined in the Glossary.